Analog Circuits and Signal Processing

The *Analog Circuits and Signal Processing* book series, formerly known as the *Kluwer International Series in Engineering and Computer Science*, is a high level academic and professional series publishing research on the design and applications of analog integrated circuits and signal processing circuits and systems. Typically per year we publish between 5-15 research monographs, professional books, handbooks, and edited volumes with worldwide distribution to engineers, researchers, educators, and libraries. The book series promotes and expedites the dissemination of new research results and tutorial views in the analog field. There is an exciting and large volume of research activity in the field worldwide. Researchers are striving to bridge the gap between classical analog work and recent advances in very large scale integration (VLSI) technologies with improved analog capabilities. Analog VLSI has been recognized as a major technology for future information processing. Analog work is showing signs of dramatic changes with emphasis on interdisciplinary research efforts combining device/circuit/technology issues. Consequently, new design concepts, strategies and design tools are being unveiled.Topics of interest include: Analog Interface Circuits and Systems; Data converters; Active-RC, switched-capacitor and continuous-time integrated filters; Mixed analog/digital VLSI; Simulation and modeling, mixed-mode simulation; Analog nonlinear and computational circuits and signal processing; Analog Artificial Neural Networks/ Artificial Intelligence; Current-mode Signal Processing; Computer-Aided Design (CAD) tools; Analog Design in emerging technologies (Scalable CMOS, BiCMOS, GaAs, heterojunction and floating gate technologies, etc.); Analog Design for Test; Integrated sensors and actuators; Analog Design Automation/Knowledge-based Systems; Analog VLSI cell libraries; Analog product development; RF Front ends, Wireless communications and Microwave Circuits; Analog behavioral modeling, Analog HDL.

More information about this series at https://link.springer.com/bookseries/7381

Hakan Kuntman • Deniz Özenli

Trends in Circuit Design for Analog Signal Processing

Springer

Hakan Kuntman
Electronics and Communication Engineering
Istanbul Technical University
Istanbul, Turkey

Deniz Özenli
Electronics Engineering
National Defence University, Turkish Air Force Academy
Istanbul, Turkey

ISSN 1872-082X ISSN 2197-1854 (electronic)
Analog Circuits and Signal Processing
ISBN 978-3-030-96835-9 ISBN 978-3-030-96836-6 (eBook)
https://doi.org/10.1007/978-3-030-96836-6

This Springer imprint is published by the registered company Springer Nature Switzerland AG
The registered company address is: Gewerbestrasse 11, 6330 Cham, Switzerland

Preface

Analog circuits are fundamentally necessary in many complex and high-performance systems, although digital signal processing is becoming increasingly more powerful and many types of signal processing have indeed moved to digital domain due to the advances in IC technology which provides compact and efficient implementation of these algorithms in silicon. This is caused by the reality that naturally occurring signals are analog. In other words, analog circuits act as a bridge between the real world and digital systems. In analog signal processing, many circuit topologies, including active filters, oscillators, and immittance simulators, have been proposed in the literature. Today, process technologies such as CMOS and BiCMOS have advanced. Now, it is easier to realize this type of topologies for analog signal processing; as a result, it is not complete to propose only this type of circuits without applying the aforementioned considerations.

In analog circuit design, the current trend is to realize the proposed circuit by showing its performance limitations and adding application examples for the proposed circuit to be used in the real world. It is known that application areas of analog signal processing are wide and range from very low frequencies at several Hz levels of biomedical signals to RF applications operating at GHz level, from EEG signals to cognitive radio and encrypted communications or low-noise amplifiers in wireless communications. In this respect, this book offers to readers a strong overview and new ideas in order to catch the recent trends and advanced applications.

This book discusses new possibilities in analog circuit design, including applications on communication, measurement, and RF systems, combining the main features for circuit design with actual circuit realizations and demonstrating several performance limitations on chosen circuit examples.

This book covers illustrations, tables, and new proposals derived from authors' published papers. In this respect, the book makes a strong overview of the recent advances in the last couple of decades.

It is known that application areas of analog signal processing are wide and range from very low frequencies at several Hz levels of biomedical signals to RF applications operating at GHz level, from EEG signals to cognitive radio and encrypted communications or low-noise amplifiers in wireless communications. Therefore, the designer should know the limitations that affect the performance of their circuit before the actual realization. Consequently, the performed work is not sufficient without demonstrating the circuit characteristics in view of microelectronics and fabrication technology. In this manner, readers get a huge opportunity to improve themselves with the help of this book covering new advances and possibilities in the related research area including application on communication, measurement, and RF systems.

The subject of our book is fully derived from published papers by Kuntman (and some parts by Ozenli). The main subject is based on the concept "Alternative Active Elements to Operational Amplifiers for Analog IC Design." In the last two decades, Kuntman et al. worked on this subject, derived several topologies, and the results were published in the literature. Our book presents a survey of these works. We think that this book will also be helpful for high-level MSc and PhD students as well as researchers from industry working in the area of analog IC design.

Istanbul, Turkey

Hakan Kuntman
Deniz Özenli

Contents

1 **Introduction** 1
Basic Concept: New Possibilities and Trends in Circuit Design for Analog Signal Processing 1
Alternative Active Elements to Operational Amplifiers for Analog IC Design 2
Limitations 5
References 7

2 **Operational Transconductance Amplifiers (OTAs) and Their Applications** 11
Basic Concept 11
Limitations for Input Signal 13
Design Example 15
Application Example: Design of EEG Filters for Biomedical Applications 17
DO-OTA: Dual-Output Operational Transconductance Amplifier 19
High Output Impedance Stages 24
CMOS Circuits Employing the High-Performance Input and Output Stages 27
OTA-C Oscillators 31
Realization of DO-OTA-C Oscillators 34
Concluding Remarks 34
References 36

3 **Current Conveyors, Variants, and Applications** 39
Basic Concept 39
Nonlinear Behavior of the Current Conveyors 42
Determination of the Maximum Input Signal Amplitude 42
Limitations on Input Signal Level in Current-Mode Active-RC Filters Using CCIIs 47

Wide Dynamic Range High Output Impedance Current-Mode Multifunction Filters with Dual-Output Current Conveyors 49
Dual-Output Current Conveyor 49
Introduced Filter Topologies 50
Simulation Results and Verification (Tables 3.2 and 3.3) 53
Realization of nth-Order Current Transfer Function Employing ECCIIs and Application Examples 56
The ECCII and Its CMOS Implementation 57
Second-Order Filter Using ECCII 60
CCCII, Current-Controlled Conveyors 65
Electronically Tunable, Active-Only Floating Inductance Simulation 68
A New CCII-Based Sinusoidal Oscillator Employing Grounded Resistors and Capacitors 73
Simulations 74
Concluding Remarks 74
Differential Difference Current Conveyor (DDCC) 76
DVCC, Differential Voltage Current Conveyor 77
Design Example: A 22.5 MHz Current-Mode KHN Biquad Using Differential Voltage Current Conveyor and Grounded Passive Elements 78
KHN Biquad 78
Third-Generation Current Conveyor, CCIII 80
Fully Differential Current Conveyor (FDCCII) 86
Application Examples 88
Current-Mode Single-Input Four-Outputs (SIFO) Biquad Filter 88
Current-Mode MISO Biquad Filters 90
Simulation Results 92
Concluding Remarks 93
References 93

4 The Operational Transresistance Amplifier (OTRA) 99
Basic Concept 99
Single-Output OTRA 100
Design Example: Two OTRA-Based Grounded Immittance Simulator Topologies 102
Application Example and Simulation Results 104
Low-Voltage CMOS Differential OTRA for Submicron Technologies 107
Universal MOS-C Filter Design Example and Simulation Results 113
References 117

5 FTFN: Four-Terminal Floating Nullor . . . 119
Basic Concept . . . 119
Simulation Results . . . 123
Universal Series and Parallel Immittance Simulators Using Four-Terminal Floating Nullors . . . 125
Concluding Remarks . . . 131
A New Four-Terminal Floating Nullor-Based Single-Input Three-Output Current-Mode Multifunction Filter . . . 131
Circuit Description . . . 131
Realization of Inductorless Chua's Circuit Using FTFN-Based Nonlinear Resistor and Inductance Simulator . . . 133
Circuit Description of FTFN-Based Chua's Circuit . . . 134
Concluding Remarks . . . 139
References . . . 141

6 Current Operational Amplifiers . . . 143
Basic Concept . . . 143
COA-Based Filter Realizations . . . 147
Simulation Results . . . 148
Simulation Results of the COA-Based Filter Shown in Fig. 6.4 . . . 150
Concluding Remarks . . . 150
A High-Drive Fully Differential Current-Mode Operational Amplifier Providing High Output Impedance with a Filter Application . . . 151
COA-Based Notch Filter Realization . . . 154
Simulation Results of the Described COA . . . 154
Simulation Results of the COA-Based Filter . . . 156
Concluding Remarks . . . 157
References . . . 158

7 Current Feedback Operational Amplifiers, CFOA, and Applications . . . 161
Basic Concept . . . 161
The Current Conveyor and the Current Feedback Op-amp . . . 163
The Realization Circuits, Application Examples . . . 164
Frequency Stability . . . 167
Experimental Results and Discussion . . . 169
CFOA-Based Lossless and Lossy Inductance Simulators . . . 171
Applications and Simulation Results . . . 172
Concluding Remarks . . . 177
References . . . 177

8 Active-Only Circuits . . . 181
Basic Concept: Realizations of Current-Mode and Voltage-Mode Multifunction Filters Without External Passive Elements . . . 181
The OTA and Op-Amp Models . . . 183

The Proposed Active-Only Multifunction Filters . . . 185
Effects of the Non-idealities of the Op-Amps . . . 187
Simulation Results and Discussion . . . 187
Concluding Remarks . . . 190
Further Filter Applications Employing Active-Only Circuits . . . 191
Simulation Results, Discussion, and Design Example . . . 194
Concluding Remarks . . . 198
References . . . 198

9 MOS-Only and MOS-C Circuits . . . 199
Basic Concept: Fundamentals of MOS-Only and MOS-C Circuits . . . 199
Exemplary Circuit I . . . 206
Exemplary Circuit II . . . 211
Exemplary Circuit III . . . 214
Exemplary Circuit IV . . . 219
Concluding Remarks . . . 222
References . . . 224

10 Versatile Active Elements, Structure, and Applications . . . 225
Basic Concept . . . 225
Current Differencing Buffered Amplifier (CDBA) . . . 226
Filter Topology Employing CDBA . . . 227
Influence of the Non-idealities . . . 231
Experimental Verification . . . 232
Concluding Remarks . . . 235
Current Differencing Transconductance Amplifier (CDTA): A New, Improved CMOS Realization of the CDTA and Its Filter Applications . . . 235
CMOS Realization of the CDTA and Simulation Results . . . 237
Biquad Filter and Simulation Results . . . 243
Application Example and Simulation Results . . . 244
Concluding Remarks . . . 246
The Voltage Differencing Transconductance Amplifier (VDTA) . . . 247
Application Example of the VDTA-Based Filter (Fig. 10.24) . . . 248
Application Example: Realization of Frequency Agile Filters for Encrypted Communications and Multi-standard Transceivers . . . 251
Method Designed with Feedback Technique . . . 255
VDTA-Based Design . . . 259
References . . . 265

11 DTMOS-Based Ultra-Low-Voltage Low-Power Circuit Design . . . 267
Basic Concept . . . 267
DTMOS Transistor . . . 268

DTMOS-Based 0.4 V Ultra-Low-Voltage Low-Power VDTA Design and Its Application to EEG Data Processing 270
Processing of the Measured EEG Data 274
A Very Compact, 0.4 V DTMOS CCII Employed in an Audio-Frequency Filter 277
Concluding Remarks 282
References 282

12 High-Precision Current-Mode Circuits Based on the MOS Translinear Principle 285
Basic Concept 285
Simulations 290
Concluding Remarks 292
High-Precision Current-Mode Multiplier/Divider 294
Current-Mode Multiplier/Divider Circuit 295
Simulation Results 297
Concluding Remarks 299
References 301

Index 303

About the Authors

Hakan Kuntman received his BSc, MSc, and PhD degrees from Istanbul Technical University in 1974, 1977, and 1982, respectively. In 1974, he joined the Department of Electronics and Communication Engineering at Istanbul Technical University. In 1993, he became Professor of Electronics in the same department (retired, 2016). His research interest includes design of electronic circuits, modeling of electron devices and electronic systems, active filters, and design of analog IC topologies. Dr. Kuntman has authored many publications on modeling and simulation of electron devices and electronic circuits for computer-aided design, analog VLSI design, and active circuit design. He is the author or the co-author of 129 journal papers published or accepted for publishing in international journals, 179 conference papers presented or accepted for presentation in international conferences, 161 Turkish conference papers presented in national conferences, and 10 books related to the above-mentioned areas. (h-index 37). He advised and completed the work of 16 PhD students and 44 MSc students. Dr. Kuntman is a member of the Chamber of Turkish Electrical Engineers (EMO).

From 2001 to 2004, he acted as the head of the Department of Electronics and Communication Engineering, and from 2004 to 2010, he was the dean of the Electrical and Electronics Engineering Faculty at Istanbul Technical University. Furthermore, Dr. Kuntman is one of the founders of the ELECO conferences and acted as the Conference Chairman several times.

Deniz Özenli received his BSc degree from Istanbul University in electrical and electronics engineering in 2009 and MS as well as PhD degrees from Istanbul Technical University in 2011 and 2018, respectively. He is now an assistant professor in the Department of Electronics Engineering at the Turkish Air Force Academy. He was also visiting assistant professor in the Department of Electrical and Electronics Engineering at Marmara University for the VLSI lectures. His main research interests are analog filters, low-voltage current and voltage mode circuits, computer-aided analog circuit design, VLSI design, as well as image and video processing. During his PhD, he carried out different analog filter applications in the basis of

MOSFET-only and MOSFET-C building blocks. Dr. Ozenli worked in the Department of Electrical and Electronics Engineering at Marmara University as research and teaching assistant between 2012 and 2019. He also worked with VLSI research group in Istanbul Technical University-VLSI Labs as laboratory assistant between 2012 and 2015.

Chapter 1
Introduction

Basic Concept: New Possibilities and Trends in Circuit Design for Analog Signal Processing

Operational amplifiers are well known and mostly used building blocks for analog circuit design. However, due to their limited performance, researchers look for better alternatives and other active blocks (Allen and Holberg, 2002; Razavi, 2000; Grey et al. 2001; Toumazou, 1990; Ferri and Guerrini, 2003). As a result, there is significant amount of past and ongoing research about new current-mode active building blocks such as operational transconductance amplifiers (OTAs), second-generation current conveyors (CCIIs), current feedback op-amps (CFOAs), four-terminal floating nullors (FTFNs), differential voltage current conveyor (DVCC), differential difference current conveyor (DDCC), third-generation current conveyor (CCIII), dual X current conveyors (DXCCII), current-controlled current conveyors (CCCIIs), current differencing transconductance amplifiers (CDTAs), etc (Acar et al. 1993; Acar and Kuntman, 1996; Kuntman and Özpınar, 1998; Düzenli et al. 1999; Acar and Özoğuz 1999; Zeki and Kuntman, 1999; Biolek et al, 2003, 2008; Altun and Kuntman, 2008; Çam et al., 2000, 2003; Ergün and Kuntman, 2000, 2001, 2005; Ibrahim et al. 2003, 2005, 2006; Kılıç et al. 2002, 2004; Minaei et al. 2001, 2003, 2006; Kuntman and Kuntman, 2018, 2019). Using these new active elements for analog design and implementing them in CMOS technology, designers have acquired new possibilities to solve classic op-amp-based problems such as bandwidth, slew rate, etc. Moreover, active-only and MOS-only topologies provide further solutions for analog circuit design (Erdoğan et al. 2004; Arslan et al. 2011; Ozenli and Kuntman 2016, 2018; Alaybeyoğlu and Kuntman, 2017, 2018, 2019). Additionally, usage of FGMOS transistors in analog circuits is an alternative approach whose efficiency has been shown recently (Keleş and Kuntman, 2011, 2018). This book discusses these new possibilities in analog circuit design including applications on communication, measurement, and RF systems, combining the main

H. Kuntman, D. Özenli, *Trends in Circuit Design for Analog Signal Processing*,
Analog Circuits and Signal Processing,
https://doi.org/10.1007/978-3-030-96836-6_1

features for the circuit design with actual circuit realizations and demonstrating several performance limitations on chosen circuit examples.

Analog circuits are fundamentally necessary in many complex and high-performance systems although digital signal processing is becoming increasingly more powerful and many types of signal processing have indeed moved to digital domain due to the advances in IC technology which provides compact and efficient implementation of these algorithms in silicon. This is caused by the reality that naturally occurring signals are analog. In other words, analog circuits act as a bridge between the real world and digital systems. In analog signal processing, many circuit topologies including active filters, oscillators, immittance simulators, etc. have been proposed in the literature. Initially, almost 20 years ago, there were circuit proposals without any actual realizations and application examples. Therefore, the main requirement was how to realize these circuits and how to use them in practical applications. Another important missing factor was the limitations of realization of topologies and the employed active element characteristics such as bandwidth, slew rate, and input and output impedances. Today, process technologies such as CMOS, BiCMOS, etc. have advanced. Now, it is easier to realize this type of topologies for analog signal processing; as a result, it is not complete to propose only this type of circuits without applying the considerations. In analog circuit design, the trend is now to realize the proposed circuit by showing its performance limitations and adding application examples for it to be used in practical world. It is known that application areas of analog signal processing are wide and ranging from very low frequencies at several Hz levels of biomedical signals to RF applications operating at GHz level, from EEG signals to cognitive radio and encrypted communications or low-noise amplifiers in wireless communications. Therefore, the designer should know the limitations that affect the performance of his circuit before the actual realization. Consequently, the performed work is not sufficient without demonstrating the circuit characteristics in view of microelectronics and fabrication technology.

Alternative Active Elements to Operational Amplifiers for Analog IC Design

Basic amplifier types used in analog IC design with their gain functions are given in Table 1.1. Besides these basic active elements, there are alternative active elements to operational amplifiers such as transconductance amplifier (OTA)

Table 1.1 Basic active elements

Class	Gain function	Operational property	Name
V–V	$V_o = A_v.(V_1-V_2)$	$A_v \rightarrow \infty$	Operational amplifier
V–I	$I_o = G_m.(V_1-V_2)$	–	Operational transconductance amplifier
I–I	$I_o = A_i.(I_1-I_2)$	$A_i \rightarrow \infty$	Current operational amplifier
I–V	$V_o = R_m.(I_1-I_2)$	–	Operational transresistance amplifier

Table 1.2 Other popular active elements

Name	Symbol	Definition equations
CC (current conveyor)	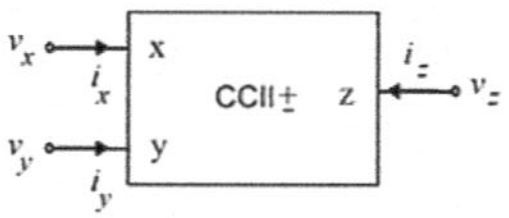	$i_y = a.i_x$ $V_x = V_y$ $i_z = \pm i_x$
DOCCII (dual-output current conveyor second generation)	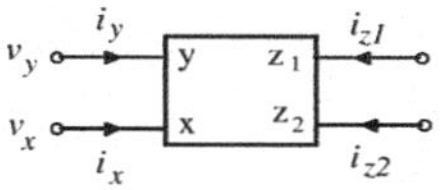	$\begin{bmatrix} v_x \\ i_y \\ i_{z1} \\ i_{z2} \end{bmatrix} = \begin{bmatrix} 0 & 1 & 0 & 0 \\ 0 & 0 & 0 & 0 \\ 1 & 0 & 0 & 0 \\ \pm 1 & 0 & 0 & 0 \end{bmatrix} \begin{bmatrix} i_x \\ v_y \\ v_{z1} \\ v_{z2} \end{bmatrix}$
CCCII (current-controlled current conveyor)	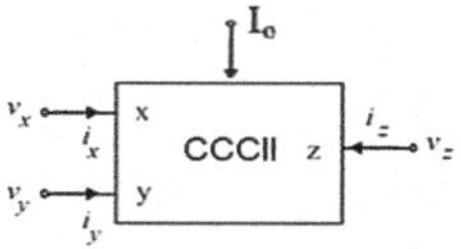	$\begin{bmatrix} I_y \\ V_x \\ I_z \end{bmatrix} = \begin{bmatrix} 0 & 0 & 0 \\ 1 & R_x & 0 \\ 0 & \pm 1 & 0 \end{bmatrix} \begin{bmatrix} V_y \\ I_x \\ V_z \end{bmatrix}, R_x = \frac{V_T}{2I_o}$
DVCCII (differential voltage current conveyor)	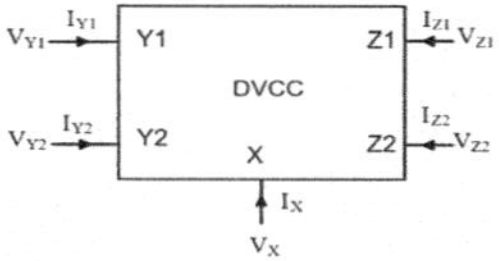	$\begin{bmatrix} V_X \\ I_{Y1} \\ I_{Y2} \\ I_{Z1} \\ I_{Z2} \end{bmatrix} = \begin{bmatrix} 0 & 1 & -1 & 0 & 0 \\ 0 & 0 & 0 & 0 & 0 \\ 0 & 0 & 0 & 0 & 0 \\ 1 & 0 & 0 & 0 & 0 \\ -1 & 0 & 0 & 0 & 0 \end{bmatrix} \begin{bmatrix} I_X \\ V_{Y1} \\ V_{Y2} \\ V_{Z1} \\ V_{Z2} \end{bmatrix}$
FTFN (four-terminal floating nullor)	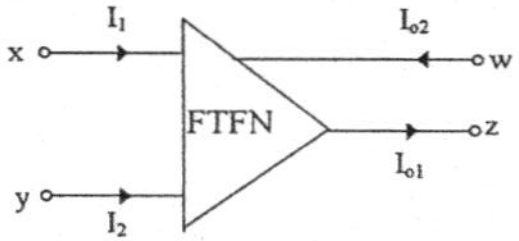	$I_1 = I_2 = 0 \quad I_{o1} = I_{o2} \quad V_x = V_y$

(continued)

Table 1.2 (continued)

Name	Symbol	Definition equations
CDBA (current differencing buffered amplifier)	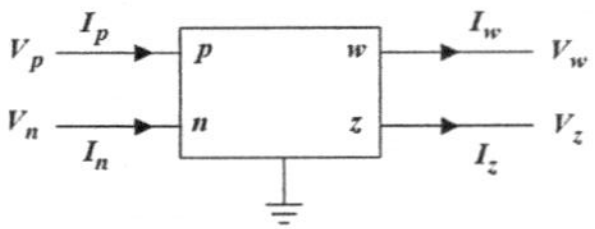	$\begin{bmatrix} i_z \\ v_w \\ v_p \\ v_n \end{bmatrix} = \begin{bmatrix} 0 & 0 & 1 & -1 \\ 1 & 0 & 0 & 0 \\ 0 & 0 & 0 & 0 \\ 0 & 0 & 0 & 0 \end{bmatrix} \begin{bmatrix} v_z \\ i_w \\ i_p \\ i_n \end{bmatrix}$
CDTA (current differencing transconductance amplifier)	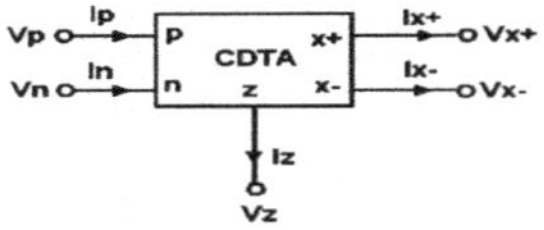	$V_p = V_n = 0$ $I_z = \alpha_p I_p - \alpha_n I_n$ $I_{x+} = gV_z$ $I_{x-} = -gV_z$
DXCCII (dual X second-generation current conveyor)	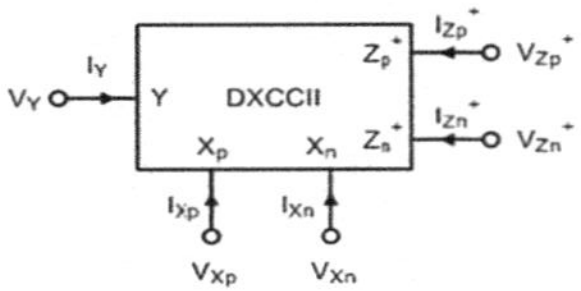	$V_{Xp} = \beta_1 V_Y \quad V_{Xn} = -\beta_2 V_Y \quad I_Y = 0$ $I_{Zn} = \alpha_n I_{Xn} \quad I_{Zp} = \alpha_p I_{Xp}$
CFOA (current feedback operational amplifier)	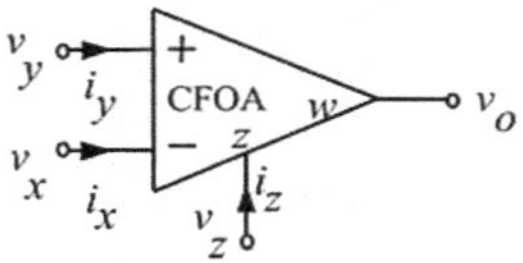	$\begin{bmatrix} i_y \\ v_x \\ i_z \\ v_o \end{bmatrix} = \begin{bmatrix} 0 & 0 & 0 \\ 1 & 0 & 0 \\ 0 & 1 & 0 \\ 0 & 0 & 1 \end{bmatrix} \begin{bmatrix} v_y \\ i_x \\ v_z \end{bmatrix}$

(Allen and Holberg, 2002; Acar et al., 1993; Bhaskar et al., 1993; Düzenli et al. 2002), four-terminal floating nullor (FTFN) (Çam and Kuntman, 1999; Çam et al., 2000a, 2000b, 2001; Saygıner et al., 2008; Kılıç et al. 2002, 2004), operational transresistance amplifier (OTRA) (Duruk and Kuntman, 2005; Duruk et al., 2007; Kacar et al. 2004, 2009b, 2009c), current feedback amplifier (CFOA) (Kacar et al. 2009a; Kacar and Kuntman 2009a), current conveyor (CC) and its derivatives (Fabre et al. 1995; Cicekoglu and Kuntman, 1998; Cicekoglu et al. 1999a, 1999b, 2001, 2002; Kuntman et al. 2000, 2002; Minaei et al. 2001, 2006; Kaçar et al. 2008, 2009c, 2010), current differencing transconductance amplifier (CDTA) (Kaçar et al. 2009b), etc. Other types of active elements are given with their symbols and definitions in Table 1.2.

Current-mode circuits have received considerable attention due to their potential advantages, such as their inherently wide bandwidth, higher slew rate, greater linearity, wider dynamic range, simple circuitry, and low power consumption. The active devices that have been used to realize current-mode circuits include current conveyors (CCIIs), current feedback op-amps (CFOAs), operational transconductance amplifiers (OTAs), and four-terminal floating nullors (FTFNs).

At the beginning, operational amplifiers were the unavoided building blocks for analog circuit design. Unfortunately, their limited performance such as bandwidth, slew rate, etc. leads the analog designer to search for other possibilities and other building blocks. As a result, new current-mode active building blocks such as operational transconductance amplifiers (OTAs), second-generation current conveyors (CCIIs), current feedback op-amps (CFOAs), four-terminal floating nullors (FTFNs), differential voltage current conveyor (DVCC), differential difference current conveyor (DDCC), third-generation current conveyor (CCIII), dual X current conveyors (DXCCII), current-controlled current conveyors (CCCIIs), etc. received considerable attention due to their larger dynamic range and wider bandwidth. Employing these new active elements for analog design and using CMOS technology for implementation, the circuit designers obtained new possibilities to solve their problems. This book covers new advances and possibilities in the related research area including application on communication, measurement, and RF systems. The main problems of using these types of active elements in active circuits are the linear operating range limitation and the limited impedances of the high output impedance.

Limitations

The designers have an assumption of linearity of the active components in the realization of active filters. However, these elements behave linear only under certain conditions which depend on their design parameters, and violation of these conditions results in nonlinear distortion in filters. Therefore, linear operation conditions of the filter to be realized should have been known in advance. Linear operation conditions for several active components such as op-amps, OTAs, CCIs, and CCIIs are given in the literature by utilizing macromodel approach; however, linear operation of active filters is not much investigated. In the implementation of an active filter, the designer should take into consideration of the input signal because active elements do not operate linearly if the terminal signals exceed certain limits. In the case of OTA-C filters, if the output voltage of any OTA saturates, a clipped waveform has been obtained. If the output current saturates, a sawtooth form has been acquired which is known as the slew-rate limitation.

Similarly, in the case of current conveyor-based filters, there are saturations in x terminal and z terminal voltages and currents which cause clipped or sawtooth waveforms limiting the performance of the filter topology.

(A) *Linearity of input differential stage, limitation caused by output resistance:* The finite output resistance, limited bandwidth, and limited input linearity range are important drawbacks in the realization of active elements such as OTA and DO-OTA, CCIIs, etc. The finite output resistance is the cause of a lossy integration because it becomes parallel to the load capacitor and leads to filtering errors. Therefore, output stages that have very high output impedance are required both to enable filtering at low frequencies and to reduce filtering errors. Another important drawback is the limited input voltage range. This limitation is caused by the input differential amplifier which affects the device linearity seriously. Substantial amount of work was performed in the last 20 years to solve these problems, and new alternative topologies were proposed to replace the conventional differential input stages.

(B) *Input and output stages*: To achieve a linear relationship between input differential voltage and output current, linearization techniques for input cells such as Nedungadi–Visvanathan input cell, cross-coupled input circuit, and Krummenacher's input circuit are available in the literature. Furthermore, to increase the output impedance and to extend the circuit performance, several output stage configurations such as employing active-feedback cascode current mirrors, are also introduced and can be found in the available literature. Using high-output-impedance current mirrors, high output impedance stage can be realized employing active-feedback cascode current mirrors. Output stages are designed in this approach; while keeping mirroring precision and GBW high, they additionally achieve much larger R_{out} and thus a much larger DC gain in comparison with classical cascode counterpart. Using this type of input and output stages in active analog blocks, high-performance OTAs and CCIIs can be realized without much difficulty.

This book is organized as follows. After a short introduction of the problem in this chapter, different structures based on OTA (operational transconductance amplifier) have been given in Chap. 2. Also, performance limitations of these circuits are underscored to the designers. Chapter 3 describes current conveyors, variants, and applications. Chapter 4 is on the operational transresistance amplifier (OTRA). Chapter 5 discusses the FTFN: the four-terminal floating nullor and its applications. Chapter 6 is on the COA: the current operational amplifier. Chapter 7 describes CFOA: the current feedback operational amplifier. Chapter 8 is on active-only circuit design. Chapter 9 describes MOS-only and MOS-C circuits. In

Chap. 10, recent advances are presented on the basis of versatile active elements. DTMOS-based ultra-low-voltage low-power circuit design is discussed in Chap. 11. Chapter 12 describes high-precision current-mode circuits based on the MOS translinear principle. Concluding remarks are given to the designers at the end of the chapters.

References

Acar, C. Anday F. and Kuntman, H.: On the realization of OTA-C filters, International Journal of Circuit Theory and Applications, Vol.21, pp.331–341, 1993.

Acar, C., Kuntman, H., "Limitations on input signal level in current-mode active-RC filters using 142 CCIIs", Electronics Letters, Vol.32, 16, pp.1461–1462, 1996.

Acar, C., Ozoguz, S., "A new versatile building block: current differencing buffered amplifier suitable for analog signal-processing filters", Microelectronics Journal Vol. 30, 157–160, 1999a.

Alaybeyoğlu, E., Kuntman, H., "A new implementation of the reconfigurable analog baseband low pass filter with cell-based variable transconductance amplifier", Analog Integrated Circuits and Signal Processing, Vol.97, Issue 1, pp.87–96, 2018a.

Alaybeyoğlu, E., Kuntman, H., "Capacitor multiplier with high multiplication factor for integrated 150 low pass filter of biomedical applications using DTMOS technique", AEU - International Journal of Electronics and Communications, Vol. 107, pp. 291–297, 2019.

Alaybeyoğlu, E., Kuntman, H., "New Realization Methods of Frequency Agile Filters for Encrypted 153 Communications and Multi-Standard Transceivers" (Invited Talk), Proc. of ICTUS'2017: International Conference on Infocom Technologies and Unmanned Systems, pp.99–112, Amity University, Dubai, 18–20 December 2017a.

Allen, P.E. and Holberg, D.R., CMOS analog circuit design (Second Edition), Oxford University 157 Press, New York Oxford, 2002.

Altun, M., Kuntman, H., 'Design of a Fully Differential Current Mode Operational Amplifier with Improved Input-Output Impedances and Its Filter Applications', AEU: International Journal of Electronics and Communications, Vol.62, NO. 3, 239–244, 2008.

Altun, M., Kuntman, H., Minaei S., Sayın, O. K., "Realization of nth-order current transfer function employing ECCIIs and application examples", IJE: International Journal of Electronics, Vol.96, No.11, 1115 –1126, 2009.

Arslan, E., Metin, B., Kuntman, H. and Cicekoglu, O., "MOS Only Second Order Current-Mode 165 LP/BP Filter", Proc. of ELECO 2011: The 7th International Conference on Electrical and 166 Electronics Engineering, pp.89–91, Bursa, TURKEY, 2011.

Bhashkar, D. R., Tripath, M., and Senani, R., Systematic derivation of all possible canonic OTA-C sinusoidal oscillators. Journal of the Franklin Institute, 330, 885-903, 1993.

Biolek, D., CDTA-building block for current-mode analog signal processing. Proceedings of the 170 ECCTD'03, Cracow, Poland, p. 397–400, 2003.

Biolek, D., Senani, R., Biolkova, V., & Kolka, Z. Active elements for analog signal processing: classification, review, and new proposals. Radioengineering, 17(4), 15–32, 2008a.

Çam, U, Çiçekoğlu, O., Kuntman, H., "A new four terminal floating nullor based single-input three output current-mode multifunction filter", Microelectronics Journal, Vol.30, No.2, pp.115-118, 1999.

Çam U. and Kuntman, H., "A new CMOS realisation of four terminal floating nullor (FTFN)", International Journal of Electronics, Vol. 87, No.7, pp. 809-817, 2000a.

Çam, U., Toker, A., Kuntman, H. , "Novel CMOS FTFN realisation based on translinear cells", Electronics Letters, Vol.36, No.15, pp.1255-1256, 2000b.

Çam U., Toker A., Çiçekoğlu O., Kuntman H, "Current-mode high output impedance multifunction filters employing minimum number of FTFN", Analog Integrated Circuits and Signal Processing, vol. 28, pp. 299–307, 2001.

Çiçekoğlu, O., Kuntman, H., "On the design of CCII+ based relaxation oscillator employing single grounded passive element for linear period control", Microelectronics Journal, Vol.29, No. 12, pp.983–989, 1998.

Cicekoglu, O., Kuntman, H., Berk, S., "All-pass Filters using a single current conveyor", International Journal of Electronics, No.8, pp.947–955,1999a

Cicekoglu, O., Özcan, S., Kuntman, H., Insensitive Multifunction Filter Implemented with Current Conveyors and Only Grounded Passive Elements, Frequenz, 53, N0.7–8,pp.158–160, 1999b.

Cicekoglu, O., Tarım, N., Kuntman, H., Wide Dynamic Range High output impedance current mode multifunction filters with dual-output current conveyors providing wide dynamic range, AEÜ: International Journal of Electronics and Communications, 56, No.1, pp. 55–60, 2002.

Cicekoglu, O., Toker, A., Kuntman, H., 'Universal immittance function simulators using current conveyors', Computers and Electrical Engineering, Vol. 27, pp. 227–238, 2001.

Duruk, A., Güneş, E. O., Kuntman, H., 'A new low voltage CMOS differential OTRA for sub-micron technologies', AEU: International Journal of Electronics and Communications, Vol.61, 291–299, 2007.

Duruk, A., Kuntman, H., "A new CMOS Differential OTRA Design for the Low Voltage Power Supplies in the Sub-Micron Technologies", Turkish Journal of Electrical Engineering and Computer Sciences (ELEKTRIK), Vol. 13, No.1, 23–37, 2005.

Düzenli, G., Kılıç, Y., Kuntman H. and Ataman, A.: On the design of low-frequency filters using CMOS OTAs operating in the subthreshold region, Microelectronics Journal, Vol.30, No. 1, pp.4554, 1999.

Erdogan, E. S., Topaloglu, R. O., Çiçekoglu, O. and Kuntman, H., "New Current-mode Special Function Continuous Time Active Filters Employing Only OTAS and Opamps", International Journal of Electronics, Volume 91, Number 6, 345–359, 2004.

Ergün, B.S., Kuntman, H., Realization of a high output impedance CMOS DO-OTA with extended linearity range, Proceedings of ELECO'2001: The 2nd International Conference on Electrical and Electronics Engineering (Electronics), pp.73–77, Bursa, 7–11 November 2001.

Ergün, B.S., Kuntman, H., "Yüksek Lineerlikte DO-OTA Gerçekleştirilmesi" Proceedings of ELECO 2000: The 1st National Symposium of Electrical, Electronics and Computer Engineering, pp 77–81, Chamber of Turkish Electrical Eng., Bursa section, Bursa Turkey, 2000.

Ergün, B.S., Kuntman, H., On the design of new CMOS DOOTA topologies providing high output impedance and extended linearity range, Journal of Electrical & Electronics Engineering, Engineering Faculty, Istanbul University, Vol.5, No.2, pp.1449–1461, 2005.

Fabre, A., Saaid, O., Wiest, F., Boucheron, C., Current controlled bandpass filter based on translinear conveyors. Electronics Letters, Vol. 31, pp.1727–1728, 1995.

Ferri, G. and Guerrini, N. C., "Low-voltage Low-power CMOS Current Conveyors", Kluwer Academic Publishers, AH Dordrecht, Netherlands, 2003.

Gray, R., Hurst, P.J., Lewis, S.H., Meyer, R.G., "Analysis and design of analog integrated circuits", John Wiley & Sons, Inc., 2001.

Ibrahim, M. A., Kuntman H., Cicekoglu O., Canonical Biquadratic All-Pass and Notch Filters Employing Differential Difference Current Conveyor, Frequenz, vol. 57, Nr. 7–8, pp. 162–165, 2003.

Ibrahim, M. A., Kuntman H., High Linearity CMOS Differential Difference Current Conveyor (DDCC), Proc. of ICM'2002: the 14th International Conference on Microelectronics, pp.6–9, Beirut, Lebanon, 2002.

Ibrahim, M. A., Kuntman H., Cicekoglu O., "Single DDCC biquads with high input impedance and 246 minimum number of passive elements", Analog Integrated Circuit and Signal Processing, 247 Vol.43, 71–79, 2005a.

Ibrahim, M. A., Minaei S. and Kuntman H., "A 22 MHz current-mode KHN-biquad using differential voltage current conveyor and grounded passive elements", AEU: International Journal of Electronics and Communications, Volume 59, 311–318, 2005b.

Ibrahim, M. A., Minaei S. and Kuntman H., DVCC based differential-mode all-pass and notch filters with high CMRR, International Journal of Electronics, Volume 93, No.4, 231–240, 2006.

Kacar, F., Kuntman H., "On the Realization of the FDNR Simulators Using Only a Single Current Feedback Operational Amplifier", Proc. of ELECO'2009: The 6th International Conference on Electrical and Electronics Engineering, Vol.2, pp.223–226, 5–8, Bursa, Turkey November 2009a.

Kacar, F., Kuntman H., "A new CMOS current differencing transconductance amplifier (CDTA) and its biquad filter application", Proceedings of EUROCON'2009 (CD-ROM), pp.208–215, St. Petersburg, RUSSIA, May 18–23, 2009b.

Kacar, F., Kuntman H., "Novel Electronically Tunable FDNR Simulator Employing Single FDCCII", Proceedings of ECCTD'09: the 19th European Conference on Circuit Theory & Design (CDROM), pp.21–24, Antalya, Turkey, August 23–27, 2009c.

Kacar, F., Metin B., Kuntman H., "A New Dual-X CMOS Second Generation Current Conveyor (DXCCII) with a FDNR Circuit Application", accepted for publication in AEU: International Journal of Electronics and Communications 2008 15th IEEE International Conference on Electronics, Circuits and Systems. IEEE, 2008.

Kacar, F., Metin B., Kuntman H., Cicekoglu, O., "A New High Performance CMOS FDCCII with Application Example of Biquad Filter Realization", accepted for publication in IJE: International Journal of Electronic 97(5), 499–510, 2010.

Kacar, F., Çam, U., Cicekoglu, O., Kuntman, H., Kuntman, A., "Novel Two Otra-Based Grounded Parallel Immittance Simulator Topologies", Analog Integrated Circuit and Signal Processing, 278 Vol.39, 169–175, 2004.

Keleş, S. and Kuntman, H., "Four quadrant FGMOS analog multiplier", TJEECS: Turkish Journal of Electrical Engineering & Computer Sciences, vol.: 19(2), pp. 291–301, 2011.

Keleş, S., Keleş, F., Kuntman, H.," Square root circuit using FGMOS translinear principle", Analog Integrated Circuits and Signal Processing, 2018.

Kılıç, R., Çam, U., Alçı, M., Kuntman, H. and Uzunhisarcıklı, E., "Realization of Inductorless Chua's Circuit Using FTFN-Based Nonlinear Resistor and Inductance Simulator", FREQUENZ, Vol.58, 1–4, 2004.

Kılıç, R., Çam, U., Alçı, M., Kuntman, H., "Improved Realisation of Mixed-mode chaotic circuit", International Journal of Bifurcation and Chaos, Vol. 12, No. 6, 1429–1435, 2002.

Kuntman, H., Özpınar, A., "On the realization of DO-OTA-C oscillators", Microelectronics Journal, Vol.29, No. 12, pp.991–997, 1998.

Kuntman, A., Kuntman, H., "Circuit Model for Statistical Method Based Reliability Estimation of MOS Transistors and Analog CMOS Circuits" (Invited Talk), Proc. of ICRITO'2018: International Conference on Reliability, Infocom Technologies and Optimization, pp.97–107, Amity University Uttar Pradesh, Noida, India, 29–31 August 2018.

Kuntman, A., Kuntman, H., "On the Reliability Estimation of Analog CMOS Circuits Based on Statistical Methods (Invited Talk), Proc. of ELECO'2019: International Conference on Electrical and Electronics Engineering, Bursa, Turkey 28–30 November, 2019.

Kuntman, H. Gülsoy, M., Çiçekoğlu, O., "Actively simulated grounded lossy inductors using third generation current conveyors", Microelectronics Journal, Vol.31, pp.245–250, 2000.

Kuntman, H., Çiçekoğlu, O., Özcan, S., "Realization of current-mode third order butterworth filters employing equal valued passive elements and unity gain buffers" Analog Integrated Circuit and Signal Processing, Vol.30, pp.253–256, 2002.

Kuntman, H., Çiçekoğlu, O., Özoğuz, S., 'A modified third generation current conveyor, its characterization and applications' FREQUENZ, Vol.56, pp.47–54, 2002.

Minaei, S., Cicekoglu, O., Kuntman, H., Türköz, S., "Electronically Tunable Active Only Floating Inductance simulation", International Journal of Electronics, Vol.89, No. 12, pp. 905–912, 2003.

Minaei, S., Cicekoglu, O., Kuntman, H, Türköz, S., "High Output Impedance Current-Mode Lowpass, Bandpass and Highpass Filters Using Current Controlled Conveyors". International Journal of Electronics., 88 (8), 915–922, 2001.

Minaei, S., Sayın O. K., Kuntman, H., "A New CMOS Electronically Tunable Current Conveyor and Its Application to Current-Mode Filters", IEEE Transactions on Circuits and Systems I, TCAS-I, Volume 53, No.7, 1448–1457, 2006.

Özenli, D., Kuntman, H., "MOS-Only Circuit Design Automation", Proc. of LASCAS 2016: 7th IEEE Latin American Symposium on Circuits and Systems, Florianopolis, Brazil, from February 28 to March 2, 2016.

Ozenli,D, Kuntman, H. , "A novel low power MOSFET-C band pass filter for low frequency applications with subthreshold models based on polynomial regression", Analog Integrated Circuits and Signal Processing, Vol.97, Issue 1, pp.97-105, October 2018.

Saygıner, M., Altun, M., Kuntman, H., "A New CMOS FTFN Realization and Grounded Inductance Simulation", Proceedings of MELECON'08: The 14th IEEE Mediterranean Electrotechnical Conference, pp. 421–424, 5–7 May, Ajaccio, Corsica, France, 2008.

Razavi, B., Design of Analog CMOS Integrated Circuits, McGraw Hill, 2000.

Toumazou, C., Lidgey, F.J. and Haigh, D.G., Analogue IC Design: The Current-Mode Approach, Peter Peregrinus, London, UK, 1990.

Zeki, A., Kuntman, H., "High-output-impedance CMOS dual-output OTA suitable for wide-range continuous-time filtering applications", Electronics Letters, 35,No.16, pp.1295–1296, 1999.

Chapter 2
Operational Transconductance Amplifiers (OTAs) and Their Applications

Basic Concept

In contrary to operational amplifiers exhibiting the behavior of a voltage-controlled voltage source where the gain is described by

$$K_{\mathrm{V}} = \frac{V_{\mathrm{O}}}{V_{I1} - V_{I2}} \tag{2.1}$$

the operational transconductance amplifier (OTA) behaves as a voltage-controlled current source defined by

$$G = \frac{I_O}{V_{I1} - V_{I2}} \tag{2.2}$$

OTA circuit symbol is illustrated in Fig. 2.1. In addition, some CMOS realization OTA topologies are shown in Fig. 2.2.

OTA-C structures have attracted considerable attention in recent years because they offer several advantages over conventional op-amp-based circuits as well as providing the evaluation of fully integrated circuits in VLSI design with CMOS technology. It is well known that OTAs provide highly linear electronic tunability of their transconductance (g_{m}) and require just a few or even no resistors for their internal circuitry and have more reliable high frequency performance because of the current-mode operation which has been established as an important topic in analog signal processing owing to its advantage over the voltage mode, particularly for higher frequency of operation.

Because of these features, the OTAs are increasingly replacing operational amplifiers, and in the past few years, a number of OTA-C-based filters and oscillators have been reported (Bhashkar et al. 1993; Çam et al. 1998; Düzenli et al. 1999, Erdogan et al. 2004; Ergün et al. 2001; Gönüleren et al. 1995). If a given transfer

H. Kuntman, D. Özenli, *Trends in Circuit Design for Analog Signal Processing*, Analog Circuits and Signal Processing,
https://doi.org/10.1007/978-3-030-96836-6_2

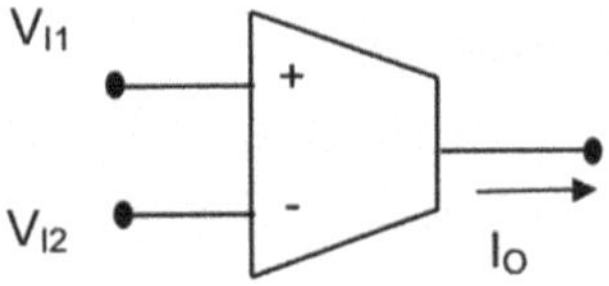

Fig. 2.1 OTA circuit symbol

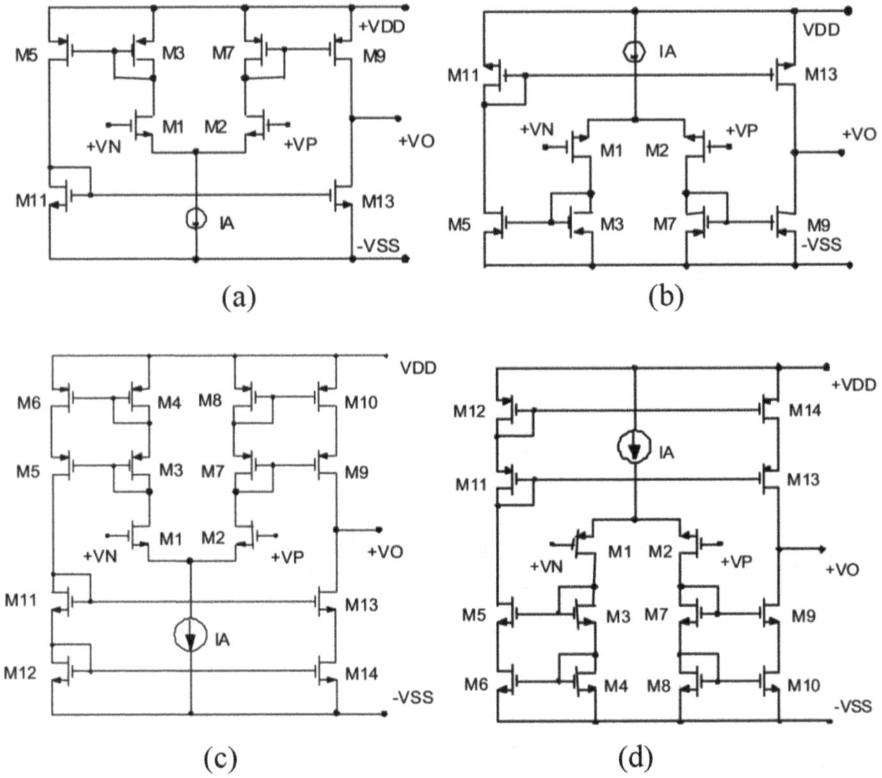

Fig. 2.2 CMOS OTA realization examples. (**a**) Symmetrical CMOS OTA with NMOS input and (**b**) with PMOS input, (**c**) symmetrical CMOS cascode OTA with NMOS input, and (**d**) symmetrical CMOS cascode OTA with PMOS input. (Allen and Holberg 2002; Gray et al. 2001)

function can be represented by the signal flow graph shown in Fig. 2.3a, the circuit realization can be given as shown in Fig. 2.3b. The aim is to present a synthesis procedure for the realization of biquadratic active filters using a minimum number of OTAs and grounded capacitors (Acar et al. 1993).

$$G(s) = \frac{a_2 s^2 + a_1 s + a_0}{s^2 + b_1 s + b_0} \tag{2.3}$$

where the relations between the transfer function coefficients and OTA transconductances and capacitances are given by

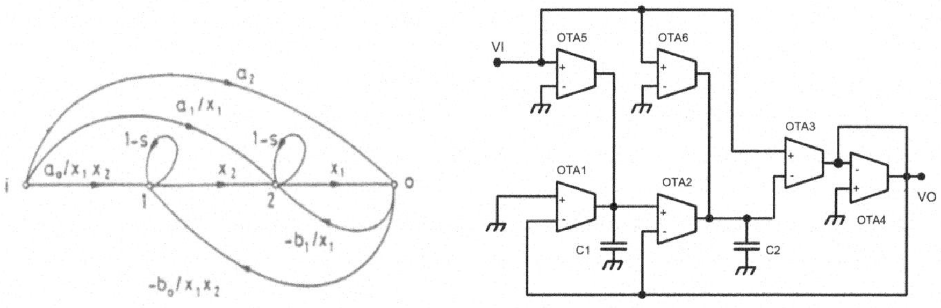

(a) (b)

Fig. 2.3 (**a**) Signal flowchart. (**b**) Realization circuit. (Acar et al. 1993)

Table 2.1 Transfer functions and basic relations of the filters illustrated in Fig. 2.4

Filter	Transfer function	Basic relations
Figure 2.4a Low-pass	$\frac{a_0}{s^2+b_1s+b_0}$	$\frac{g_{m1}}{C_1}=\frac{b_0}{b_1}$ $\frac{g_{m2}}{C_2}=b_1$ $\frac{g_{m3}}{C_1}=\frac{a_0}{b_1}$
Figure 2.4b High-pass	$\frac{a_0}{s^2+b_1s+b_0}$	$a_0=b_0$ $\frac{g_{m1}}{C_1}=\frac{b_0}{b_1}$ $\frac{g_{m2}}{C_2}=b_1$
Figure 2.4c Band-pass	$\frac{a_1s}{s^2+b_1s+b_0}$	$\frac{g_{m1}}{C_1}=\frac{b_0}{b_1}$ $\frac{g_{m2}}{C_2}=b_1$ $\frac{g_{m3}}{C_2}=a_1$
Figure 2.4d High-pass	$\frac{a_2s^2}{s^2+b_1s+b_0}$	$\frac{g_{m1}}{C_1}=\frac{b_0}{b_1}$ $\frac{g_{m3}}{g_{m4}}=a_2$ $\frac{g_{m2}}{C_2}=\frac{b_1}{a_2}$
Figure 2.4e Band-stop	$\frac{a_2s^2+a_0}{s^2+b_1s+b_0}$	$\frac{g_{m1}}{C_1}=\frac{b_0}{b_1}$ $\frac{g_{m2}}{C_2}=\frac{b_1}{a_2}$ $a_0=b_0$ $\frac{g_{m3}}{g_{m4}}=a_2$
Figure 2.4f Band-stop	$\frac{a_2s^2+a_0}{s^2+b_1s+b_0}$	$\frac{g_{m1}}{C_1}=\frac{b_0}{b_1}$ $\frac{g_{m2}}{C_2}=\frac{b_1}{a_2}$ $\frac{g_{m5}}{C_1}=\frac{a_0}{b_1}$ $\frac{g_{m3}}{g_{m4}}=a_2$
Figure 2.4g All-pass	$\frac{s^2-b_1s+b_0}{s^2+b_1s+b_0}$	$\frac{g_{m1}}{C_1}=\frac{b_0}{b_1},\ \frac{g_{m2}}{C_2}=\frac{b_1}{a_2},\ \frac{g_{m5}}{C_2}=b_1,\ \frac{g_{m3}}{g_{m4}}=1$

$$\frac{g_{m1}}{C_1}=\frac{b_0}{b_1},\ \frac{g_{m2}}{C_2}=\frac{b_1}{a_2},\ \frac{g_{m3}}{g_{m4}}=a_2,\ \frac{g_{m5}}{C_1}=\frac{a_0}{b_1},\ \frac{g_{m6}}{C_2}=\frac{a_1}{a_2} \tag{2.4}$$

Various second-order OTA-C filters with a minimum number of OTAs can be derived from the general circuit. These filters realize different filter characteristics such as LP, HP, BP, AP, and BS functions. These filters are illustrated in the following figures with the transfer functions given in Table 2.1.

Limitations for Input Signal

In the implementation of an OTA-C filter, the designer must be careful in the determination of the input signal. Actually, active components are nonlinear and they behave linear only under certain conditions. These conditions depend on design parameters of active components. Violation of these conditions causes nonlinear distortion in filters.

In practice OTAs do not operate linearly if their output signals exceed certain limits. If the output voltage of any OTA saturates, we get a clipped waveform. If the output current saturates, we get a sawtooth form, which is known as the slew-rate limiting problem.

In the following the maximum input signal level not causing clipping and slew-rate limiting is studied. For linear operation the input signal level must be adjusted so that

$$\begin{aligned} ||V_k| &\leq V_{ks}\ , \quad k = 1, 2, \ldots, n \\ I_k| &\leq I_{ks}\ , \quad k = 1, 2, \ldots, n \end{aligned} \tag{2.5}$$

Be simultaneously satisfied for the designer specified frequency band $\omega \in [\omega_1, \omega_2]$. Here n is the total number of OTAs used in the design, and $V_k = V_k(j\omega)$ and $I_k = I_k(j\omega)$ are the phasor voltage and current, respectively, at the output of kth =OTA. V_{ks} and I_{ks} are bounds of linear region and we call them the saturation voltage and current, respectively, of the kth OTA. In the case

$$\begin{aligned} V_{1s} &= \ldots = V_{ns} = V_s \\ I_{1s} &= \ldots = I_{ns} = I_s \end{aligned} \tag{2.6}$$

where the OTAs are identical, these bounds are equal.

In terms of the input voltage these conditions can be written as

$$\begin{aligned} |V_i| \cdot |H_k| &\leq V_{ks} = V_s\ , \quad k = 1, 2, .., n \\ |V_i| \cdot |Y_k| &\leq I_{ks} = I_s\ , \quad k = 1, 2, .., n \end{aligned} \tag{2.7}$$

where V_i is the amplitude of the filter's input voltage, $|H_k|$ is the voltage transfer function which is defined as the ratio of the kth OTA's phasor voltage to phasor input voltage, and $|Y_k|$ is the transfer admittance function which is defined as the ratio of the kth OTA's phasor output current to phasor input voltage.

There exist 2n inequalities. They put the following constraints on the input voltage amplitude for $\omega \in [\omega_1, \omega_2]$:

$$\begin{aligned} |V_i| &\leq = \frac{V_s}{|H_k|}\ , \quad k = 1, 2, .., n \\ |V_i| &\leq \frac{I_s}{|Y_k|}\ , \quad k = 1, 2, .., n \end{aligned} \tag{2.8}$$

The common solution of these inequalities which gives the maximum value of the input voltage amplitude not causing clipping and slew-rate limiting can be expressed as

$$|V_i|_{\text{maks}} = \min\left(\frac{V_s}{|H_k(j\omega)|_{\text{maks}}}, \frac{I_s}{|Y_k(j\omega)|_{\text{maks}}}\right),\quad (2.9)$$
$$k = 1..n$$

where $H_{k\text{maks}}$ and $Y_{k\text{maks}}$ are maximum values of H_k and Y_k, respectively, for the designer specified band.

Design Example

The method is demonstrated on a 3 MHz Butterworth LP filter design illustrated in Fig 2.4b as follows:

The OTAs are designed using the simple symmetrical CMOS OTA structure shown in Fig. 2.5. To get a transconductance of 1.33 mA/V, the amplifier bias current is chosen as $I_B = 336$ μA. The limits are specified as $V_S = 3.27$ v, $I_S = 560$ μA; the capacitances are determined as $C_1 = 100$ pF, $C_2 = 50$ pF. The related voltage transfer functions and the admittance functions are

$$\begin{aligned}
H_1 &= \frac{V_1}{I_1} = \left(1 + j\omega\frac{Q_p}{\omega_p}\right).H_2(j\omega) \\
H_2 &= \frac{V_2}{I_2} = \frac{{\omega_p}^2}{(j\omega)^2 + \frac{Q_p}{\omega_p}(j\omega) + {\omega_p}^2} \\
Y_1 &= \frac{I_1}{V_1} = g_m(1 - H_2) \\
Y_2 &= \frac{I_2}{V_2} = g_m(H_1 - H_2)
\end{aligned}\quad (2.10)$$

Using these equations and the linearity boundaries of active elements, we obtain

$$\begin{aligned}
|H_1(j\omega)|_{maks} &= 1.029 \\
|H_2(j\omega)|_{maks} &= 1 \\
|Y_1(j\omega)|_{maks} &= 1632\,\mu A/V \\
|Y_2(j\omega)|_{maks} &= 666\,\mu A/V
\end{aligned}$$

The maximum input voltage value of the filter can be calculated as

$$|V_I|_{maks} = 0.34$$

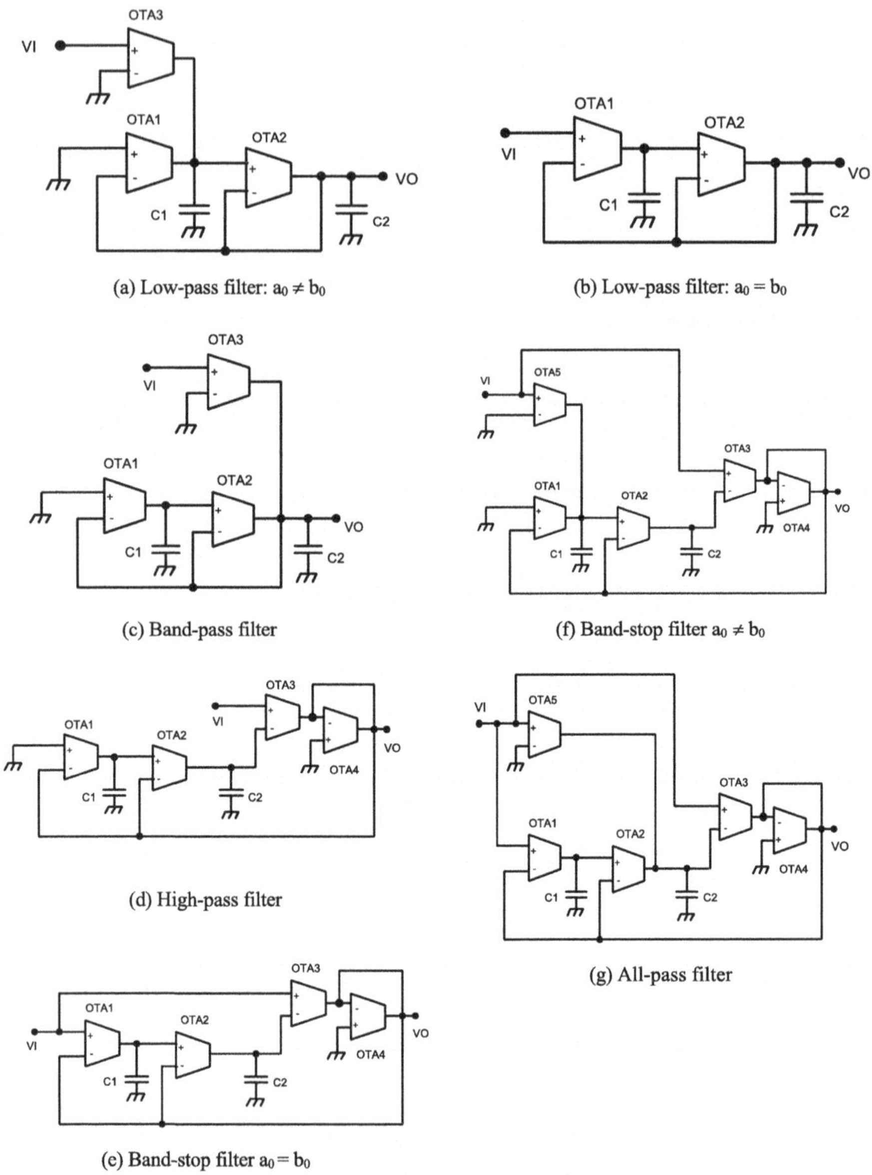

(a) Low-pass filter: $a_0 \neq b_0$

(b) Low-pass filter: $a_0 = b_0$

(c) Band-pass filter

(f) Band-stop filter $a_0 \neq b_0$

(d) High-pass filter

(g) All-pass filter

(e) Band-stop filter $a_0 = b_0$

Fig. 2.4 OTA-C-based low-pass, band-pass, high-pass, band-stop, and all-pass active-filter structures. (Acar et al. 1993)

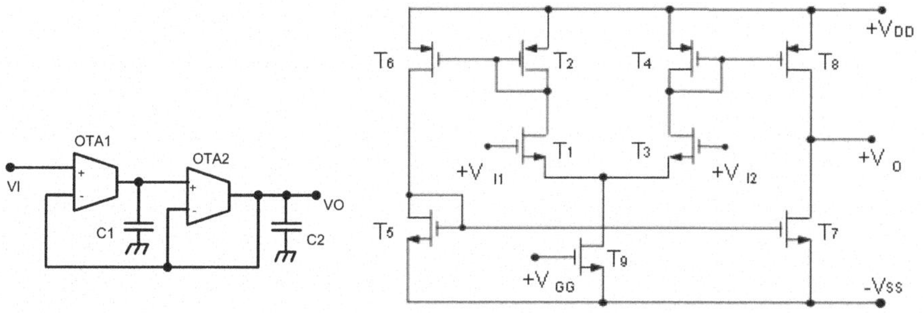

Fig 2.5 LP filter derived from Fig. 2.4b and simple symmetrical CMOS OTA structure. (Acar et al. 1993)

Application Example: Design of EEG Filters for Biomedical Applications

The rapid increasing use of battery-operated portable equipment in application areas such as telecommunications and medical electronics imposes the use of low-power and small-sized circuits realized with VLSI (very large-scale integrated) technologies. CMOS (complementary metal–oxide semiconductor) circuits operating in the subthreshold (weak inversion) region introduce a versatile solution for the realization of low-power VLSI building blocks.

Circuits needed for processing of biological signals are a typical and good example of low-power and small-sized building blocks. While carrying out various functions, certain body systems generate their own monitoring signals. These signals convey useful information about their activities and are called biological signals. The main features of biological signals are their low amplitude and low frequency range. In many cases these signals are modified by the simultaneous actions of other organs. Hence, they may not contain direct messages about the parameters of the system under consideration. In order to obtain useful information, these signals should be processed.

The human electroencephalogram (EEG), which provides a rich picture of the electrical activities of the brain, is one of the most important biological signals (Clark 1992). The voltage amplitudes of EEG signals range from about 1 to 100 mV peak to peak at low frequencies (0.5–100 Hz) at the cranial surface.

It is possible to realize low-frequency OTA-C active filters with small capacitance values of the order of 25–400 pF. The circuit technique described is applied to the α (8–12 Hz), β (13–40 Hz), θ (4–8 Hz), and δ (1–4 Hz) band filters for EEG signals. Because of small capacitance values, the filter circuit is suitable for realization on a single VLSI chip using the CMOS technology and enables the user to implement the circuit on implantable biotelemetric applications.

The filter chip is fabricated in Turkish Scientific and Technological Council (TUBITAK) laboratory. The realized filter topology, filter frequency responses, CMOS OTA structure, chip layout, capacitance values, biasing currents, and OTA

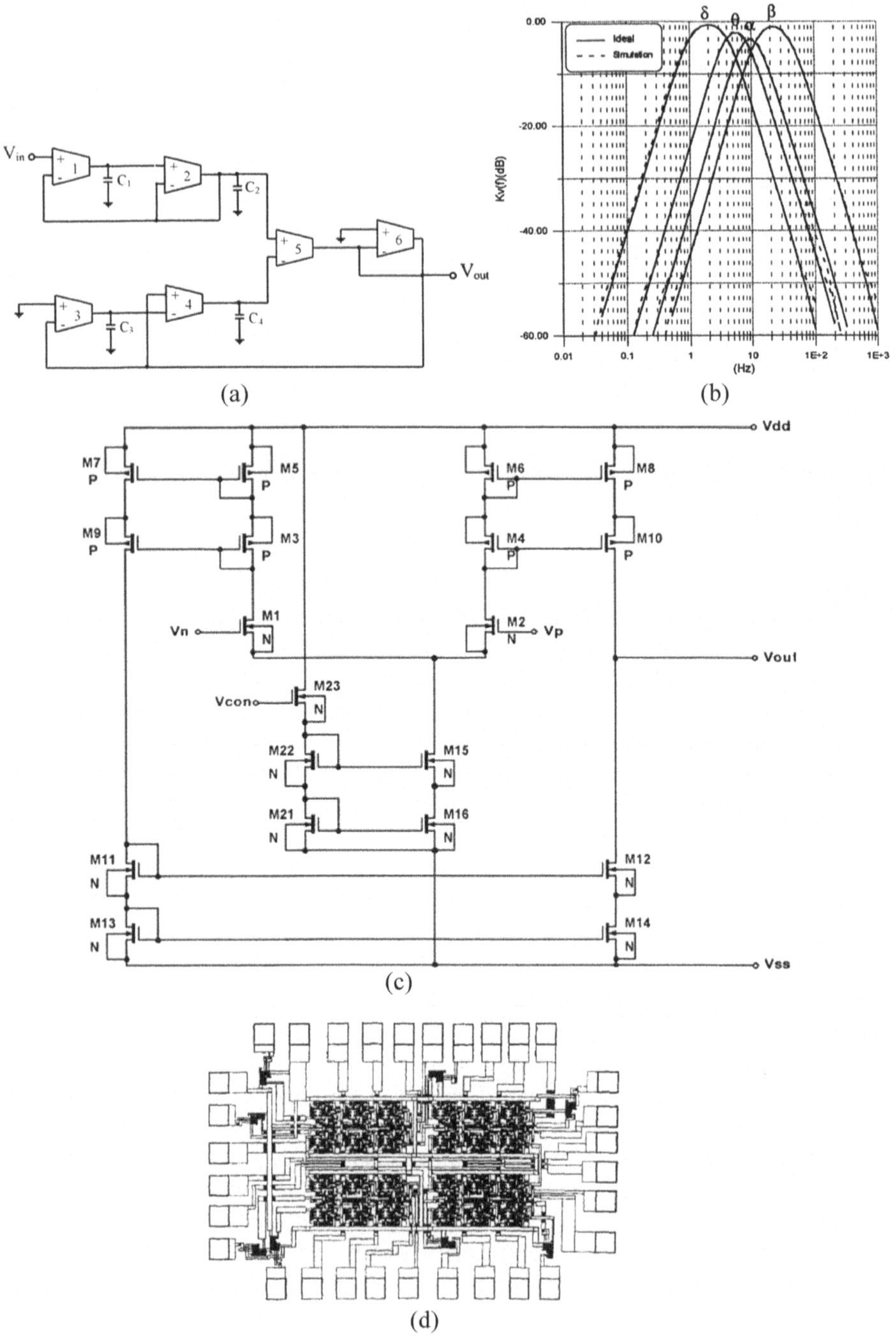

Fig. 2.6 (**a**) Fourth-order OTA-C-based EEG filter, (**b**) frequency responses, (**c**) CMOS OTA, (**d**) and filter layout. (Düzenli et al. 1999)

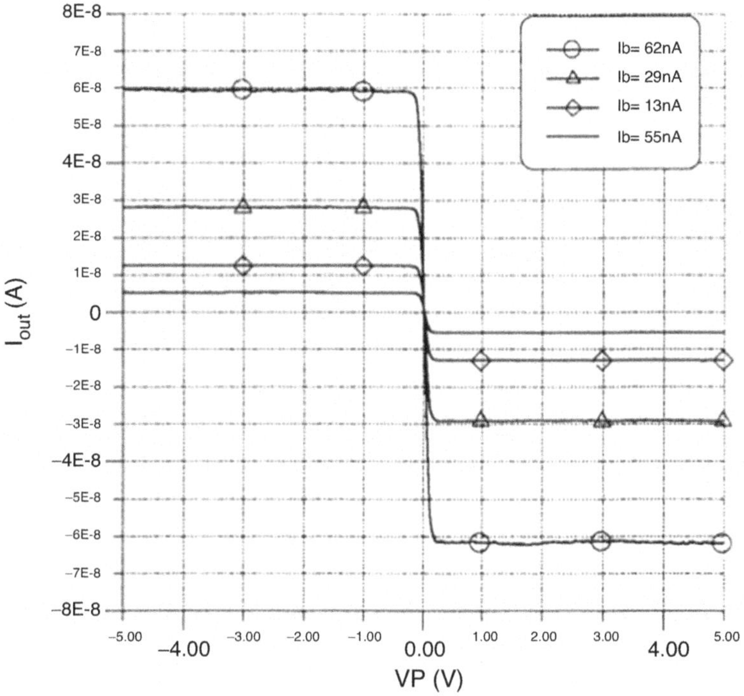

Fig. 2.7 Dependence of OTA output current on input differential voltage for several biasing current values in the subthreshold region, obtained by measurements with parameter analyzer HP 4145A. (Düzenli et al. 1999)

transconductances are shown in Fig. 2.6 (Düzenli et al. 1999). Furthermore, the measured characteristics of the OTA designed are illustrated in Figs. 2.7, 2.8, 2.9, 2.10, and 2.11 (Table 2.2).

Measured variations of the output current against input voltage for the subthreshold regions are illustrated in Fig. 2.7. The measured variations of OTA transconductance, G_m, with the control voltage, V_{con}, and with the biasing current, I_B, are given in Fig. 2.8.

The measured frequency response of OTA transconductance operating in the subthreshold region is given in Fig. 2.9. The measured frequency response of band-pass OTA-C filter operating in the subthreshold region is illustrated in Fig. 2.10. Experimental output waveforms of the filter for $V_{in} = V_{inmax} = 80$ mV are given in Fig. 2.11.

The most important result obtained in this work is the possibility of constructing the EEG band filters with small capacitance values of the order of 25–400 pF, which cannot be obtained by conventional active-filter techniques where capacitance values of the order of several uF are necessary for this frequency range. This is achieved by the use of CMOS OTA structures operating in the subthreshold region. Owing to the small capacitance values, the complete filter circuit is realizable on a single VLSI chip using the CMOS technology. The IC realization of the complete filter will enable the user to implement the circuit on implantable biotelemetric applications where low power consumption and small-sized capacitors are required.

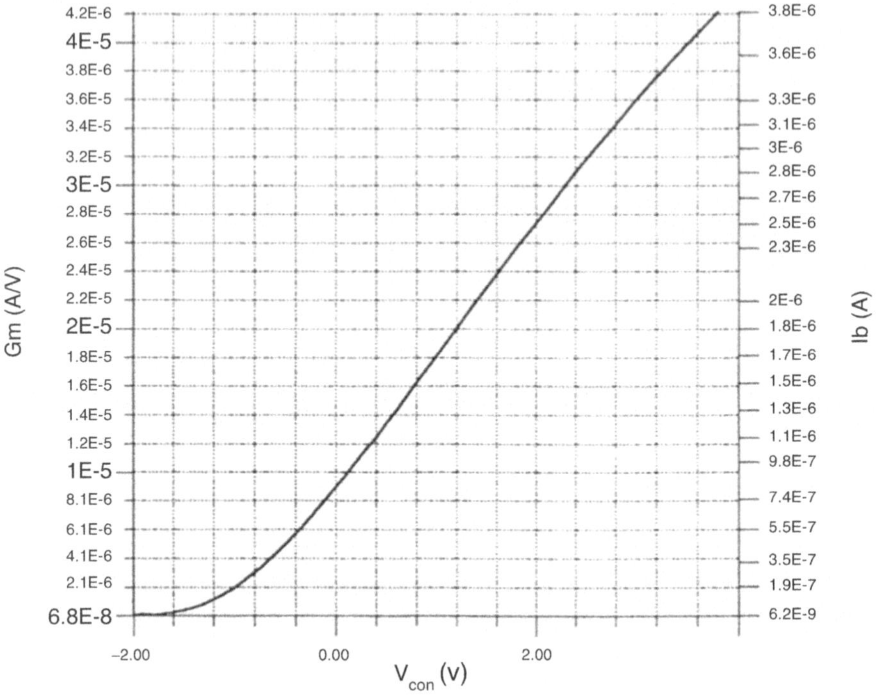

Fig. 2.8 Measured variation of OTA transconductance G_{m} with control voltage Vcon and biasing current. (Düzenli et al. 1999)

DO-OTA: Dual-Output Operational Transconductance Amplifier

The circuit symbol of the DO-OTA (dual-output operational transconductance amplifier) is given in Fig. 2.11a. Ideally, DO-OTA is assumed as an ideal voltage-controlled current source and can be described by the following equations:

$$I_0{}^{+} = g_{m1}(V^{+} - V^{-}) \quad I_0{}^{-} = g_{m2}(V^{-} - V^{+}) \tag{2.11}$$

Generally, the transconductances of the DO-OTAs are chosen as $g_{m1} = g_{m2}$ similar to the topologies described in this work. For some applications unequal $g_{m1} = g_{m2}$ values are obtained by taking different width and length values for MOS transistors of the output stages. Figure 2.12b illustrates the realization of the DO-OTA. From Fig. 2.12b it can be easily observed that the dual-output operational transconductance amplifier consists of two main blocks, namely, the differential input stage and the output stage. A simple CMOS realization circuit is shown in Fig. 2.12c.

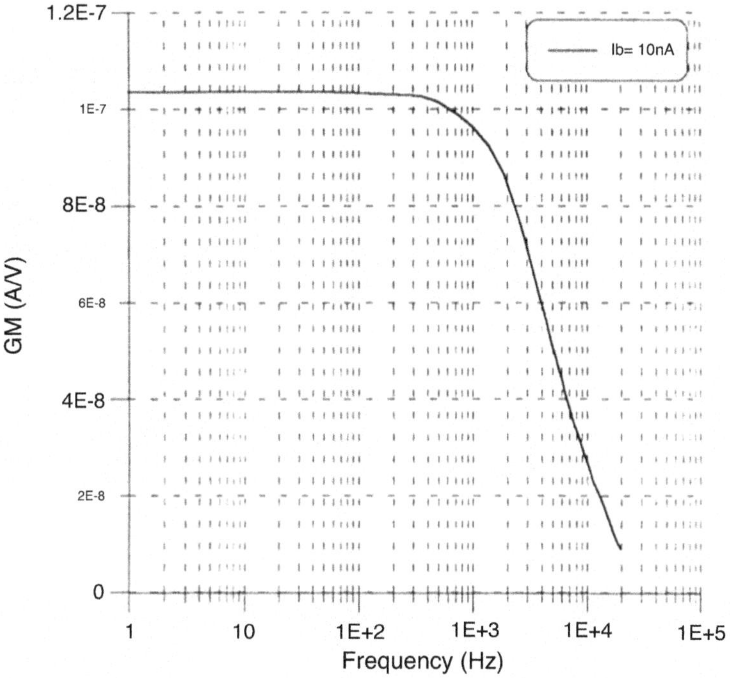

Fig. 2.9 Measured frequency response of OTA transconductance in the subthreshold region. (Düzenli et al. 1999)

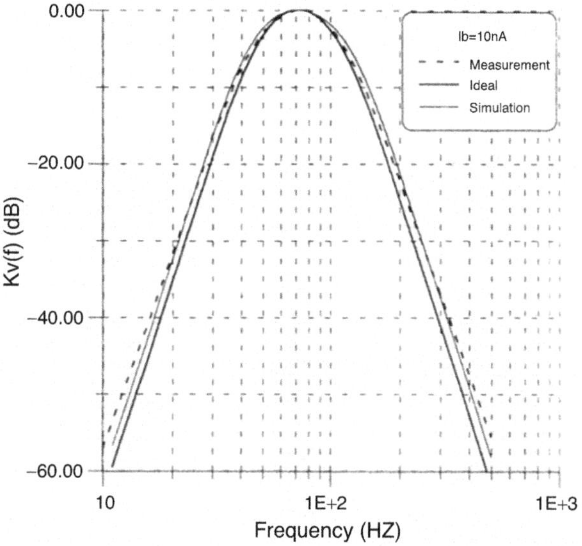

Fig. 2.10 Measured frequency response of band-pass OTA-C filter operating in the subthreshold region. (Düzenli et al. 1999)

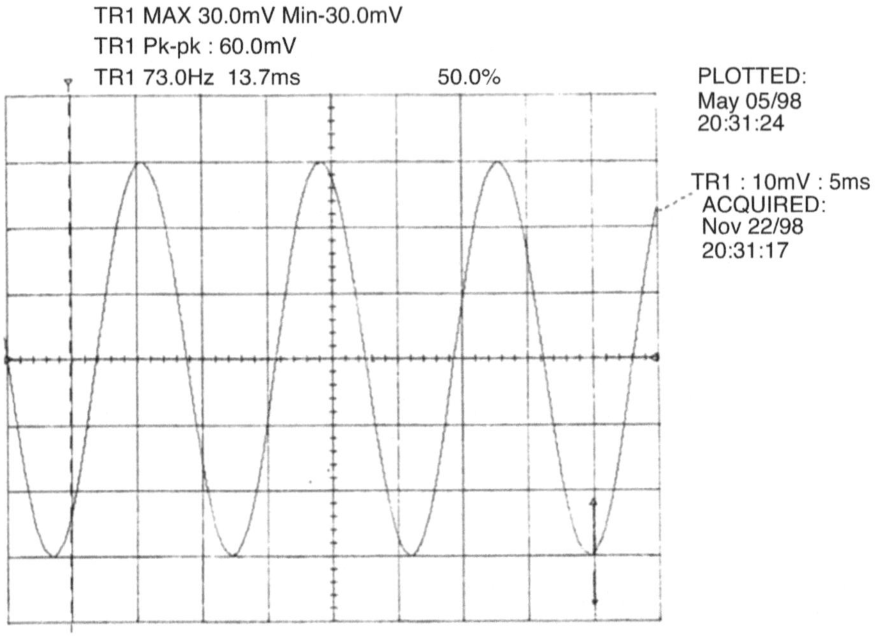

Fig. 2.11 Experimental output waveforms: (a) for $V_{in} = V_{inmax} = 80$ mV. (Düzenli et al. 1999)

Table 2.2 Capacitance values, biasing currents, and OTA transconductances in Fig. 2.5

Band	f_{P1} (Hz)	f_{P2} (Hz)	C_1 (pF)	C_2 (pF)	C_3 (pF)	C_4 (pF)	I_B (nA)	$G_{mi}(nA/V)$
A	12	8	166	81	250	122	0.69	8.8
B	40	13	81	40	250	122	1.125	14.3
Θ	8	4	125	61	250	122	0.347	4.4
Δ	4	1	62.55	30.6	250	122	0.1	1.1

Input Stages

To compare DO-OTA linearized input stages, one classical input stage and three other linearized input stages are given at this work. The difference between these circuits is given by total harmonic distortion (THD) analysis.

The main problems of using these types of active elements in active circuits are:

- Linear operating range limitation.
- Limited impedances of the high output impedance.
- Active components behave linear only under certain conditions.
- Violation of these conditions results in nonlinear distortion in filters.

Important drawbacks in the realization of active elements such as OTA and DO-OTA, CCIIs, etc. are:

- The finite output resistance.
- Limited bandwidth.

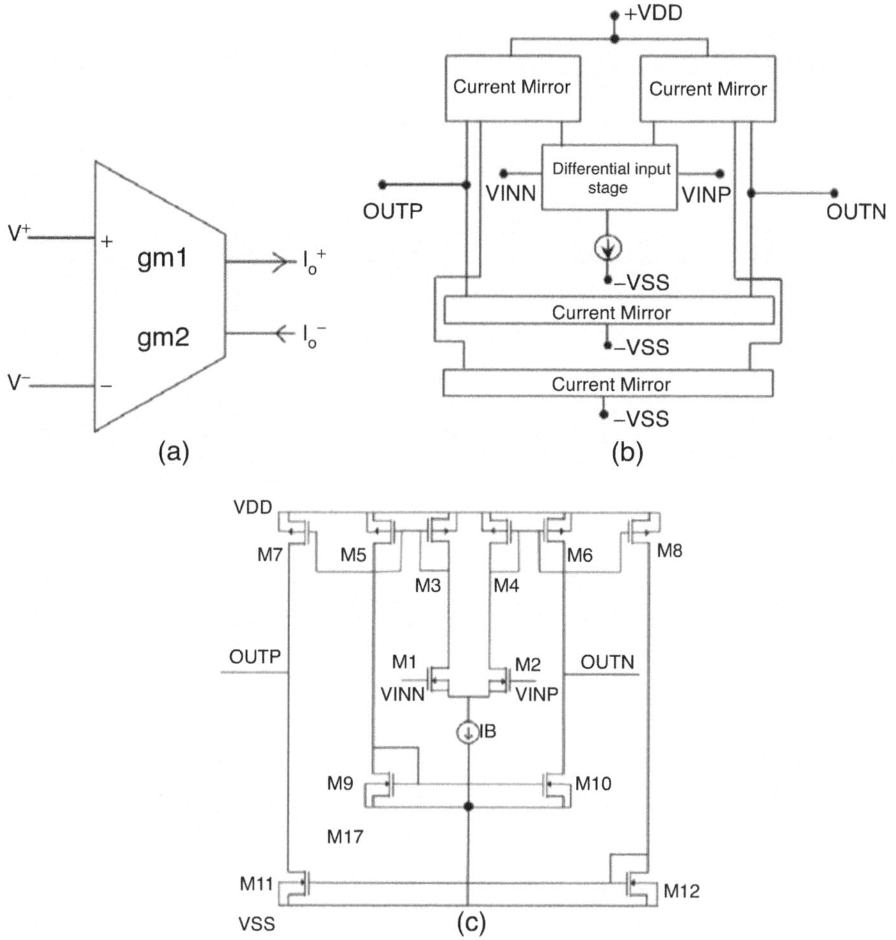

Fig. 2.12 (**a**) Circuit symbol of the DO-OTA, (**b**) realization of DO-OTA, and (**c**) simple CMOS realization: symmetrical CMOS DO-OTA. (Ergün and Kuntman 2005)

Fig. 2.13 OTA-C-based integrator. (Zeki and Kuntman 1996, 1999)

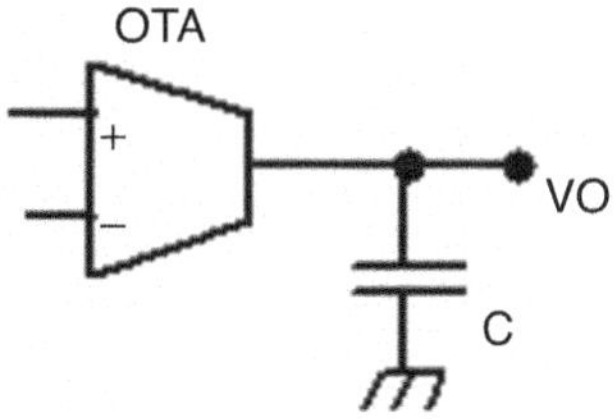

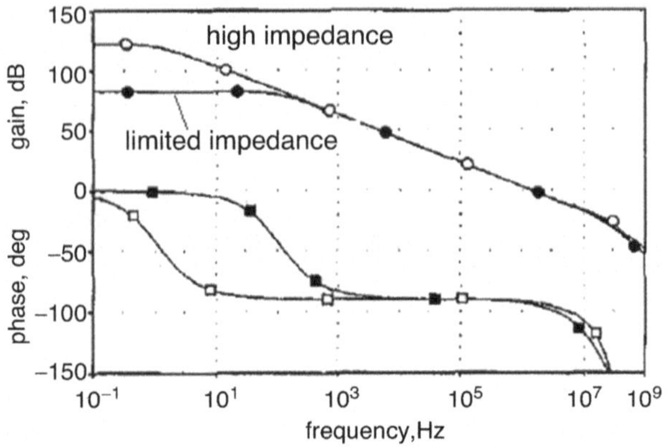

Fig. 2.14 Influence of output impedance on integrator characteristics. (Zeki and Kuntman 1996, 1999)

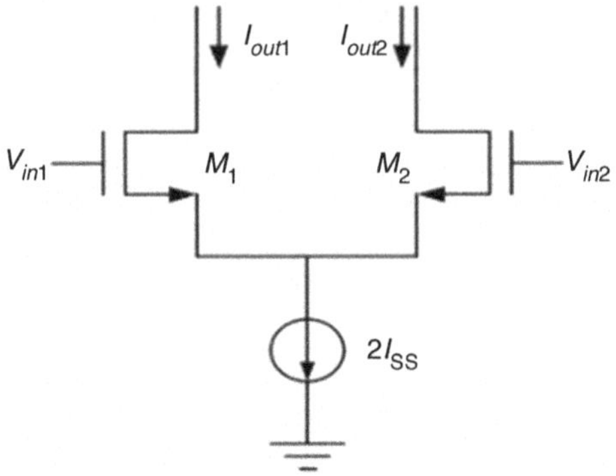

Fig. 2.15 Conventional MOS differential pair exhibiting limited input linearity range

- Limited input linearity range.
- The finite output resistance is the cause of a lossy integration.
- Output stages exhibiting very high output impedance are needed to enable filtering at low frequencies and to reduce filtering errors. As an example, OTA-C-based integrator and the influence of output impedance on integrator characteristics are illustrated in Figs. 2.13 and 2.14.

The conventional MOS differential pair in Fig. 2.15 exhibits limited input linearity range. Linearization techniques for input cells with extended linearity range are:

- Nedungadi–Visvanathan input circuit
- Cross-coupled input circuit
- Krummenacher's input circuit

The aspect ratios of M_3 and M_4 are n times larger than M_1 and M_2. Comparison of equations shows clearly the improvement obtained (Figs. 2.16 and 2.17).

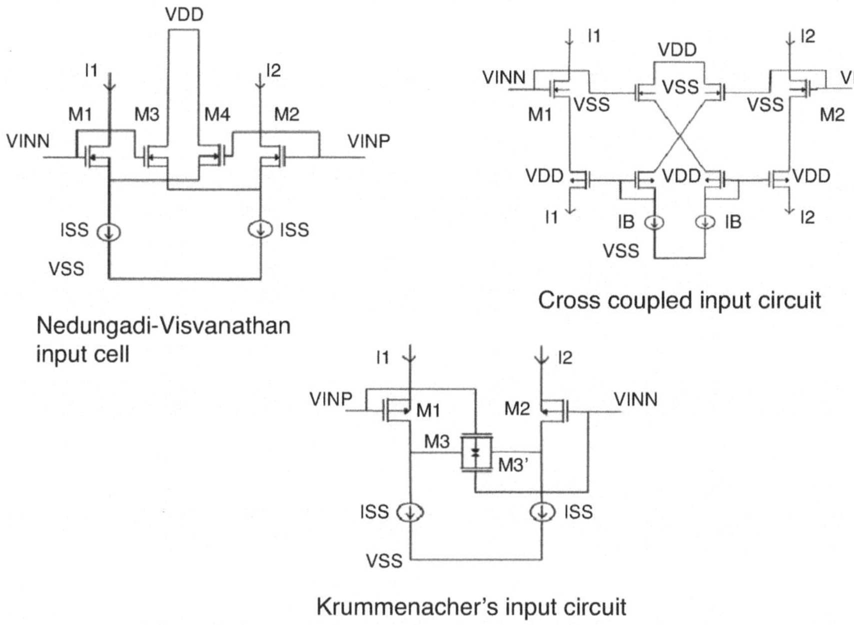

Fig. 2.16 Linearized input stages. (Nedungadi and Viswanathan 1984; Krummenacher and Joehl 1988; Seevinck and Wassenaar 1987)

High Output Impedance Stages

The use of current-output-based active devices (COBADs), such as OTAs, current conveyors, etc. in continuous-time filter design, has been attracting a large amount of interest, especially because of their wider bandwidths with respect to those of equivalent op-amps. An important drawback of COBAD-based continuous-time filters is the finite output resistance (R_{out}) of a COBAD, which, in a basic integrator structure, is in parallel with the load capacitor $\boldsymbol{C}_{\mathbf{L}}$ causing a lossy integration, thus generating filtering errors (Khorramabadi and Gray 1984). Classical cascode current output stages cannot handle a load capacitor of the order of several picofarads at low frequencies (e.g., for $f < 1$ kHz): therefore, integration cannot be performed beyond this limit (Zeki and Kuntman 1997). Choosing large C values (e.g., several nanofarads) is not a reasonable solution, since this requires very large areas on a chip. Therefore, very high output impedance current output stages are required, both to enable filtering at low frequencies and to reduce filtering errors (Fig. 2.18).

- The conventional active-feedback cascode current mirror (CAFCCM) achieves a much larger output impedance and a relatively wider output voltage swing than those of the classical cascode current mirror.
- High output impedance is achieved by the active negative feedback through the amplifier stage I_K-M_K and the source follower M_3.

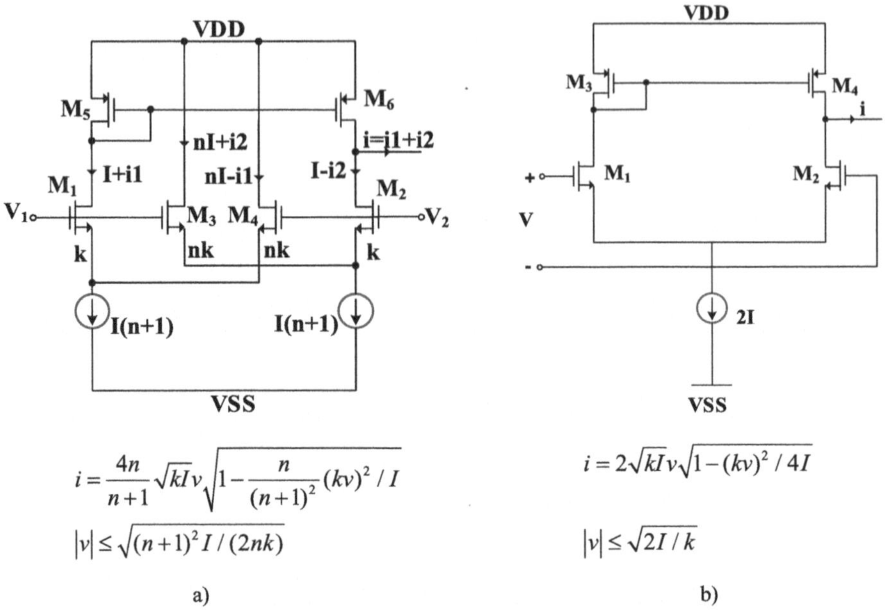

Fig. 2.17 Comparison of Nedungadi–Visvanathan input cell with conventional differential input cell. (**a**) Nedungadi–Visvanathan input cell. (**b**) Conventional differential input cell. (Nedungadi and Viswanathan 1984)

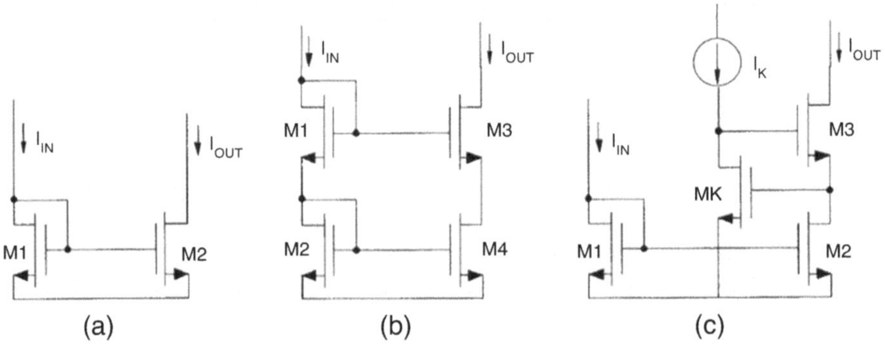

Fig. 2.18 (**a**) Simple current mirror, (**b**) classical cascode current mirror, and (**c**) conventional active-feedback cascode current mirror. (Zeki and Kuntman 1998)

- I_K and MK must be chosen such that $V_{DS2} = V_{GS2}$ is satisfied; otherwise transfer accuracy is rather lower than that of a classical cascode current mirror, due to the channel length modulation effect.

If the input current I_{IN} is not fixed, then two problems arise:

(a) $V_{DS2} = V_{GS2}$ *cannot be achieved, except in a single case (i.e., when* $V_{GSK} = V_{GS1}$*) degrading transfer accuracy.*

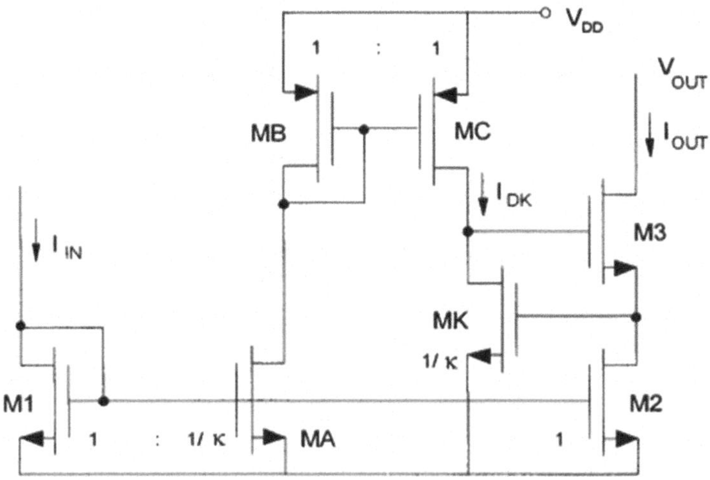

Fig. 2.19 Improved active-feedback cascode current mirror. (Zeki and Kuntman 1996, 1999)

(b) If I_{IN} *(thus* V_{GS2}*) is large enough to drive M2 into the triode region (because* of fixed V_{DS2}*), then the current mirror does not operate properly.*

- The simplest method to achieve $V_{DS2} = V_{GS2}$ is to choose M_K matched with M_1 and set $I_K = I_{IN}$.
- This is provided by improved active-feedback cascode current mirror (Fig. 2.19).

The transfer error of the circuit can be kept very low, *if* $I_{DK} = I_{D1}$ *is achieved satisfactorily.*

- The improved active-feedback cascode current mirror achieves a very high output impedance over active negative feedback through amplifier M_K-M_C and source follower M_3.
- Precision is maintained by making I_K dependent on I_{IN} to achieve $V_{GSK} = V_{GS1}$ equality (thus, $V_{DS2} = V_{DS1}$ equality), for any input current level.
- To save power and area, M_A and M_K can be relatively smaller than M_1, i.e., $(W/L)_A = (W/L)_K = (W/L)_1/\kappa$, where $\kappa > 1$ (Fig. 2.20).

CMOS Circuits Employing the High-Performance Input and Output Stages

Combining high-performance input and output stages and dual-output operational transconductance amplifier and differential difference current conveyor (DDCC) is constructed as examples.

The performances of the CMOS DO-OTA topologies are demonstrated by SPICE simulations. For the simulations AMS 0.8 μm MOS models are used. The supply

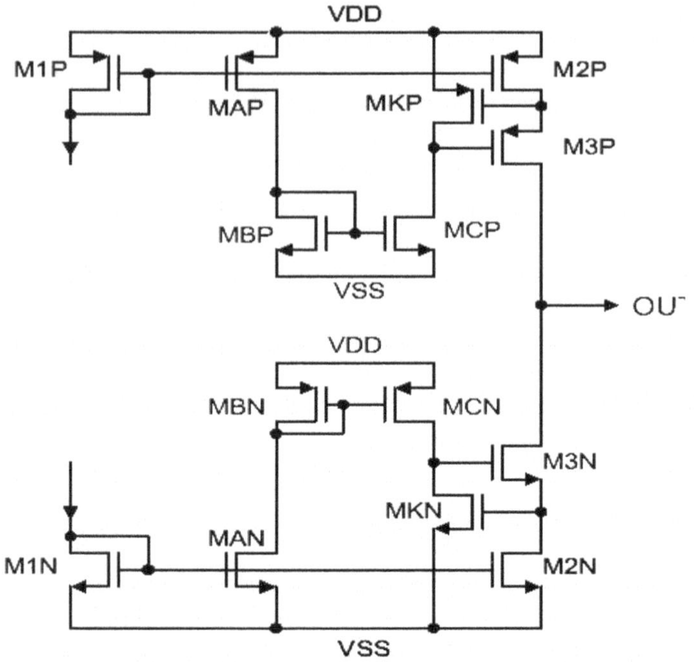

Fig. 2.20 Output stage employing active-feedback cascode current mirrors. (Zeki and Kuntman 1996, 1999)

voltages are taken as 2.5 V and –2.5 V. SPICE simulations were performed for circuit characterizations and for a design example of a BP active filter. The results obtained for the circuits are compared with each other. The simulated characteristic plots of DO-OTA structures employing conventional and Nedungadi input differential pairs and the output stage of Fig. 2.21 are given in Fig. 2.22.

To compare the linearity regions of DO-OTA circuits proposed, the integrator of Fig. 2.23 is used as test circuit. Capacitors are taken 1 nF for all tests. Connecting the inverting input to the reference node a sinusoidal input voltage of 1 kHz is applied to the noninverting circuit and the signal amplitude is varied in the range of 0–2 V. For each step the total harmonic distortion at the outputs is calculated and given in Fig. 2.24. From Fig. 2.24 it can be easily observed that, compared to the conventional differential pair, the Nedungadi–Visvanathan, Krummenacher, and cross-coupled input stages exhibit lower harmonic distortion at the output where the Nedungadi–Visvanathan circuit exhibits the best performance among the four circuits.

The performance of the CMOS DO-OTA circuits proposed is also demonstrated on a design example of a current-mode BP filter shown in Fig. 2.25. The test filter is a current-mode version of the voltage-mode OTA-C filter given previously in the literature (Ergün 2001; Zeki 1997).

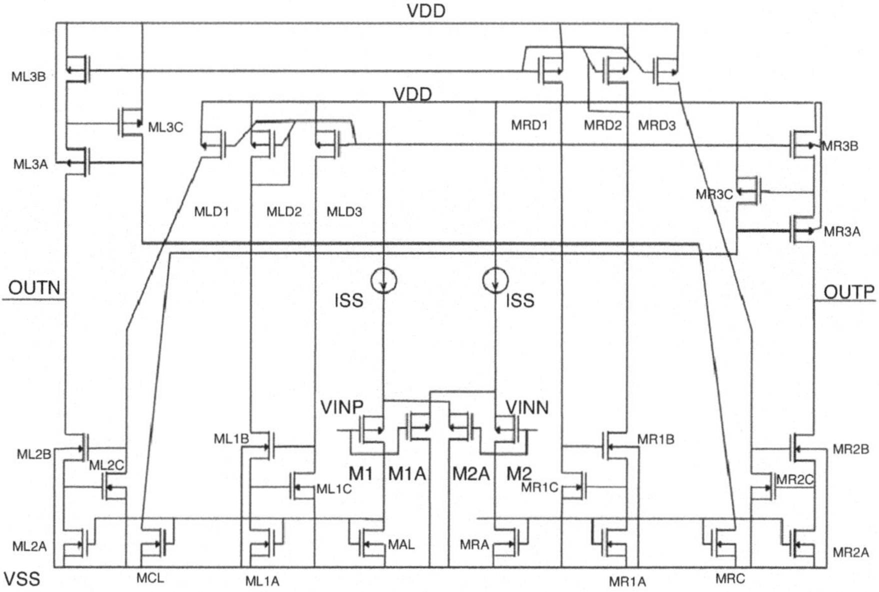

Fig. 2.21 DO-OTA circuit employing the active-feedback output stage and Nedungadi–Visvanathan input stage. (Ergün and Kuntman 2000, 2001, 2005)

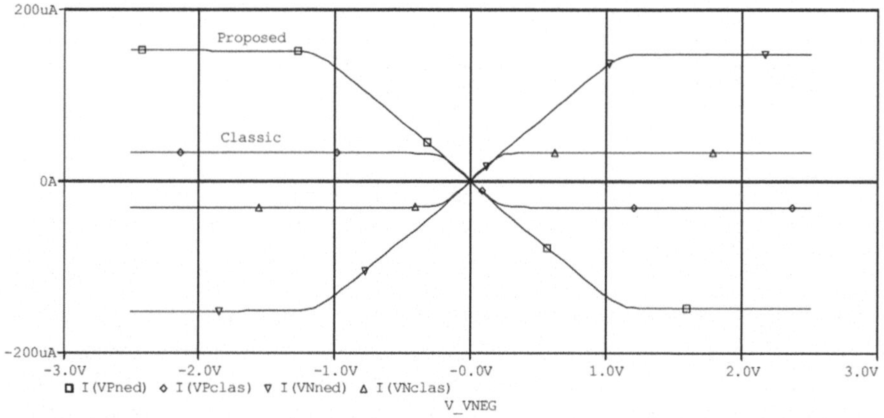

Fig. 2.22 Input linearity range of conventional and high-performance OTA. (Ergün and Kuntman 2000, 2001, 2005)

SPICE simulations were performed by using the DO-OTAs employing the output stage illustrated in Fig. 2.20. Besides the conventional differential input stage, Nedungadi–Visvanathan, Krummenacher, and cross-coupled input stages are also used for simulations. The resulting frequency responses of BP filters are investigated. Each of the filters realizes Butterworth response with a pole frequency of

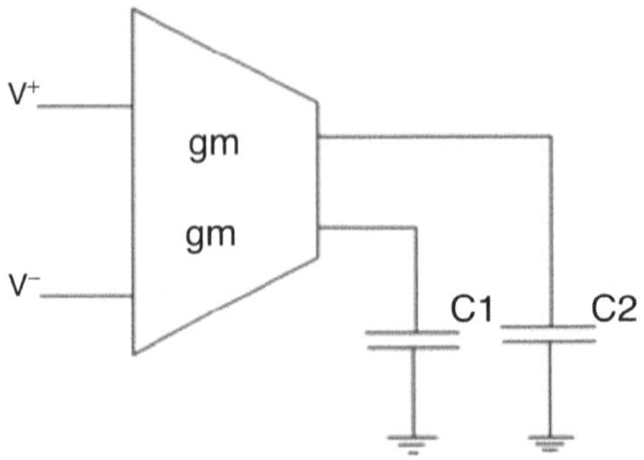

Fig. 2.23 Lossy integrator employing DOTA. (Ergün and Kuntman 2000, 2001, 2005)

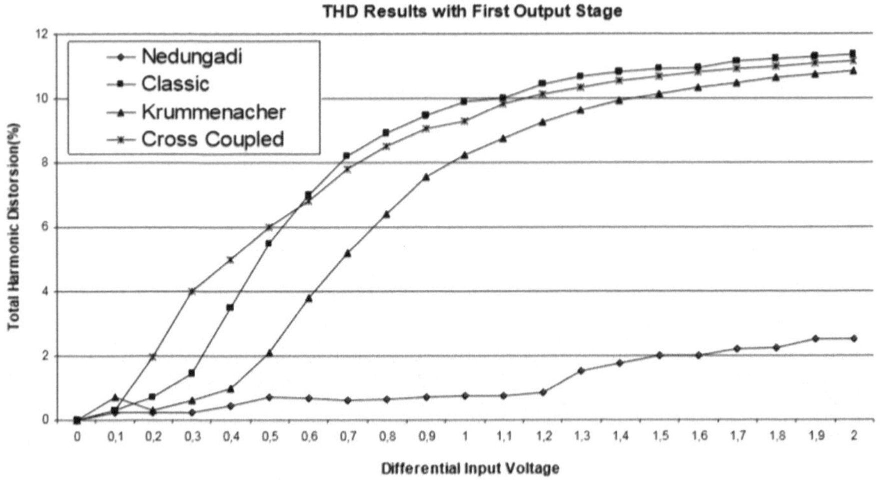

Fig. 2.24 Dependence of total harmonic distortion (THD) on input voltage obtained for conventional, Nedungadi–Visvanathan, Krummenacher, and cross-coupled input stages. (Ergün and Kuntman 2000, 2001, 2005)

1 MHz. For simulations the basic quantities are chosen as $g_m = 150$ μA/V, $C_1 = 30$ pF, and $C_2 = 15$ pF. Figure 2.26 illustrates the frequency responses of BP filters realized with conventional symmetrical CMOS-DOTA in Figure 2.12b and the CMOS DO-OTA in Fig. 2.21. To provide simplicity, only the results of one new CMOS-DO-OTA realization are shown in Fig. 2.26. The other CMOS-DO-OTA realizations yield similar improved results. It is observed that all of the filter characteristics agree with the ideal filter responses in a wide frequency range. The deviations at high frequencies can be considered the result of limited DO-OTA bandwidth. The deviations at low frequencies in BP filter characteristics are caused by the limited DO-OTA output resistances. However, these deviations appear below noise level and have no importance from the point of view of filter function. It can be easily observed that the topologies proposed provide an extended improvement due to the increased output resistance.

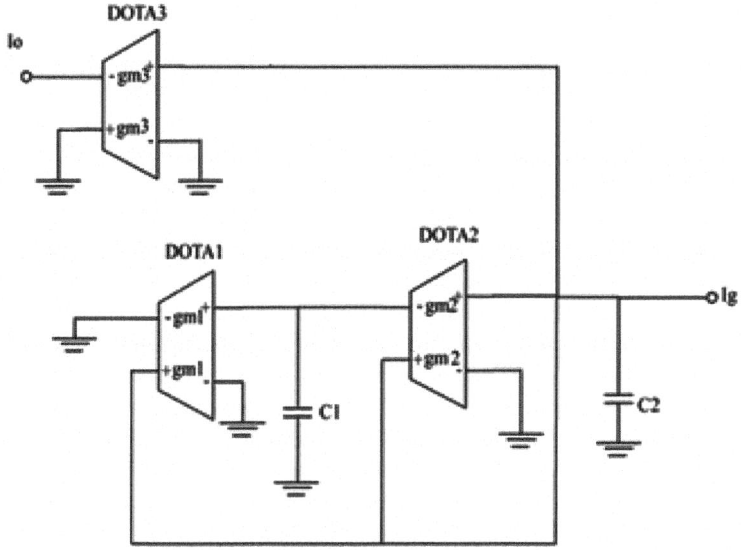

$$H(s) = \frac{a_1 s}{s^2 + b_1 s + b_0}$$

where

$$\frac{g_{m1}}{C_1} = \frac{b_0}{b_1}, \qquad \frac{g_{m2}}{C_2} = b_1, \qquad \frac{g_{m3}}{C_2} = a_1$$

$$b_1 = \frac{\omega_p}{Q_p} \qquad b_0 = \omega_p{}^2$$

$$\omega_p = \frac{\sqrt{g_{m1}} \cdot \sqrt{g_{m2}}}{\sqrt{C_1} \cdot \sqrt{C_2}} \qquad Q_p = \frac{\sqrt{g_{m1}} \cdot \sqrt{C_2}}{\sqrt{g_{m2}}\sqrt{C_1}}$$

Fig. 2.25 Current-mode BP filter. (Ergün and Kuntman 2000, 2001, 2005)

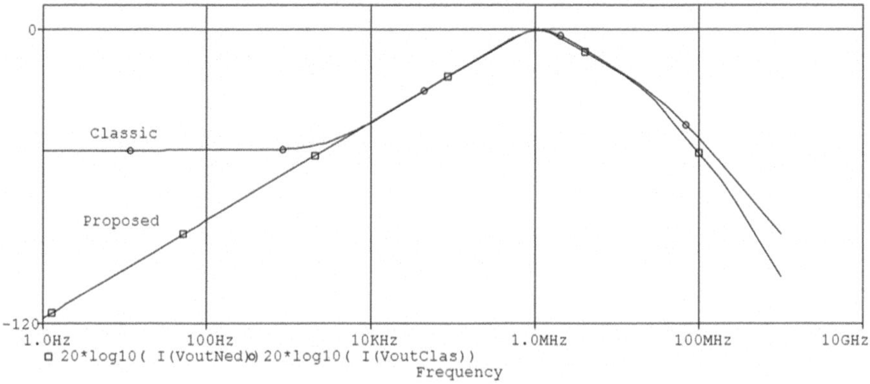

Fig. 2.26 Simulated frequency responses of BPF circuits constructed with classical and proposed DO-OTA structures. (Ergün and Kuntman 2000, 2001, 2005)

OTA-C Oscillators

Oscillators have a wide range of use in the areas of telecommunications, control systems, signal processing, and measurement systems. Up to now, for the design of a tunable sinusoidal oscillator, a variety of active op-amp-RC networks have been reported. However, these networks do not operate at high frequencies and effectively do not have the frequency tuning parameter. On the other hand, OTA-C oscillator networks have some advantages such as open-loop tunability of OTA, compatibility to VLSI, and adjustment ability of the oscillation frequency by transconductance of the OTA. Since transconductance of an OTA (operational transconductance amplifier) is a function of its tail current, the oscillation frequency can be easily adjusted via this current. Therefore, OTAs are increasingly being used in place of operational amplifiers as active devices in analog circuit design (Sanchez-Sinencio et al. 1988; Linares-Barronco et al. 1992; Acar et al. 1993; Bashkar and Senani 1994; Gonuleren et al. 1995; Cam 1996). A general sinusoidal oscillator circuit has two conjugate poles on the imaginary axis and can be described by a second-order characteristic equation, as follows:

$$\left(s^2 + bs + V_0{}^2\right)V_{out} = 0 \tag{2.12}$$

where $b = 0$ is the oscillation condition and V_0 is the oscillation frequency. The basic aim of OTA-C oscillator design is to achieve noninteractive control of band V_0 with a minimum number of components.

The general biquadratic transfer function is given as follows:

$$H(s) = \frac{V_0(s)}{V_i(s)} = \frac{a_2 s^2 + a_1 s + a_0}{s^2 + b_1 s + b_0} \tag{2.13}$$

There are two possible methods of obtaining the sinusoidal oscillator from this transfer function. In the first method the characteristic equation of the oscillator is obtained by equating the input voltage of the filter Vin(s) to zero. In this case the following oscillator characteristic equation is obtained:

$$\left(s^2 + b_1 s + b_0\right)V_0(s) = 0 \tag{2.14}$$

The second method is based on connecting the output terminal of a filter to the input terminal (Senani et al. 1991). By doing so, the resulting oscillator characteristic equation is expressed as

$$\left(s^2 + \left(\frac{b_1 - a_1}{1 - a_2}\right)s + \left(\frac{b_0 - a_0}{1 - a_2}\right)\right)V_0(s) = 0 \tag{2.15}$$

If the oscillation condition is satisfied by equating the coefficient of s to zero, these two equations yield undamped oscillators. In this study, six OTA-C sinusoidal oscillator structures, two of them known from the literature (Sanchez-Sinencio et al. 1989; Linares-Barronco et al. 1992), are obtained by converting filters (Acar et al. 1993; Sanchez-Sinencio et al. 1988) into oscillators. The oscillation condition of an oscillator must include both positive and negative terms to obtain stable oscillation, which can be achieved by equating the oscillation condition term to zero. In order to generate oscillators from the filters (Acar et al. 1993; Sanchez-Sinencio et al. 1988), negative and positive resistors, implemented with CMOS OTAs, are added to the oscillator networks (Geiger et al. 1985). These configurations have oscillation frequencies controlled by the transconductance gain without affecting the oscillation condition and the capability of operation at high frequencies. All of the proposed topologies are very attractive in both monolithic integrated technology (IC) and thin film fabrication due to their inclusion of only grounded capacitors. Furthermore, the parasitic capacitances can be easily accounted for and tuned since these capacitances are in parallel with grounded capacitors (Abuelma'atti and Almaskati 1988). The proposed topologies are shown in Figure 2.27. The oscillator circuits illustrated in Fig. 2.27b, d, e, f are newly introduced networks. The configurations shown in Fig. 2.27a, c may be seen in the literature (Rodrigez-Vasqez et al. 1990; Sanchez-Sinencio et al. 1988; Abuelma'atti and Almaskati 1988; Senani et al. 1991).

For the oscillator topologies given in Fig. 2.27, nodal analysis yields the oscillation condition (b = 0) and oscillation frequency (omega 0) expressions according to (1), by using the ideal model of OTA. The obtained b and V_0 expressions are given in Table 2.3.

Realization of DO-OTA-C Oscillators

The basic aim of this section is to give design considerations of current-mode high-frequency DO-OTA-C oscillator topologies achieving noninteractive control of b and fo with a minimum number of components. Starting from DO-OTA-C (grounded capacitor) filter topologies, reported in the literature and employing a minimum number of components (Sun and Findler 1996, 1997), novel DO-OTA-C oscillator topologies are generated by converting filters into oscillators (Kuntman and Özpınar 1998). Furthermore, the influence of the OTA non-idealities on oscillator performance is investigated by including the finite input and output impedances and transconductance frequency dependencies of DO-OTAs into derived equations (Fig. 2.28 and Table 2.4).

Concluding Remarks

In this section, design considerations of current-mode high-frequency DO-OTA-C oscillator topologies are given, achieving noninteractive control of oscillation condition and oscillation frequency with a minimum number of components. Starting

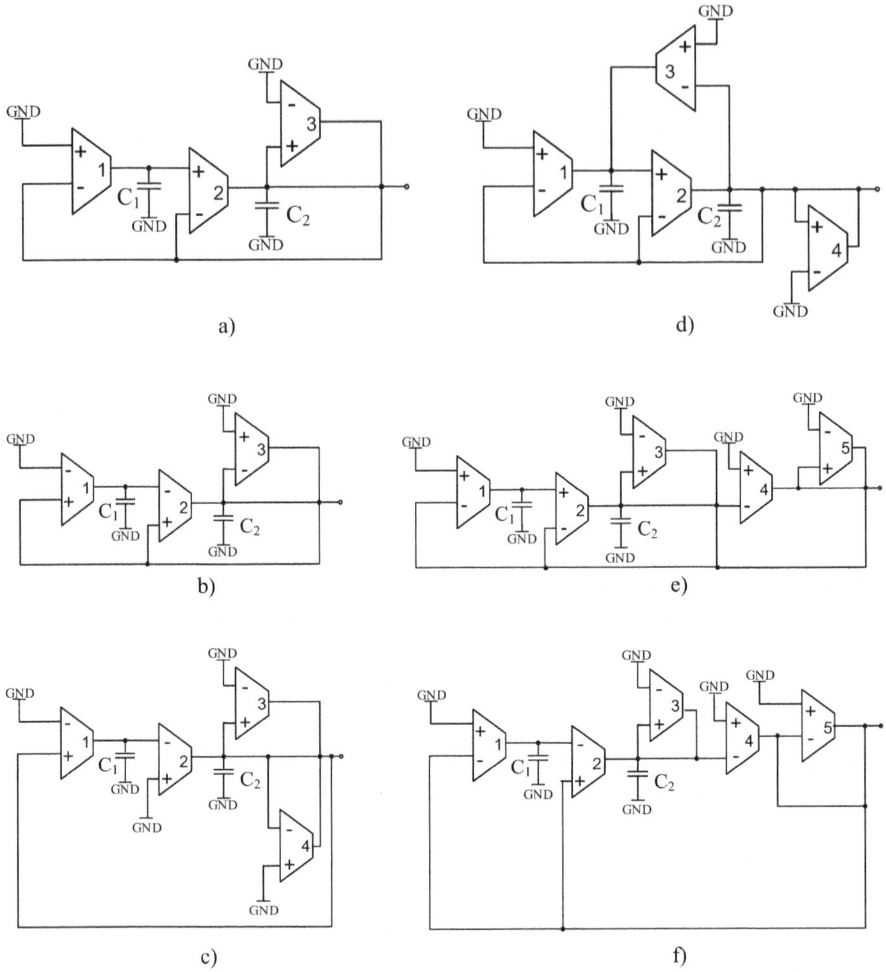

Fig. 2.27 (**a**) 3OTA-2C I oscillator circuit, (**b**) 3OTA-2C II oscillator circuit, (**c**) 4OTA-2C I oscillator circuit, (**d**) 4OTA-2CII oscillator circuit, (**e**) 5OTA-2CI oscillator circuit, and (**f**) 5OTA-2C II oscillator circuit. (Çam et al. 1998)

Table 2.3 Expressions of b and Ω0 for the oscillator topologies in Figure 1 constructed with ideal OTAs

Topology	**b**	Ω_0^2
3OTA-2C I	$\frac{g_{m2}-g_{m3}}{C_2}$	$\frac{g_{m1}g_{m2}}{C_1C_2}$
3OTA-2C II	$\frac{g_{m3}-g_{m2}}{C_2}$	$\frac{g_{m1}g_{m2}}{C_1C_2}$
4OTA-2C I	$\frac{g_{m4}-g_{m3}}{C_2}$	$\frac{g_{m1}g_{m2}}{C_1C_2}$
4OTA-2C II	$\frac{g_{m2}-g_{m4}}{C_2}$	$\frac{g_{m2}(g_{m1}+g_{m3})}{C_1C_2}$
5OTA-2C I	$\frac{\frac{g_{m2}g_{m4}}{g_{m5}}-g_{m3}}{C_2}$	$\frac{g_{m1}g_{m2}g_{m4}}{C_1C_2g_{m5}}$
5OTA-2C II	$\frac{\frac{g_{m2}g_{m4}}{g_{m5}}-g_{m3}}{C_2}$	$\frac{g_{m1}g_{m2}g_{m4}}{C_1C_2g_{m5}}$

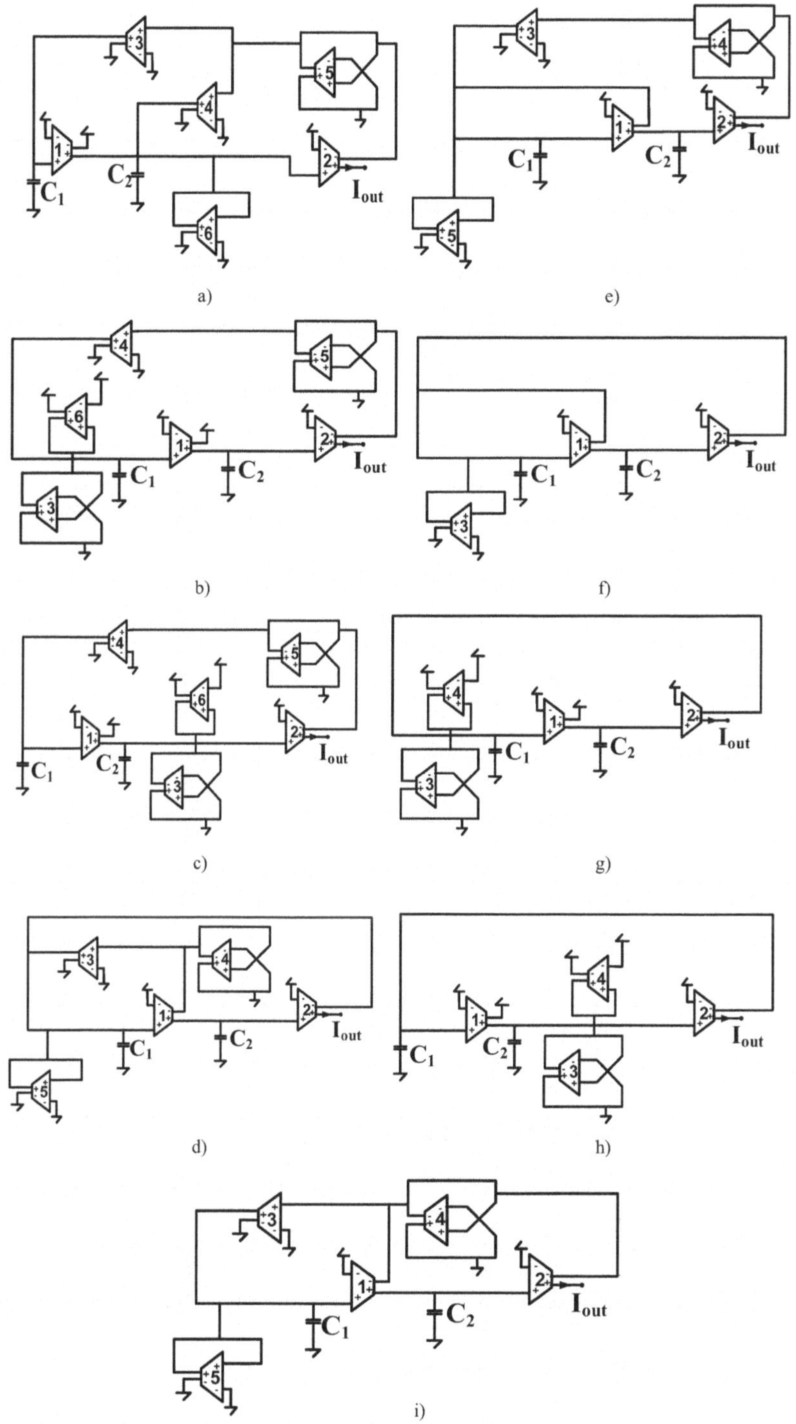

Fig. 2.28 Proposed DO-OTA-C oscillator topologies. (Kuntman and Özpınar 1998)

Table 2.4 Expressions for oscillation conditions and oscillator frequencies of topologies illustrated in Fig. 2.24, derived assuming ideal OTAs

Topology	b	Ω_0^2
Figure 2.28a	$\frac{\frac{g_{m2}g_{m4}}{g_{m5}}-g_{m6}}{C_2}$	$\frac{g_{m1}g_{m2}g_{m3}}{C_1C_2g_{m5}}$
Figure 2.28b	$\frac{g_{m3}-g_{m6}}{C_1}$	$\frac{g_{m1}g_{m2}g_{m4}}{C_1C_2g_{m5}}$
Figure 2.28c	$\frac{g_{m3}-g_{m6}}{C_2}$	$\frac{g_{m1}g_{m2}g_{m4}}{C_1C_2g_{m5}}$
Figure 2.28d	$\frac{\frac{g_{m1}g_{m3}}{g_{m4}}-g_{m5}}{C_1}$	$\frac{g_{m1}g_{m2}}{C_1C_2}$
Figure 2.28e	$\frac{g_{m1}-g_{m5}}{C_1}$	$\frac{g_{m1}g_{m2}g_{m3}}{C_1C_2g_{m4}}$
Figure 2.28f	$\frac{g_{m1}-g_{m3}}{C_1}$	$\frac{g_{m1}g_{m2}}{C_1C_2}$
Figure 2.28g	$\frac{g_{m3}-g_{m4}}{C_1}$	$\frac{g_{m1}g_{m2}}{C_1C_2}$
Figure 2.28h	$\frac{g_{m3}-g_{m4}}{C_2}$	$\frac{g_{m1}g_{m2}}{C_1C_2}$
Figure 2.28i	$\frac{\frac{g_{m1}g_{m3}}{g_{m4}}-g_{m5}}{C_1}$	$\frac{g_{m1}g_{m2}g_{m3}}{C_1C_2g_{m4}}$

from DO-OTA-C (grounded capacitor) filter topologies reported in the literature and employing a minimum number of components, novel DO-OTA-C oscillator topologies are generated. Furthermore, the main advantage provided by the circuits proposed is the high output voltage amplitudes of the order of several volts obtained by choosing proper values for the load resistors because of current-mode operation. It was recently demonstrated that in conventional single-output OTA-C topologies the output signal level is limited primarily by the slew-rate effect which seriously decreases the amplitude to a level of several hundred millivolts at high frequencies. The topologies proposed eliminate this limitation because of resistive load which makes these circuits advantageous compared to the well-known single-output OTA-C circuits. Furthermore, the oscillator configurations enable frequency control by changing the OTA transconductances with biasing currents without affecting oscillation condition and are very suitable for VLSI design since they are composed only of DO-OTAs and grounded capacitors. Therefore, the circuits obtained provide new possibilities for the circuit designer.

References

Acar, C. Anday F. and Kuntman, H.: On the realization of OTA-C filters, International Journal of Circuit Theory and Applications, Vol.21, pp.331–341, 1993.

Allen, P.E. and Holberg, D.R., CMOS analog circuit design (Second Edition), Oxford University Press, New York Oxford, 2002.

Bhashkar, D. R., Tripath, M., and Senani, R., Systematic derivation of all possible canonic OTA-C sinusoidal oscillators. Journal of the Franklin Institute, 330, 885±903, 1993.

Clark, J.W. Jr., The origin of biopotentials, in: J.G. Webster (Ed.), Medical Instrumentation: Application and Design, Houghton Mifflin Co., Boston, MA, 1992.

Çam, U., Kuntman, H., Acar, C., "On the realization of OTA-C oscillators", International Journal of Electronics, Vol.85, No.3, pp.313–326, 1998.

Düzenli, G., Kılıç, Y., Kuntman H. and Ataman, A.: On the design of low-frequency filters using CMOS OTAs operating in the subthreshold region, Microelectronics Journal, Vol.30, No. 1, pp.4554, 1999.

Erdogan, E. S., Topaloglu, R. O., Çiçekoglu, O. and Kuntman, H., "New Current-mode Special Function Continuous Time Active Filters Employing Only OTAS and Opamps", International Journal of Electronics, Volume 91, Number 6, 345–359, 2004.

Ergün, B.S., Kuntman, H., and Özcan, S., "Realization of a high performance CMOS DOTA with extended linearity range", Proc. of SCS 2001: International Symposium on Signals Circuits and Systems, pp. 81–84, Iasi, Romania, 2001.

Ergün, B.S., Kuntman, H., Realization of a high output impedance CMOS DO-OTA with extended linearity range, Proceedings of ELECO'2001: The 2nd International Conference on Electrical and Electronics Engineering (Electronics), pp.73–77, Bursa, 7–11 November 2001.

Ergün, B.S., Kuntman, H., "Yüksek Lineerlikte DO-OTA Gerçekleştirilmesi" Proceedings of ELECO 2000: The 1st National Symposium of Electrical, Electronics and Computer Engineering, pp 77–81, Chamber of Turkish Electrical Eng., Bursa section, Bursa Turkey, 2000.

Ergün, B.S., Kuntman, H., On the design of new CMOS DO-OTA topologies providing high output impedance and extended linearity range, Journal of Electrical & Electronics Engineering, Engineering Faculty, Istanbul University, Vol.5, No.2, pp.1449–1461, 2005.

Ergün B. S.,"Yüksek Lineerlikte CMOS DO-OTA Gerçekleştirilmesi", MSc Thesis, I.T.U, Institute of Science and Technology, 2001.

Gray, R., Hurst, P.J., Lewis, S.H., Meyer, R.G., "Analysis and design of analog integrated circuits", John Wiley & Sons, Inc., 2001.

Gönüleren, A.N., Köprü, R. and Kuntman H., "Multiloop feedback bandpass OTA-C filters using quads", Proc. 12th European Conference on Circuit Theory and Design, Vol.2, pp.607–610, 27–31 August, İstanbul, 1995.

Khorramabadi, H. and Gray, P. R. "High-frequency CMOS continuous-time filters", IEEE Journal of Solid-State Circuits, 19(6), 939–948, 1984.

Krummenacher, F. and Joehl, N. A., "4 MHz CMOS continuous-time filter on-chip automatic tuning", IEEE J. Solid-State Circuits, Vol. 23, pp. 750–757,. 1988.

Kuntman, H., Özpınar, A., "On the realization of DO-OTA-C oscillators", Microelectronics Journal, Vol.29, No. 12, pp.991–997, 1998.

Linares-Barranco, B., Rodríguez-Vázquez, A., Sanchez-Sinencio, E., and Huertas, J. L. "Generation, design and tuning of OTA-C high-frequency sinusoidal oscillators", IEE Proceedings G (Circuits, Devices and Systems), 139(5), 557–568, 1992.

Nedungadi, A. and Viswanathan, T. R., "Design of Linear CMOS Transconductance Elements" IEEE Trans. Circuits Syst., vol CAS31, pp 891–89, 1984.

Sánchez-Sinencio, E., Ramirez-Angulo, J., Linares-Barranco, B. and Rodriguez-Vázquez, A. "Operational transconductance amplifier-based nonlinear function syntheses", IEEE Journal of Solid-State Circuits, 24(6), 1576–1586, 1989.

Sanchez-Sinencio, E., Geiger, R. L. and Nevarez-Lozano, H., "Generation of continuous-time two integrator loop OTA filter structures", IEEE Transactions on Circuits and Systems, 35(8), 936–946, 1988.

Seevinck, E. A. and Wassenaar, R.W., "A versatile CMOS linear transconductor/square-law function circuit", IEEE J. Solid-State Circuits, Vol. SC-22, pp. 366–377, 1987.

Sun, Y. and Fidler, J. K. "Current-mode multiple-loop feedback filters using dual-output OTAs and grounded capacitors", International Journal of Circuit Theory and Applications, 25(2), 69–80, 1997.

Sun, Y. and Fidler, J. K. "Structure generation of current-mode two integrator loop dual output-OTA grounded capacitor filters", IEEE Transactions on Circuits and Systems II: Analog and Digital Signal Processing, 43(9), 659–663, 1996.

Zeki, A. and Kuntman, H. "A CMOS CCII suitable for continuous-time active networks" Proc. Int. Syrnp. On circuits and Systems (SCS '97), Iasi, Romania, 2, 393–396, 1997.

Zeki, A. and Kuntman, H., "High-output-impedance CMOS dual-output OTA suitable for wide-range continuous-time filtering applications",- Electronics Letters, 35,No.16, pp.1295–1296, 1999.

Zeki, A. and Kuntman, H., "A novel CMOS OTA structure suitable for OTA-C filters, Proc. of the 8th International Conference on Microelectronics (ICM'96), pp.7–10, December 16–18, Cairo, Egypt, 1996.

Zeki. A., and Kuntman. H. "Accurate active-feedback CMOS cascode current mirror with improved output swing", International Journal of Electronics, 84, (4). pp. 335–343, 1998.

Zeki A., "Novel High Performance OTA Structures Suitable for Continuous Time OTA-C Filters", PhD Thesis, October, I.T.U, Institute of Science and Technology, 1997.

Chapter 3
Current Conveyors, Variants, and Applications

Basic Concept

The current conveyor is an active element where the current is transferred between terminals with different impedance levels, introduced firstly by Sedra and Smith in 1970. The general circuit symbol and the definition matrix equation are illustrated in Fig. 3.1.

Furthermore, current conveyors can be classified as follows:

- $a = 1$, first-generation current conveyor, CCI
- $a = 0$, second-generation current conveyor, CCII
- $a = -1$, third-generation current conveyor, CCIII
- For CCI, if $b = 1$, noninverting first-generation current conveyor, CCI+
- For CCI, if $b = -1$, inverting first-generation current conveyor, CCI−
- For CCII, if $b = 1$, noninverting second-generation current conveyor, CCII+
- For CCII, if $b = -1$, inverting second-generation current conveyor, CCII−

The use of circuits constructed with current conveyors has been demonstrated to be potentially advantageous for the realization of high-frequency continuous-time monolithic analog systems. The current conveyor is capable of operating over a wide frequency range and therefore finds applications in the realization of broadband analog building blocks such as amplifiers, multipliers, oscillators, and active filters. Circuit symbols and circuit configurations for CMOS realization of CCII+ and CCII- are illustrated in Fig. 3.2, where $v_x(t)$, $v_y(t)$, $v_z(t)$, $i_x(t)$, $i_y(t)$, and $i_z(t)$ are terminal voltages and currents, respectively. In ideal case CCIIs are characterized by

H. Kuntman, D. Özenli, *Trends in Circuit Design for Analog Signal Processing*, Analog Circuits and Signal Processing,
https://doi.org/10.1007/978-3-030-96836-6_3

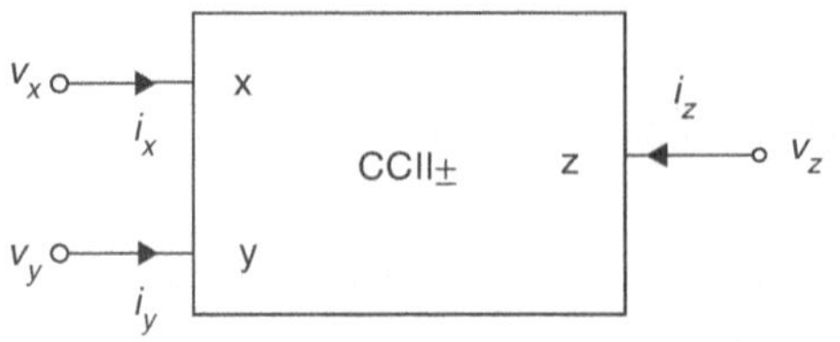

$$\begin{bmatrix} i_y \\ V_x \\ i_z \end{bmatrix} = \begin{bmatrix} 0 & a & 0 \\ 1 & 0 & 0 \\ 0 & b & 0 \end{bmatrix} \begin{bmatrix} V_y \\ i_x \\ V_z \end{bmatrix}$$

Fig. 3.1 General circuit symbol of current conveyor and definition matrix equation. (Sedra and Smith 1970)

$$\begin{aligned} i_y(\mathrm{t}) &= 0 \\ v_x(\mathrm{t}) &= v_y(\mathrm{t}) \\ i_z(\mathrm{t}) &= \pm i_x(\mathrm{t}) \end{aligned} \tag{3.1}$$

The behavior of the ideal current conveyor is characterized by in matrix form

$$\begin{pmatrix} I_Y \\ V_X \\ I_Z \end{pmatrix} = \begin{pmatrix} 0 & 0 & 0 \\ 1 & 0 & 0 \\ 0 & \pm 1 & 0 \end{pmatrix} \begin{pmatrix} V_Y \\ I_X \\ V_Z \end{pmatrix} \qquad I_X = -\frac{V_Y}{R_X} \tag{3.2}$$

where R_X is the external resistor connected to the terminal X. The behavior of an actual current conveyor differs from the ideal current conveyor and can be represented by the following equations:

$$\begin{pmatrix} I_Y \\ V_X \\ I_Z \end{pmatrix} = \begin{pmatrix} Y_Y(s) & 0 & 0 \\ A_V(s) & Z_X(s) & 0 \\ 0 & \pm A_I(s) & Y_Z(s) \end{pmatrix} \begin{pmatrix} V_Y \\ I_X \\ V_Z \end{pmatrix} \qquad I_X = -\frac{A_V(s) V_Y}{R_X + Z_X(s)} \tag{3.3}$$

Figure 3.3a depicts the non-ideal current conveyor represented by Eq. (3.3). In realization of active filters, the designers assume that active components are linear components. In fact they are nonlinear and they behave linear under certain conditions. These conditions depend on design parameters of active components. Violation of these conditions causes nonlinear distortion in filters. Hence, the designers should know in advance linear operation conditions of the filter to be realized. Linear operation conditions for several active components such as op-amps, OTAs, CCIs, and CCIIs are investigated by introducing macromodels. But, not much study is performed as to linear operation of active filters. However, limitations on input signal level in voltage-mode OTA-C filters and CCII-based current-mode active

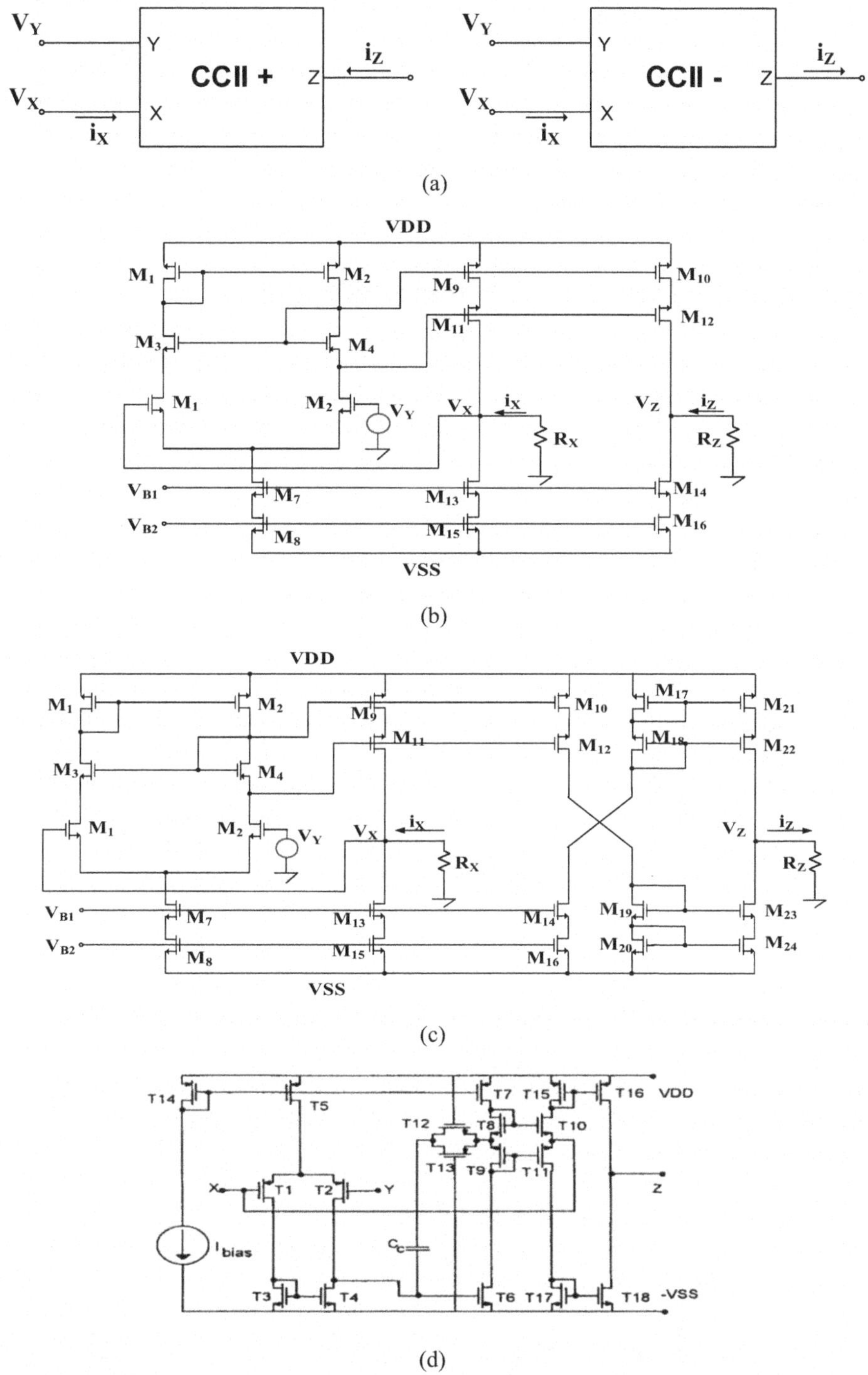

Fig. 3.2 (**a**) Circuit symbols for noninverting and inverting current conveyors and (**b**), (**c**), and (**d**) circuit examples for CMOS realization of noninverting and inverting current conveyors. (Sedra et al. 1990)

RC filters have appeared recently in literature. Although CCII-based voltage-mode filters are widely used in filter design, this problem is not taken into consideration in this type of filter design. In this section, using linear operation conditions of typical CMOS current conveyors, limitations on input signal amplitude in voltage-mode current conveyor filters are studied. A simple formula is derived for maximum input signal amplitude not causing a nonlinear operation.

Nonlinear Behavior of the Current Conveyors

The circuit representation of noninverting (CCII+) and inverting (CCII-) types of second-generation current conveyors is shown in Fig. 3.4a, b.

In actual case CCIIs are nonlinear components and behave linearly if the following conditions are satisfied:

$$\begin{aligned} V_{x\,\mathrm{m}-} &\leq v_x(\mathrm{t}) \leq V_{x\,\mathrm{m}+} \\ V_{z\,\mathrm{m}-} &\leq v_z(\mathrm{t}) \leq V_{z\,\mathrm{m}+} \\ I_{x\,\mathrm{m}-} &\leq i_x(\mathrm{t}) \leq I_{x\,\mathrm{m}+} \end{aligned} \tag{3.4}$$

where $V_{x\mathrm{m}+}$, $V_{x\mathrm{m}-}$, $V_{z\mathrm{m}+}$, $V_{z\mathrm{m}-}$, $I_{x\mathrm{m}+}$, and $I_{x\mathrm{m}-}$ represent maximum positive and negative voltage at the x terminal, maximum positive, and negative voltage at the z terminal and the maximum positive and negative current at the x terminal, respectively.

In contrary to operational amplifier and OTA applications where nonlinear macromodels of different complexity and accuracy levels are available, there are only few linear and nonlinear equivalent models for current conveyors. The purpose of this section is also to describe a simple and accurate current conveyor macromodel which is especially suitable for simulation of active filter applications.

Determination of the Maximum Input Signal Amplitude

For linear operations, the input signal level of the filter under consideration must be adjusted so that the above conditions are simultaneously satisfied for every current conveyor for designer's specified frequency band $\omega[\omega_1, \omega_2)$:

$$\begin{aligned} &|V_{xk}| \leq V_{\mathrm{s}xk} \\ &|V_{zk}| \leq V_{\mathrm{s}zk}, k = 1, 2, \ldots, n \\ &|I_{xk}| \leq I_{\mathrm{s}xk} \end{aligned} \tag{3.5}$$

where n denotes the total number of current conveyors used in the design. $V_{xk} = V_{xk}(\mathrm{j}\omega)$ and $V_{zk} = V_{zk}(j\omega)$ are respectively phasor voltages at x and z terminal

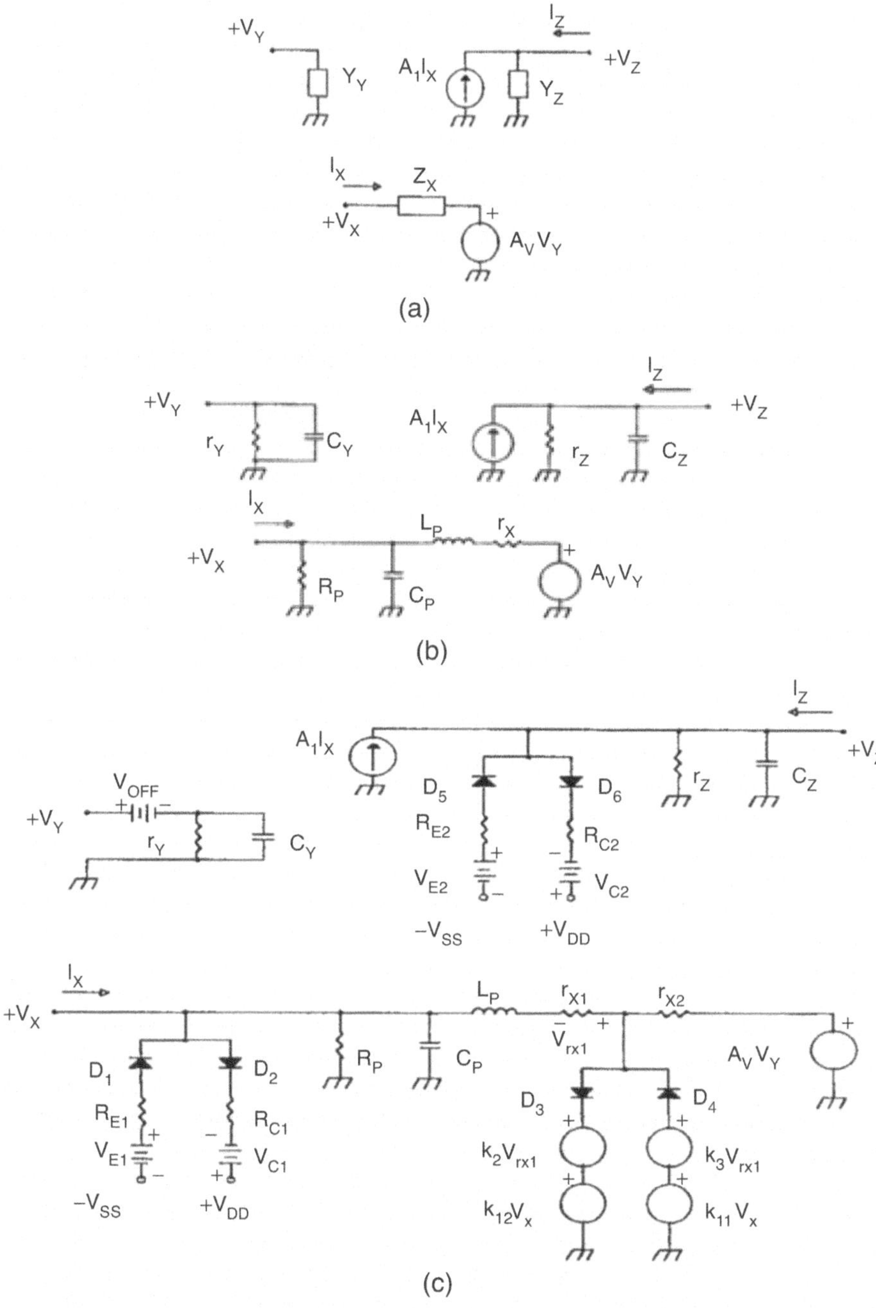

Fig. 3.3 (**a**) The non-ideal current conveyor, (**b**) a detailed circuit of the non-ideal current conveyor, and (**c**) the proposed macromodel for current conveyor. (Tarım et al. 1996)

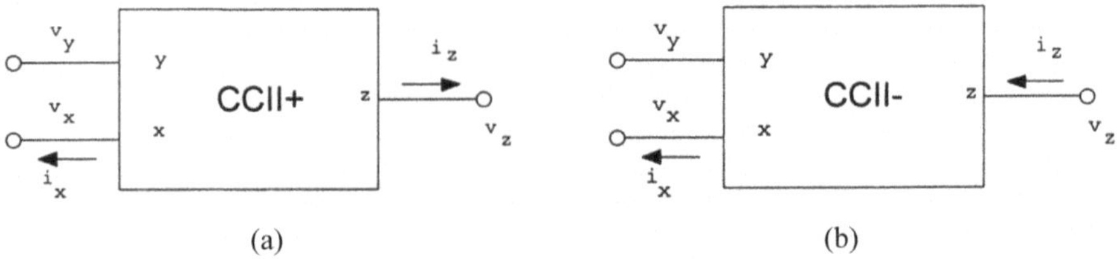

Fig. 3.4 The circuit representation of noninverting (CCII+) and inverting (CCII−) types of second-generation current conveyors. (Sedra and Smith 1970)

of the kth current conveyor. $I_{xk} = I_{xk}(j\omega)$ is also phasor current through x terminal of the kth current conveyor. V_{sxk}, V_{szk}, and I_{sxk} are the bounds of linear region and defined as

$$\begin{aligned} V_{sxk} &= \min\{|V_{x\,m+}|, |V_{x\,m-}|\} \\ V_{szk} &= \min\{|V_{z\,m+}|, |V_{z\,m-}|\} \\ I_{sxk} &= \min\{|I_{x\,m+}|, |I_{x\,m-}|\} \end{aligned} \tag{3.6}$$

We call these quantities saturation voltages and current of the kth current conveyor.

In terms of input voltage amplitude, these conditions can be written as follows:

$$\begin{aligned} &|V_i| \cdot |T_{xk}| \le V_{sxk} \\ &|V_i| \cdot |T_{zk}| \le V_{szk}, k = 1, 2, \ldots, n \\ &|V_i| \cdot |Y_{xk}| \le I_{sxk} \end{aligned} \tag{3.7}$$

where V_i is the amplitude of the filter's input voltage. $T_{xk} = T_{xk}(j\omega)$ is the voltage transfer function which is defined as the ratio of the kth current conveyor's x terminal phasor voltage to the phasor input voltage:

$$T_{xk} = \frac{V_{xk}(j\omega)}{V_i}, k = 1, 2, \ldots, n \tag{3.8}$$

$T_{zk} = T_{zk}(j\omega)$ is also the voltage transfer function defined as the ratio of the kth current conveyor's z terminal phasor voltage to the phasor input voltage:

$$T_{zk} = \frac{V_{zk}(j\omega)}{V_i}, k = 1, 2, \ldots, n \tag{3.9}$$

$Y_{xk} = Y_{xk}(j\omega)$ is the transfer admittance function which is defined as the ratio of the kth current conveyor's x terminal phasor current to the phasor input voltage:

$$Y_{xk} = \frac{I_{xk}(j\omega)}{V_{\text{i}}}, k = 1, 2, \ldots, n \tag{3.10}$$

There exist 3n inequalities and they put the following constraints on input voltage amplitude for $\omega \in (\omega_1, \omega_2)$:

$$\begin{aligned} |V_i| &\leq \frac{V_{\text{s}xk}}{|T_{xk}|} \\ |V_i| &\leq \frac{V_{\text{s}xk}}{|T_{zk}|} \\ |V_i| &\leq \frac{I_{\text{s}xk}}{|Y_{xk}|} \end{aligned} \tag{3.11}$$

The common solution of these inequalities which gives the maximum value of the input voltage's amplitude not causing nonlinear distortion can be expressed as

$$\begin{aligned} &|V_i|_{\max} \\ &= \min\left\{\frac{V_{\text{s}xk}}{|T_{xk}|_{\max}}, \frac{V_{\text{s}zk}}{|T_{zk}|_{\max}}, \frac{I_{\text{s}xk}}{|Y_{xk}|_{\max}}, k = 1, 2, \ldots, n\right\} \\ &\omega \in (\omega_1, \omega_2) \end{aligned} \tag{3.12}$$

In the following, the evaluation of the maximum input signal value is shown using the circuit of Fig. 3.5. This circuit realizes a low-pass filter characteristic whose DC gain, pole frequency, and pole quality factor are respectively $T(j0) = 1$, $\omega_P = 10^6$ rad sec^{-1}, and $Q_P = 1$, if the element values are chosen as $R_1 = R_2 = R_3 = R_4 = 10$ k, $C_1 = C_2 = 100$ pF.

The saturation voltages and current for CCII+ are specified as

$$\begin{aligned} V_{\text{s}x} &= 6.7 \text{ V} \\ V_{\text{s}z} &= 7.17 \text{ V} \\ I_{\text{s}x} &= 204\mu\text{A} \end{aligned}$$

and the saturation voltages and current of CCII− are determined as

$$\begin{aligned} V_{\text{s}x} &= 7.94 \text{ V} \\ V_{\text{s}z} &= 9.1 \text{ V} \\ I_{\text{s}x} &= 116\mu\text{A} \end{aligned}$$

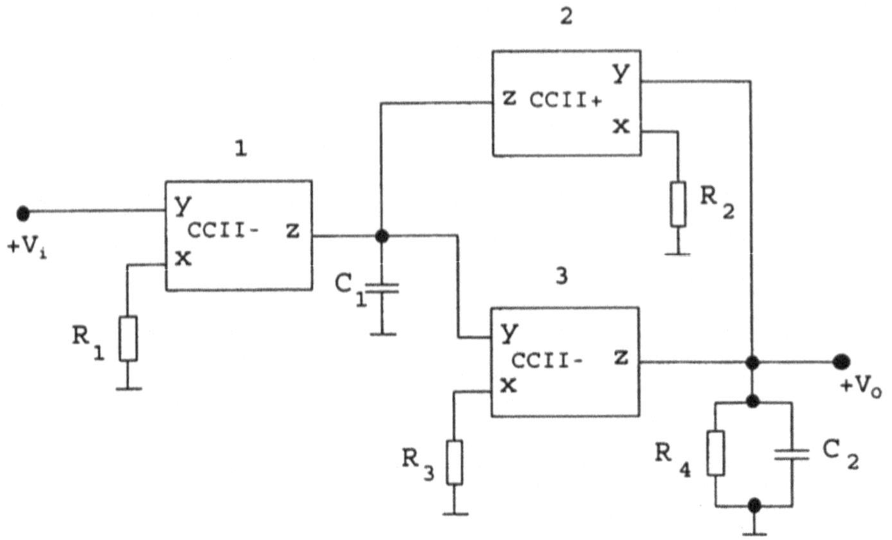

Fig. 3.5 Second-order low-pass filter using three current conveyors. (Tek and Anday 1989; Acar and Kuntman 1999)

$$T_{x1} = \frac{V_{x1}}{V_\mathrm{i}} = 1$$

$$T_{x2} = \frac{V_{x2}}{V_i} = \frac{G_1 \cdot G_3}{D(j\omega)}$$

$$T_{x3} = \frac{V_{x3}}{V_i} = -\frac{G_1 \cdot (j\omega C_2 + G_4)}{D(j\omega)}$$

$$T_{z1} = \frac{V_{z1}}{V_i} = -\frac{G_1 \cdot (j\omega C_2 + G_4)}{D(j\omega)}$$

$$T_{z2} = \frac{V_{z2}}{V_i} = -\frac{G_1 \cdot (j\omega C_2 + G_4)}{D(j\omega)}$$

$$T_{z3} = \frac{V_{z3}}{V_i} = \frac{G_1 \cdot G_3}{D(j\omega)}$$

$$Y_{x1} = \frac{I_{x1}}{V_i} = G_1$$

$$Y_{x2} = \frac{I_{x2}}{V_i} = \frac{G_1 \cdot G_2 \cdot G_3}{D(j\omega)}$$

$$Y_{x3} = \frac{I_{x3}}{V_i} = -\frac{G_1 \cdot G_3(j\omega C_2 + G_4)}{D(j\omega)}$$

where $D(j\omega)$ is the characteristic polynomial of the filter:

$$D(j\omega) = (j\omega)^2 C_1 C_2 + (j\omega) \cdot C_1 G_4 + G_2 G_3$$

Using the previous transfer function for the passband region of q ∈ (0,qp), we obtain:

$$\begin{aligned} |T_{x1}(j\omega)|_{\max} &= 1 \\ |T_{x2}(j\omega)|_{\max} &= 1.1516 \\ |T_{x3}(j\omega)|_{\max} &= 1.464 \\ |T_{z1}(j\omega)|_{\max} &= 1.464 \end{aligned}$$

$$\begin{aligned} |T_{x2}(j\omega)|_{\max} &= 1.1516 \\ |T_{x3}(j\omega)|_{\max} &= 1.464 \\ |T_{z1}(j\omega)|_{\max} &= 1.464 \\ |T_{z2}(j\omega)|_{\max} &= 1.464 \\ |T_{z3}(j\omega)|_{\max} &= 1.1516 \\ |Y_{x3}(j\omega)|_{\max} &= 146.404\mu\text{AV}^{-1} \end{aligned}$$

Using these quantities we obtain

$$\begin{aligned} |V_i|_{\max} &= \min\{7.94\text{ V}, 5.82\text{ V}, 5.42\text{ V}, 6.22\text{ V}, 4.9\text{ V}, 7.9\text{ V}, \\ &\quad 1.16\text{ V}, 1.77\text{ V}, 0.79\text{ V}\} \\ |V_i|_{\max} &= 0.79\text{ V} \end{aligned}$$

This result is verified by the SPICE simulations of the filter. Note that the output voltage is distorted for $|V_i| \geq |V_i|\max = 0.79$ V.

Limitations on Input Signal Level in Current-Mode Active-RC Filters Using CCIIs

In the realization of the current conveyor filters, we assume that current conveyors, shown in Fig. 3.1, are linear components, and in the ideal case we define them using the following equations:

$$i_y(t) = \cdot 0, v_z(t) = \cdot v_y(t), i_z(t) = \pm i_x(t)$$

In fact, CCIIs are nonlinear but they behave linearly under certain conditions, as described in the former section. For linear operation of the filter, the input signal level must be adjusted so that the following conditions are simultaneously satisfied for every current conveyor in the filter for the specified frequency band $\omega \in (\omega_1, \omega_2)$:

$$\begin{aligned} &|V_{xk}| \le V_{sxk} \quad |V_{zk}| \le V_{szk} \quad |I_{xk}| \le I_{sxk} \quad k = 1, 2, \ldots, n \\ &|I_{in}| \le \frac{V_{sxk}}{|Z_{xk}|} \quad |I_{in}| \le \frac{V_{szk}}{|Z_{in}|} \quad |I_{in}| \le \frac{I_{sxk}}{|A_{in}|} \quad k = 1, 2, \ldots, n \end{aligned} \tag{3.13}$$

$$Z_{xk} = \frac{V_{xk}(j\omega)}{I_{in}} \quad Z_{zk} = \frac{V_{zk}(j\omega)}{I_{in}} \quad A_{xk} = \frac{I_{xk}(j\omega)}{I_{in}} \quad k = 1, 2, \ldots, n \tag{3.14}$$

$$|I_{\text{in}}|_{\max} = \min\left\{\frac{V_{sxk}}{|Z_{xk}|_{\max}}, \frac{V_{szk}}{|Z_{xk}|_{\max}}, \frac{I_{sxk}}{|A_{xk}|_{\max}}, k = 1, 2, \ldots, n\right\} \omega \in (\omega_1, \omega_2) \tag{3.15}$$

In the following, as an example, the maximum input signal level is calculated for the all-pole band-pass filter shown in Fig. 3.6. This filter realizes with unity gain at a resonant frequency of $\omega_0 = 10^6$ rad/s.

$$\frac{I_{\text{out}}}{I_{\text{in}}} = \left.\frac{\sqrt{2}S^2}{S^3 + 2S^2 + 2S + 1}\right|_{S=s/\omega_p} \tag{3.16}$$

This circuit uses only CCIIs and they are designed using the CMOS configuration of Fig. 3.2. $|Z_{xk}|_{\max}$, $|Z_{zk}|_{\max}$, and $|A_{xk}|_{\max}$, $k = 1, 2, 3, 4$, which are necessary for determining $|I_{\text{in}}|_{\max}$, are obtained from the filter given topology of Fig. 3.6, by the use of related equations for the 3 dB passband region as

$$\begin{aligned} &|Z_{xi}(j\omega)|_{\max} = 0k\Omega, i = 1, 2, 3 \quad |Z_{x4}(j\omega)|_{\max} = 20k\Omega \\ &|Z_{z1}(j\omega)|_{\max} = 20.19k\Omega \\ &|Z_{z1}(j\omega)|_{\max} = 23.4k\Omega \quad |Z_{z1}(j\omega)|_{\max} = 10k\Omega \\ &|Z_{z4}(j\omega)|_{\max} = 0k\Omega \\ &|A_{x1}(j\omega)|_{\max} = 1.11 \quad |A_{x2}(j\omega)|_{\max} = 1.56 \\ &|A_{x3}(j\omega)|_{\max} = 1.17 \quad |A_{x4}(j\omega)|_{\max} = 0.69 \end{aligned} \tag{3.17}$$

Using these quantities we obtain

$$\begin{aligned} |I_i|_{\max} = \min\{&\infty, \infty, \infty, 397\,\mu\text{A}, 450\,\mu A, 388\,\mu\text{A}, \infty, \\ &104.6\,\mu\text{A}, 74.4\,\mu\text{A}, 93.3\,\mu\text{A}, 167\,\mu\text{A} = 74.4\,\mu\text{A}\} \end{aligned}$$

This result is verified by the SPICE simulations of the filter by simulating the output response of the filter to a sinusoidal input current at several signal levels of $|I_i| < 74.4$ μA and $I_i > 74.4$ μA with angular frequency of, respectively. For $|I_i| < 74.4$ μA the total harmonic distortion (THD) at the output is less than 1.1%. It is observed that THD rapidly increases with increasing input amplitude for $|I_i| > 74.4$ μA.

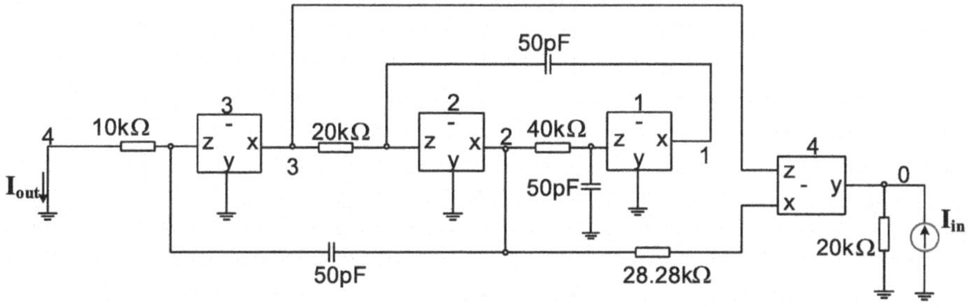

Fig. 3.6 Third-order all-pole band-pass filter using four current conveyors. (Acar and Kuntman 1996)

Wide Dynamic Range High Output Impedance Current-Mode Multifunction Filters with Dual-Output Current Conveyors

Recently there is a growing interest to current-mode multifunction filters employing current conveyors, which are accepted to have wider bandwidth and greater linearity compared to voltage-mode operational amplifiers. Multifunction filters are capable of realizing more than one basic filter function simultaneously with the same topology. A current-mode filter theoretically should exhibit high output impedance to enable easy cascadability and to enable additional filter responses by simply connecting the outputs. If the output signal on the other hand is available on a passive element, then additional active elements will be required to sense the current, and easy addition of the output signal will not be possible. This section reports five different topologies all exhibiting high output impedances, employing four dual-output current conveyors (DO-CCII) and only grounded passive elements which are advantageous from the integrated circuit implementation point of view. No component matching condition is imposed. The filters permit orthogonal adjustment of quality factor Q and resonant angular frequency ωo. The circuits presented realize low-pass, high-pass, band-pass, and notch filter functions simultaneously all at high output impedance terminals, thus permitting easy cascadability. The circuits can easily be modified to realize also a second-order all-pass function.

Dual-Output Current Conveyor

The dual-*output* current conveyor also called four-terminal pair active current conveyor shown in Fig. 3.7 is derived from the positive-type second-generation current conveyor and is defined by the following matrix equation:

$$\begin{bmatrix} v_x \\ i_y \\ i_{z1} \\ i_{z2} \end{bmatrix} = \begin{bmatrix} 0 & 1 & 0 & 0 \\ 0 & 0 & 0 & 0 \\ 1 & 0 & 0 & 0 \\ \pm 1 & 0 & 0 & 0 \end{bmatrix} \begin{bmatrix} i_x \\ v_y \\ v_{z1} \\ v_{z2} \end{bmatrix} \tag{3.18}$$

The ± sign indicates whether the element is of p (DO-CCII P) or n type (DO-CCII N). In case of non-ideal DO-CCII the definition equation converts to

$$\begin{bmatrix} v_x \\ i_y \\ i_{z1} \\ i_{z2} \end{bmatrix} = \begin{bmatrix} 0 & \beta & 0 & 0 \\ 0 & 0 & 0 & 0 \\ \alpha_1 & 0 & 0 & 0 \\ \pm \alpha_2 & 0 & 0 & 0 \end{bmatrix} \begin{bmatrix} i_x \\ v_y \\ v_{z1} \\ v_{z2} \end{bmatrix} \tag{3.19}$$

where α_1, α_2 denote the current tracking and β denotes the voltage tracking coefficients, respectively.

Introduced Filter Topologies

The filter topologies introduced are illustrated in Fig. 3.8. For the derived filter circuits the transfer functions from the input to high-pass, low-pass, and band-pass outputs are given by

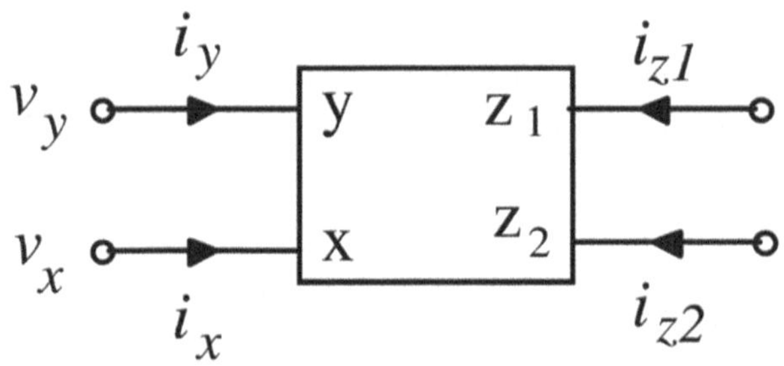

Fig. 3.7 The dual-output current conveyor symbol. (Çiçekoğlu et al. 2002)

$$\begin{aligned} H_{HP}(s) &= \frac{H_1 s^2}{s^2 + \dfrac{\omega_O}{Q} s + \omega_O^2} \\ H_{LP}(s) &= \frac{H_2 \omega_O^2}{s^2 + \dfrac{\omega_O}{Q} s + \omega_O^2} \\ H_{BP}(s) &= \frac{H_3 \dfrac{\omega_O}{Q} s}{s^2 + \dfrac{\omega_O}{Q} s + \omega_O^2} \end{aligned} \tag{3.20}$$

Expressions for pole angular frequencies (ω_o), pole quality factors (Q), and gain values (H_n) of the proposed filter topologies are illustrated in Table 3.1. The passive sensitivities are calculated for the topology given in Fig. 3.8a as

$$\begin{aligned} &S^{\omega_0}_{G_1,G_2,G_3} = 1/2, \quad S^{\omega_0}_{G_4,C_5,C_6} = -1/2, \quad S^{\omega_O}_{C_1} = 0, \\ &S^{Q}_{G_1,G_4,C_5,C_6} = 1/2, \quad S^{Q}_{G_2,G_3} = -1/2, \quad S^{Q}_{C_1} = -1, \end{aligned}$$

for Fig. 3.8b as

$$\begin{aligned} &S^{\omega_0}_{G_2,G_3,G_5} = 1/2, \quad S^{\omega_0}_{C_1,C_4,G_6} = -1/2, \quad S^{\omega_O}_{C_5} = 0, \\ &S^{O}_{C_1,C_4,G_5,G_6} = 1/2, \quad S^{Q}_{G_2,G_3} = -1/2, \quad S^{O}_{C_5} = -1, \end{aligned}$$

for Fig. 3.8c as

$$\begin{aligned} &S^{\omega_0}_{G_5,G_6} = 1/2, \quad S^{\omega_0}_{C_2,C_4} = -1/2, \quad S^{\omega_0}_{G_1,C_6} = 0 \\ &S^{Q}_{C_2,C_4,G_6} = 1/2, \quad S^{Q}_{G_5} = -1/2, \quad S^{Q}_{C_6} = -1, \quad S^{Q}_{G_1} = 0, \end{aligned}$$

for Fig. 3.8d as

$$\begin{aligned} &S^{\omega_0}_{G_2,G_3} = 1/2, \quad S^{\omega_0}_{C_1,C_4} = -1/2, \quad S^{\omega_0}_{G_1,C_5} = 0, \\ &S^{Q}_{C_1,G_2,G_3} = 1/2, \quad S^{Q}_{C^{Q}_{4}} = -1/2, \quad S^{Q}_{G_1} = -1, \quad S^{Q}_{C_5} = 0 \end{aligned}$$

for Fig. 3.8e as

$$\begin{aligned} &S^{\omega_O}_{G_2,G_3} = 1/2, \quad S^{\omega_0}_{C_1,C_5} = -1/2, \quad S^{\omega_0}_{G_1,C_4} = 0, \\ &S^{Q}_{C_1,G_2,G_3} = 1/2, \quad S^{Q}_{C_5} = -1/2, \quad S^{Q}_{C_4} = 0, \quad S^{Q}_{G_1} = -1 \end{aligned}$$

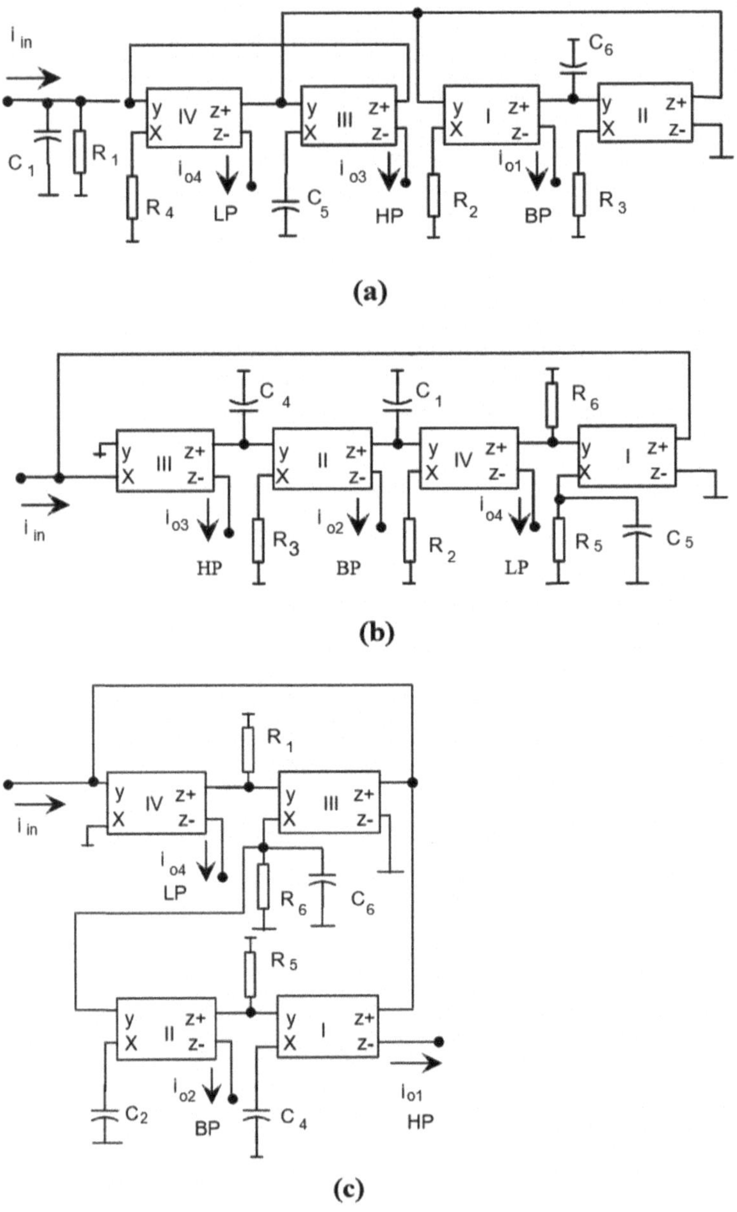

Fig. 3.8 Introduced multifunction filter topologies. (Çiçekoğlu et al. 2002)

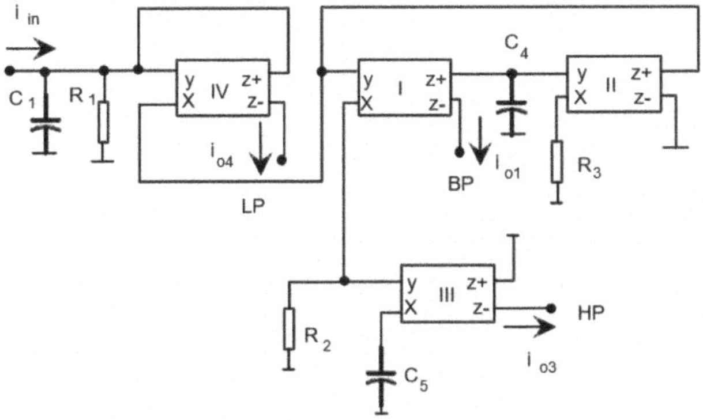

(d)

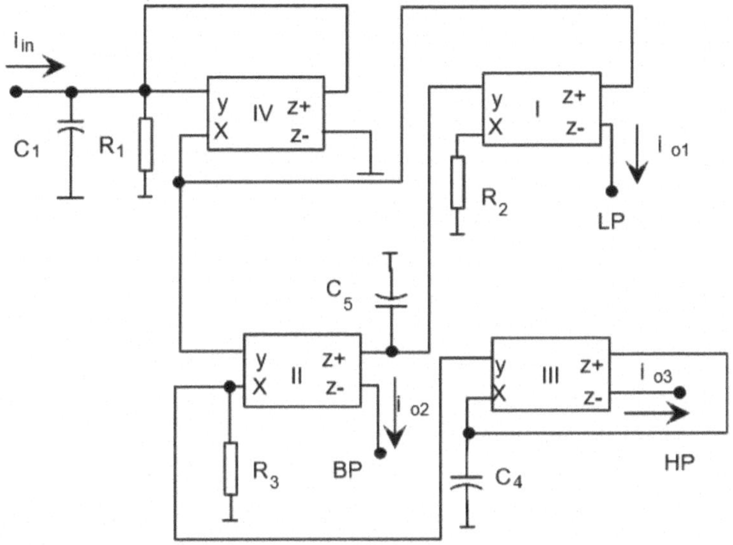

(e)

Fig. 3.8 (continued)

Simulation Results and Verification (Tables 3.2 and 3.3)

The performance of the filter topology given in Fig. 3.8a is tested on a chosen circuit example by using the CMOS realization of the DO-CCII shown in Fig. 3.9. The transistor dimensions for the DO-CCII circuit are also illustrated in Fig. 3.9. The supply voltages were chosen as $V_{DD} = 5$ V and $-V_{SS} = -5$ V. To obtain a universal filter with a pole frequency of $f_0 = 100$ kHz and quality factor of $Q = 0.707$, the element values were chosen as $R_1 = 10$k, $C_1 = 225$ pF, $R_2 = 1$k, $R_3 = 10$k,

Table 3.1 Expressions for pole frequencies, pole quality factors, and gain values of filter topologies

Topology	ω_o	Q	H_1	H_2	H_3
Figure 3.8a	$\sqrt{\frac{G_1G_2G_3}{C_5C_6G_4}}$	$\frac{1}{C_1}\sqrt{\frac{G_1G_4C_5C_6}{G_2G_3}}$	1	$-\frac{G_4}{G_1}$	$\frac{G_4C_6}{G_3C_1}$
Figure 3.8b	$\sqrt{\frac{G_2G_3G_5}{C_1C_4G_6}}$	$\frac{1}{C_5}\sqrt{\frac{G_5G_6C_1C_4}{G_2G_3}}$	1	$\frac{G_6}{G_5}$	$\frac{G_6C_1}{G_2C_3}$
Figure 3.8c	$\sqrt{\frac{G_5G_6}{C_2C_4}}$	$\frac{1}{C_6}\sqrt{\frac{G_6C_2C_4}{G_5}}$	1	$\frac{G_1}{G_6}$	$\frac{C_2}{C_6}$
Figure 3.8d	$\sqrt{\frac{G_2G_3}{C_1C_4}}$	$\frac{1}{G_1}\sqrt{\frac{G_2G_3C_1}{C_4}}$	$-\frac{C_5}{C_1}$	1	$-\frac{G_2}{G_1}$
Figure 3.8e	$\sqrt{\frac{G_2G_3}{C_1C_5}}$	$\frac{1}{G_1}\sqrt{\frac{G_2G_3C_1}{C_5}}$	$-\frac{1}{2}\frac{C_4}{C_1}$	-1	$-\frac{G_3}{G_1}$

Table 3.2 The active sensitivities of ωo and Q for the topologies under the assumption of equal αs and equal βs

Topology	$S_{\alpha}^{\omega_0}$	$S_{\beta}^{\omega_0}$	S_{α}^{Q}	S_{β}^{Q}
Figure 3.8a	0	0	0	0
Figure 3.8b	2	3/2	−2	−3/2
Figure 3.8c	−1/2	−1	1/2	1
Figure 3.8d	3/2	3/2	3/2	3/2
Figure 3.8e	1	3/2	1	3/2

Table 3.3 The sensitivities of L_{eq} and R_{eq} to passive elements

	Type	Figure	$S_{C_1}^{L_{eq}}$	$S_{R_1}^{L_{eq}}$	$S_{R_2}^{L_{eq}}$	$S_{R_3}^{L_{eq}}$	$S_{R_1}^{R_{eq}}$	$S_{R_2}^{R_{eq}}$	$S_{R_3}^{R_{eq}}$
1	L with series R	Fig. 3.1a	1	1	1	–	–	1	–
2	L with series R	Fig. 3.1b	1	1	1	–	$\frac{R_1}{R_1+R_2}$	$\frac{R_2}{R_1+R_2}$	–
3	L with series R	Fig. 3.1c	1	1	1	–	1	–	–
4	L with parallel R	Fig. 3.1d	1	1	–	1	$\frac{R_2}{2R_1+R_2}$	$\frac{-R_2}{R_1+R_2}$	1
5	L with series R	Fig. 3.1e	1	1	1	–	$\frac{R_1}{R_1+R_2}$	$\frac{R_2}{R_1+R_2}$	–

$R_4 = 10\text{k}$, $C_5 = 159$ pF, $C_6 = 159$ pF. The resulting frequency characteristics obtained from SPICE simulations are illustrated with ideal frequency responses in Fig. 3.10. Furthermore, the large signal behavior of the circuit is tested by investigating the dependence of the output harmonic distortion on the input signal amplitude. The results obtained are shown in Fig. 3.11. Note from Fig. 3.11 that harmonic distortion rapidly increases if the input signal is increased beyond the 40 μA level for the example chosen. For input signal levels lower than 40 μA, the total harmonic distortion remains in acceptable limits of the order of THD = 5%. The dependence of the output current and the output voltage on the load resistance R_L is simulated for an input signal level of $I_{in} = 20$ μA and for a signal frequency of $f_0 = 100$ kHz. The results are illustrated in Table 3.4. From Table 3.4, it can be easily observed that the output current level remains approximately constant and can be considered independent from load resistance value.

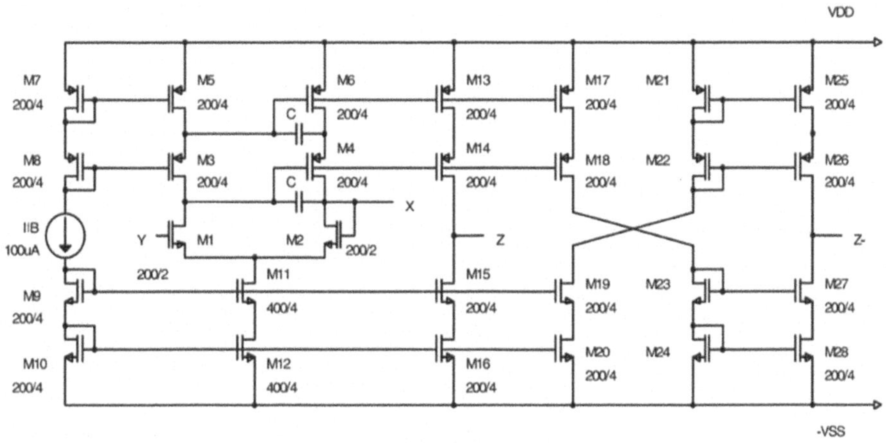

Fig. 3.9 CMOS realization of DO-CCII. (Çiçekoğlu et al. 2002)

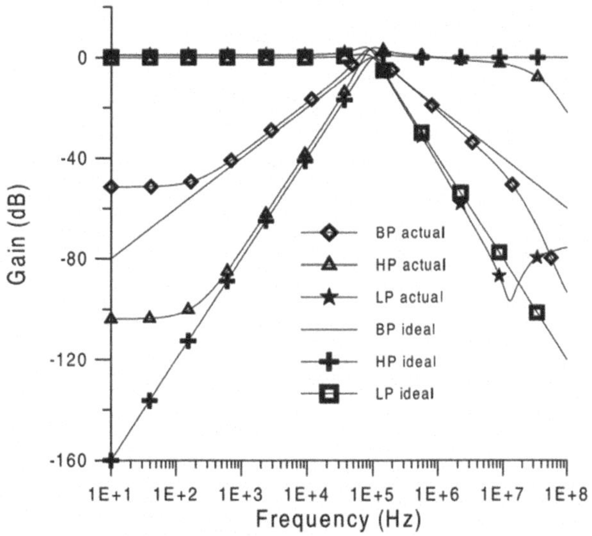

Fig. 3.10 Resulting frequency characteristics obtained from SPICE simulations. (Çiçekoğlu et al. 2002)

Furthermore, the large signal behavior of the circuit is tested by investigating the dependence of the output harmonic distortion on the input signal amplitude. Note from Fig. 3.11 that the harmonic distortion rapidly increases if the input signal is increased beyond the 40 μA level for the example chosen. For input signal levels lower than 40 μA, the total harmonic distortion remains in acceptable limits of the order of THD = 5%. The dependence of the output current and the output voltage on the load resistance RL is simulated for an input signal level of $I_{in} = 20$ μA and for a signal frequency of $f_0 = 100$ kHz.

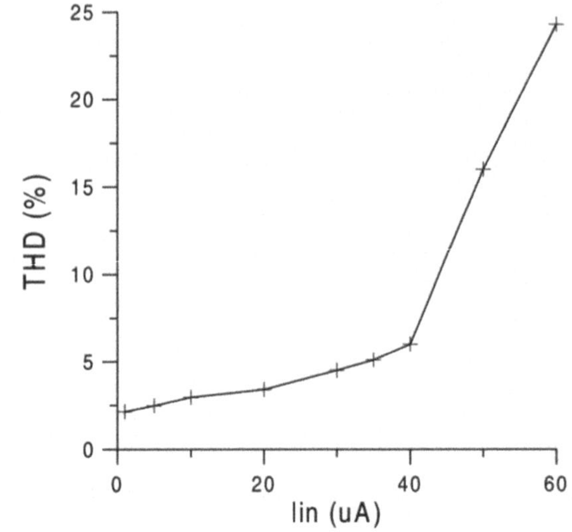

Fig. 3.11 Dependence of output harmonic distortion on input signal amplitude. (Çiçekoğlu et al. 2002)

Table 3.4 Transistor aspect ratios for the FDCCII

Transistors	W(μm)	L(μm)
M,-M_6	8.75	0.7
M_7–M_9, M_{13}, M_{17}, M_{21}, M_{27}, M_{IP}–M_{3P}	70	0.7
M_{10}–M_{12}, M_{16}, M_{18}, M_{IN}–M_{3N}	17.5	0.7
M_{14}, M_{15}, M_{19}, M_{20}	0.7	0.7
M_{22}, M_{28}, M_{AN}, M_{BN}, M_{CN}, M_{KN}	3.5	0.7
M_{23}–M_{24}, M_{29}–M_{30}	35	0.7
M_{25}–M_{26}, M_{31}–M_{32}	105	0.7
M_{AP}, M_{BP}, M_{CP}, M_{KP}	7	0.7

Realization of nth-Order Current Transfer Function Employing ECCIIs and Application Examples

In many applications, it is required to change the coefficients of the filter transfer function (TF). Then further research has been focused on current conveyors with adjustable current gain (Senani 1980a, b; Surakampontorn and Thitimajshima 1988; Surakampotorn and Kumwatchara 1992; Fabre and Mimeche 1994; Papazoglou and Karybakas 1997; Sayın 2004; Minaei et al. 2006a, b). The electronically tunable current conveyor (ECCII) which has a controllable current gain was first proposed by Senani (1980a, b) using an operational amplifier (OA) and an operational transconductance amplifier (OTA). Then the ECCII was designed and realized in both CMOS and bipolar technology (Surakampontorn and Thitimajshima 1988; Surakampotorn and Kumwatchara 1992; Fabre and Mimeche 1994; Minaei et al. 2006b). High-order filters are needed and widely used for realization of communication systems. Several nth-order voltage- and current-mode TF synthesis methods and circuits are available in the literature employing active elements such as CCIIs and current differencing buffered amplifiers (CDBAs) (Tek and Anday 1989; Chang and Chen 1991; Gunes and Anday 1995; Acar and Özoğuz 1996, 2000). However, these circuits suffer from the lack of electronic tunability which is very important for today's modern circuit design. High-order filters employing ECCIIs seem to bring a solution to this problem (Sayın 2004; Minaei et al. 2006a). All coefficients of the current TF can be tuned independently by adjusting the current gain of the ECCIIs used in the configuration. Moreover, a second-order current-mode filter which realizes low-pass, high-pass, band-pass, and notch responses is derived from the proposed configuration. Furthermore, a fourth-order video band filter is designed using second-order low-pass and low-pass notch filters connected in cascade. Simulation results using transistor level implementation for the ECCIIs are given to confirm the theoretical analysis.

The ECCII and Its CMOS Implementation

The electrical symbol of an ECCII is shown in Fig. 3.12. The terminal current–voltage relations of an ideal ECCII can be given by

$$\begin{bmatrix} i_y \\ v_x \\ i_z \end{bmatrix} = \begin{bmatrix} 0 & 0 & 0 \\ 1 & 0 & 0 \\ 0 & \pm k & 0 \end{bmatrix} \begin{bmatrix} v_y \\ i_x \\ v_z \end{bmatrix} \tag{3.21}$$

The ECCII, like a CCII, has a unity voltage gain between its Y and X terminals but unlike a CCII has a tunable current gain +k between its X and Z terminals. The plus and minus signs of k denote positive-type (ECCII+) and negative-type (ECCII−) conveyors, respectively. Recently, an improved CMOS realization of the ECCII has been introduced (Minaei et al. 2006b). This CMOS implementation of the positive-type electronically tunable current conveyor (ECCII+) is shown in Fig. 3.13.

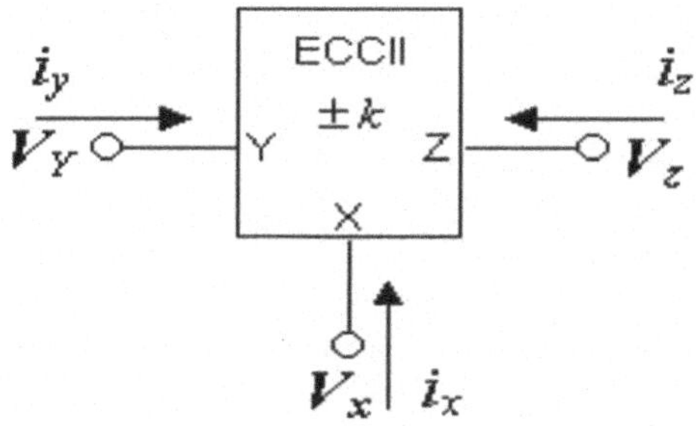

Fig. 3.12 Electrical symbol of the ECCII. (Minaei et al. 2006b)

$$i_z = nI_1 - nI_2 = n\left(\frac{I_B}{2I_A}\right) \cdot i_x = k \cdot i_x \tag{3.22}$$

where

$$k = n\left(\frac{I_B}{2I_A}\right) \tag{3.23}$$

is the small-signal current gain of the amplifier and can be controlled electronically by changing DC bias currents I_A and I_B. The parameter n is the current multiplication factor of the current mirrors used in the output stage of the ECCII+. It should be noted that to implement a negative-type electronically tunable current conveyor (ECCII−), it is sufficient to apply the currents $I_B + i_x$ and $I_B - i_x$ into terminals B and A of the small-signal current amplifier, respectively.

The ECCII can be used for realization of the nth-order current TF (Minaei et al. 2006a). An nth-order current TF can be expressed as

$$T(s) = \frac{I_{out}}{I_{in}} = \frac{a_n s^n + a_{n-1}s^{n-1} + \cdots\cdots\cdots + a_1 s + a_0}{s^n + b_{n-1}s^{n-1} + \cdots\cdots\cdots + b_1 s + b_0} \tag{3.24}$$

where I_{in} and I_{out} are the input and output currents, respectively. Related circuit is given in Fig. 3.14.

$$\begin{aligned}\frac{I_{out}}{I_{in}} = &\left[k_{a_n}s^n\left(\frac{R_1}{R_{n+2}}\right) + k_{a_{n-1}}s^{n-1}\left(\frac{R_1}{R_{n+4}R_2C_1}\right) + k_{a_{n-2}}s^{n-2}\left(\frac{R_1}{R_{n+6}R_2R_3C_1C_2}\right)\right.\\ &\left.+ .. + k_{a_0}\left(\frac{R_1}{R_{3n+2}(R_2..R_{n+1})(C_1..C_n)}\right)\right]\\ &\Big/\left[s^n + k_{b_{n-1}}s^{n-1}\left(\frac{R_1}{R_{n+3}R_2C_1}\right) + k_{b_{n-2}}s^{n-2}\left(\frac{R_1}{R_{n+5}R_2R_3C_1C_2}\right)\right.\\ &\left.+ \ldots + k_{b_0}\left(\frac{R_1}{R_{3n+1}(R_2\ldots..R_{n+1})(C_1\ldots.C_n)}\right)\right]\end{aligned} \tag{3.25}$$

The coefficients k_{ai} and k_{bj} ($i = 0...n$ and $j = 0...n-1$) are the current gains of the ECCIIs, which can be controlled electronically. Each of the ECCIIs is used to adjust an individual coefficient in the current TF. It means that each coefficient can be tuned independently by adjusting the current gain of the relevant ECCII. This is the most important advantage of the proposed circuit compared with the conventional designs, which makes it very attractive for analog designers. For the realization of an nth-order filter, $3n + 2$ active elements are necessary in general form which could be considered as a large element number. Note that this large number can be reduced for special realization purposes such as low-pass (LP), high-pass (HP), band-pass (BP), and band-stop (BS) filter functions by removing the unused sections

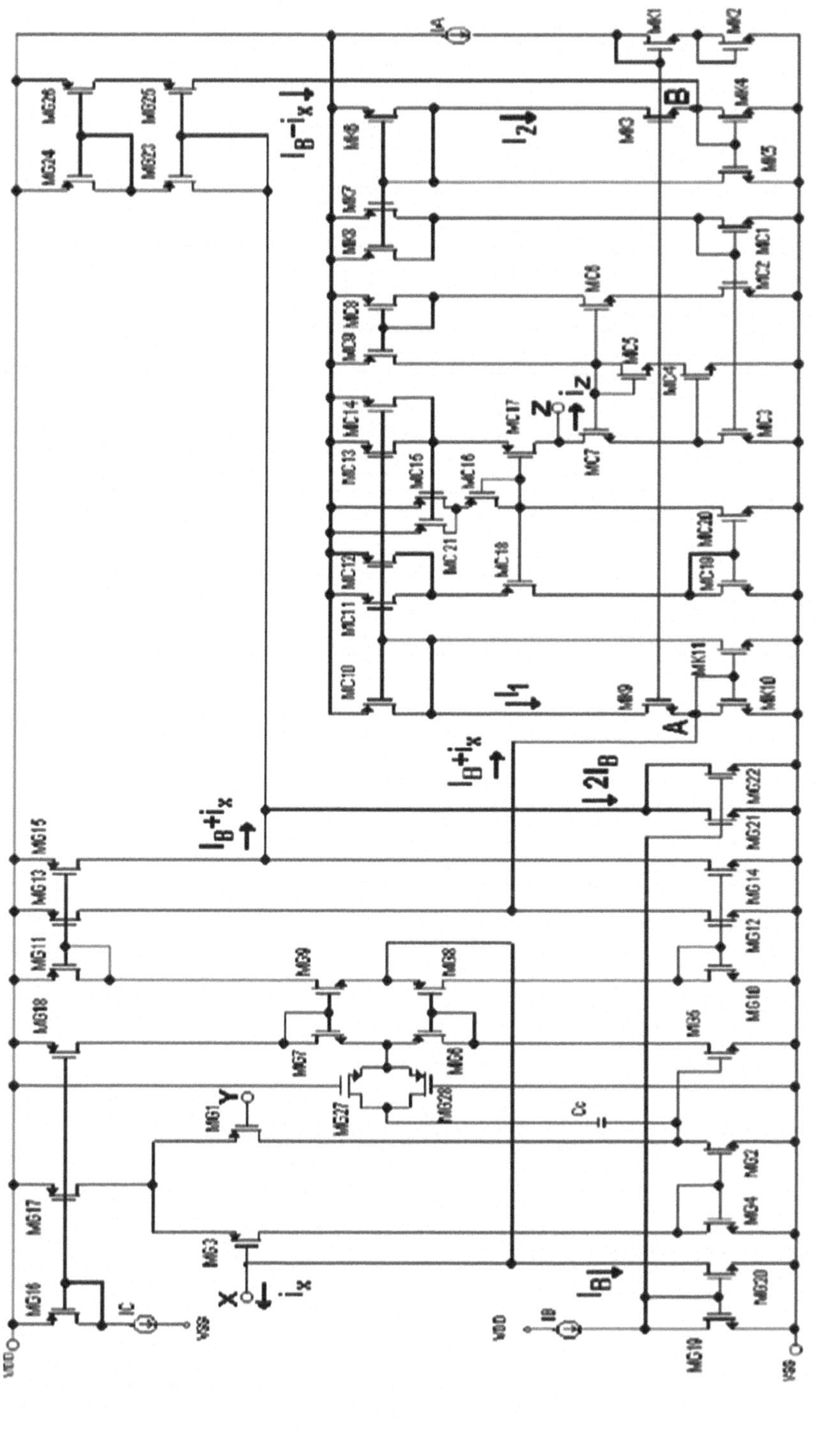

Fig. 3.13 CMOS realization of the ECCII+ (Minaei et al. 2006b) The output current of the circuit i_z can be calculated as (Surakampotorn and Kumwatchara 1992; Minaei et al. 2006b)

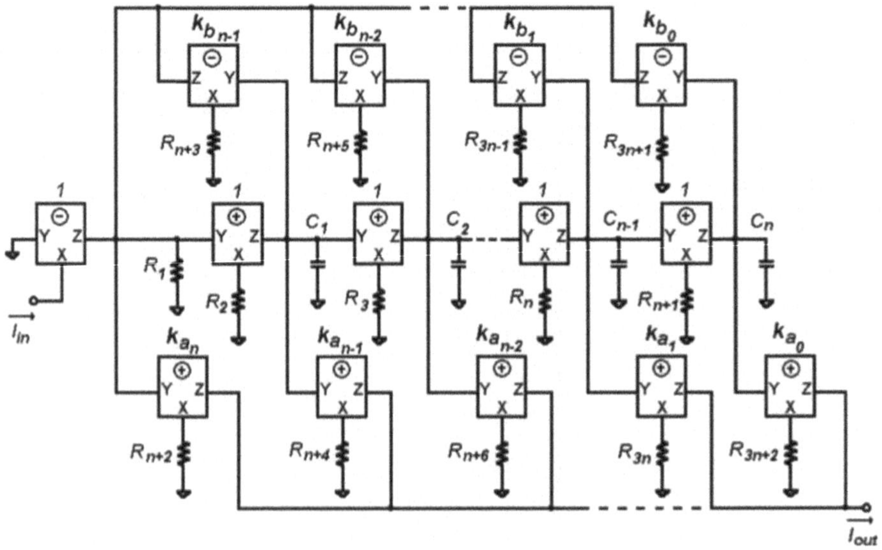

Fig. 3.14 Implementation of the nth-order current TF using ECCII (Minaei et al. 2006b)

depending on the aim of the design. The remaining part is necessary for the tuning of the filter parameters independently.

Second-Order Filter Using ECCII

Figure 3.15 shows the proposed circuit which realizes a second-order current TF. It can be considered as a filter with low-pass, high-pass, and band-pass responses at high output impedances. The sum of the LP, HP, and BP responses yields the circuit TF as

$$T(s) = \frac{I_{\text{out}}}{I_{\text{in}}} = \frac{I_{\text{HP}} + I_{\text{BP}} + I_{\text{LP}}}{I_{\text{in}}} \\ = \frac{k_{a_2}\left(\frac{R_1}{R_4}\right)s^2 + k_{a_1}\left(\frac{R_1}{R_6R_2C_1}\right)s + k_{a_0}\left(\frac{R_1}{R_8R_2R_3C_1C_2}\right)}{s^2 + k_{b_1}\left(\frac{R_1}{R_5R_2C_1}\right)s + k_{b_0}\left(\frac{R_1}{R_7R_2R_3C_1C_2}\right)} \tag{3.26}$$

The ω_0 and Q parameters of the filter are obtained as

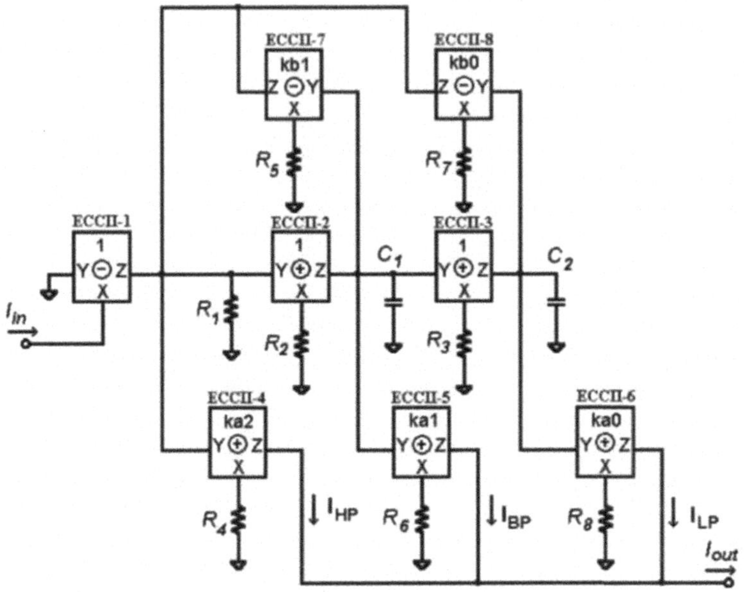

Fig. 3.15 Second-order current-mode filter realization. (Minaei et al. 2006b)

$$\omega_0 = \sqrt{\frac{k_{b0}R_1}{R_7R_2R_3C_1C_2}} \tag{3.27}$$

$$Q = \frac{R_5}{k_{b1}}\sqrt{\frac{k_{b0}R_2C_1}{R_1R_7R_3C_2}} \tag{3.28}$$

It can be seen that the parameter ω_0 can be tuned electronically by adjusting k_{b0}. By keeping the value of k_{b0} constant and varying k_{b1}, the parameter Q can also be tuned without disturbing the parameter ω_0. Also the gain factors of the HP, BP, and LP responses can be tuned electronically by changing k_{a2}, k_{a1}, and k_{ao}, respectively, without disturbing the parameters ω_0 and Q. Sensitivity analysis of this filter gives

$$\begin{aligned} S_{R_7}^{\omega_0} = S_{R_2}^{\omega_0} = S_{R_3}^{\omega_0} = S_{C_1}^{\omega_0} = S_{C_2}^{\omega_0} = -S_{k_{b0}}^{\omega_0} = -S_{R_1}^{\omega_0} = -\frac{1}{2} \\ S_{R_1}^{Q} = S_{R_7}^{Q} = S_{R_3}^{Q} = S_{C_2}^{Q} = -S_{k_{b0}}^{Q} = -S_{R_2}^{Q} = -S_{C_1}^{Q} = -\frac{1}{2} \\ S_{R_5}^{Q} = -S_{k_{b1}}^{Q} = 1 \end{aligned} \tag{3.29}$$

Related simulation results are illustrated in Figs. 3.16, 3.17, 3.18, and 3.19. Finally, the tunability of the pole natural frequency of the BP response is tested by selecting different values for k_{b0}. Different biasing currents of $I_B = 25$ μA, 50 μA, and 100 μA are selected for the ECCII-8 which results in $k_{b0} = 0.5$, 1, and 2 and $f_o =$ 0.707 MHz, 1 MHz, and 1.41 MHz, respectively, as shown in Fig. 3.19. It can be

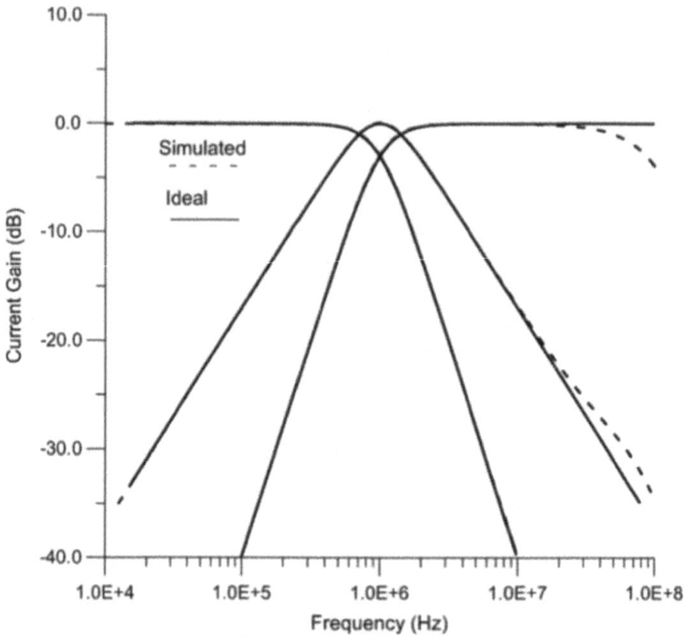

Fig. 3.16 Simulated and ideal LP, BP, and HP filter responses; $Q = 0.707, f_0 = 995$ kHz $\rightarrow$1 MHz, $R_5 = R_6 = 5.65$ kOhm, $R_i = 8$ kOhm (other resistance values), $C_i = 20$ pF, $k_i = 1$ (gain of the current conveyors). (Minaei et al. 2006b)

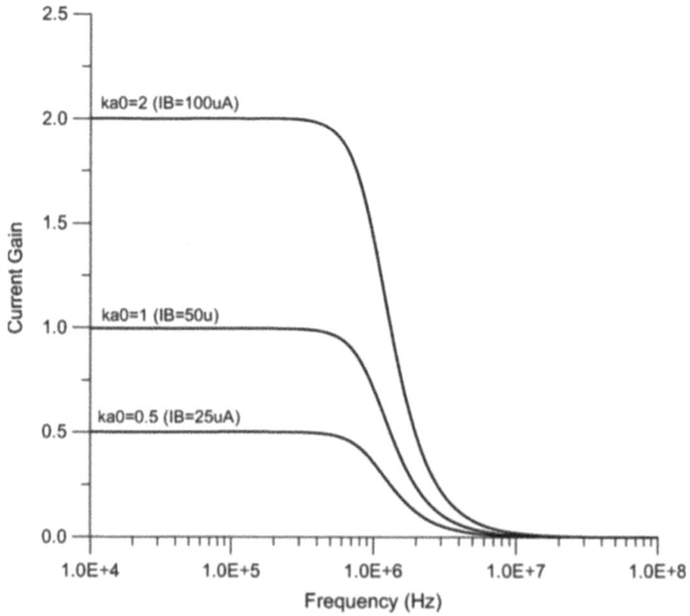

Fig. 3.17 Tuning of the gain for LP response (gain = 0.5, 1, 2). (Minaei et al. 2006b)

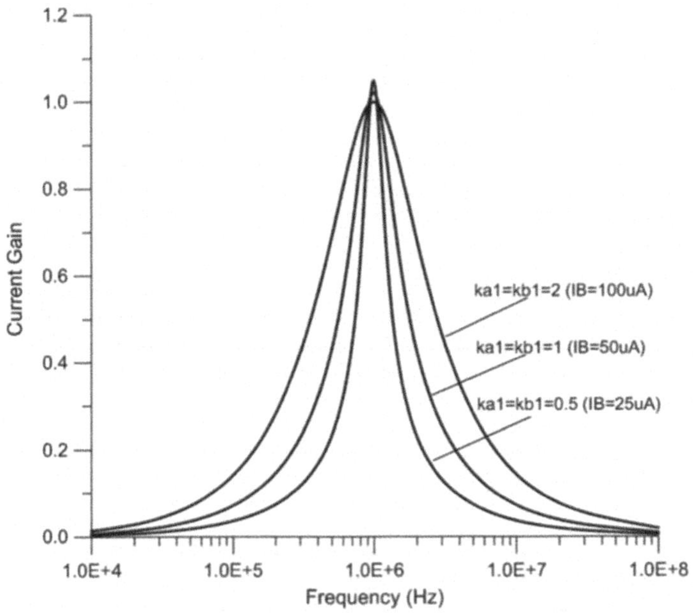

Fig. 3.18 Tuning of the quality factor of BP response ($Q = 0.707, 1.414, 2.828$) (Minaei et al. 2006b)

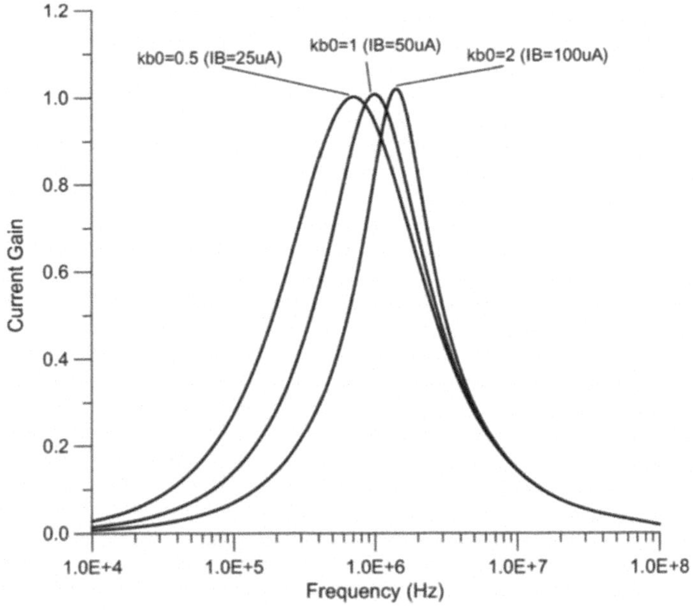

Fig. 3.19 Tuning of the natural frequency of BP response ($f_o = 0.707$ MHz, 1 MHz, 1.41 MHz). (Minaei et al. 2006b)

easily observed that the simulation results are in good agreement with the theoretical values. In the following, a design example of a fourth-order video band filter is given to illustrate the possibilities in analog circuit design provided by the topology introduced. The TF of the filter is

$$H(s) = H\frac{s^2 + w_{Z1}^2}{s^2 + \frac{w_{P1}}{Q_{P1}}s + w_{P1}^2}\frac{w_{P2}^2}{s^2 + \frac{w_{P2}}{Q_{P2}}s + w_{P2}^2} \tag{3.30}$$

The element values are chosen as $R_5 = 2.7$ kOhm, $R_i = 4$ kOhm (other resistance values), $C_i = 15$ pF, $k_i = 1$ (gain of the current conveyors) for the LP section and $R_5 = 7.9$ kOhm, $R_4 = 4.6$ kOhm, $R_i = 2.3$ kOhm (other resistance values), $C_i = 20$ pF, $k_i = 1$ (gain of the current conveyors) for the LP notch section. where

$$H = w_{P1}^2/w_{Z1}^2 \approx 2, f_{P1} = 3.46\text{MHz}, Q_{P1} = 3.42, f_{Z1} = 4.83\text{MHz},$$
$$f_{P2} = 2.65\text{MHz}, Q_{P2} = 0.675.$$

This TF is realized by cascading an LP notch filter with an LP section. Simulated and ideal fourth-order filter responses are given in Fig. 3.20.

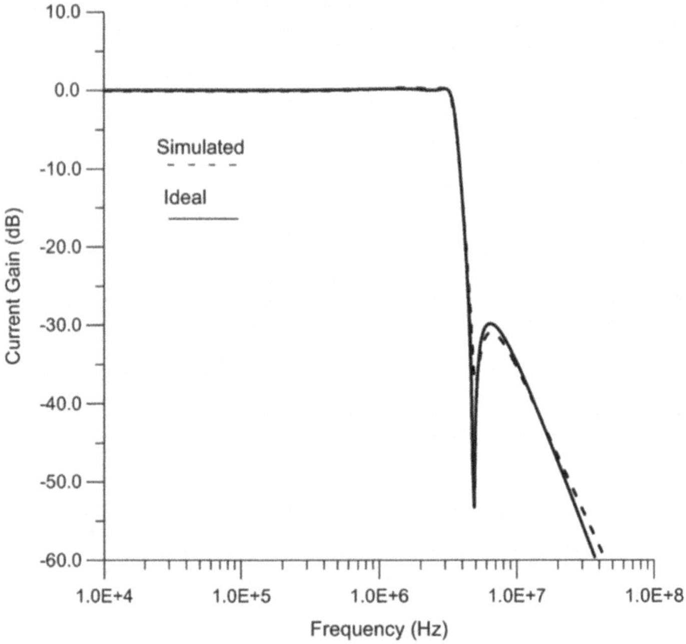

Fig. 3.20 Simulated and ideal fourth-order filter responses. (Minaei et al. 2006b)

CCCII, Current-Controlled Conveyors

At present there is a growing interest in designing current-mode current conveyor (CC)-based active filters. This is attributed to their larger signal bandwidth, greater linearity, wider dynamic range, simple circuitry, and low power consumption (Roberts and Sedra 1989). Therefore, some current-mode filters using current conveyors have been proposed (Aronhime et al. 1990; Fabre et al. 1990; Higashimura and Fukui 1990; Liu et al. 1990a, b; Alami and Fabre 1991; Chang 1991, 1993a, b; Chang et al. 1993). However, all of these filters suffer from the lack of electronic adjustability. Moreover, the circuits presented in Alami and Fabre (1991) and Chang (1993) suffer from the lack of high output impedance. A current-mode filter theoretically should exhibit high output impedance to enable easy cascadability and to enable additional filter responses by simply connecting the outputs. By using the second-generation current-controlled conveyor (CCCII) introduced in Fabre et al. (1995b), current conveyor applications can be extended to the domain of electronically adjustable functions. Electronic adjustability of the CCCII is attributed to the dependence of the parasitic resistance at port x on the bias current of the current conveyor. Therefore, in recent past, there has been great emphasis on the design of current-mode circuits using current-controlled conveyors (Fabre et al. 1995b; Abuelma'atti and Tasadduq 1998a, b; Khan and Zaidi 2000). However, these filters suffer from the use of three to six dual-output CCCIIs (Abuelma'atti and Tasadduq 1998a, b) or the outputs of the filter responses are not in high output impedance (Khan and Zaidi 2000). The band-pass filter proposed in Fabre et al. (1995b) uses two CCCIIs and grounded capacitors. However, the output of this filter also does not exhibit high output impedance. In this section, we introduce three new configurations for realizing low-pass, band-pass, and high-pass filter responses all at high output impedance using one or two current-controlled conveyors and reduced number of passive elements. The proposed circuits enjoy current control of the parameter ωo without disturbing the parameter ωo/Q. The proposed filters are simulated with PSPICE to verify the theoretical analysis.

The port relations of a CCCII as shown in Fig. 3.21 can be characterized by

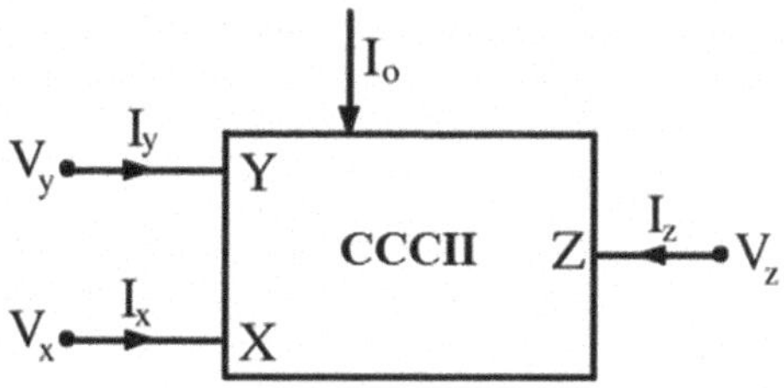

Fig. 3.21 Electrical symbol of the CCCII. (Fabre et al. 1995b)

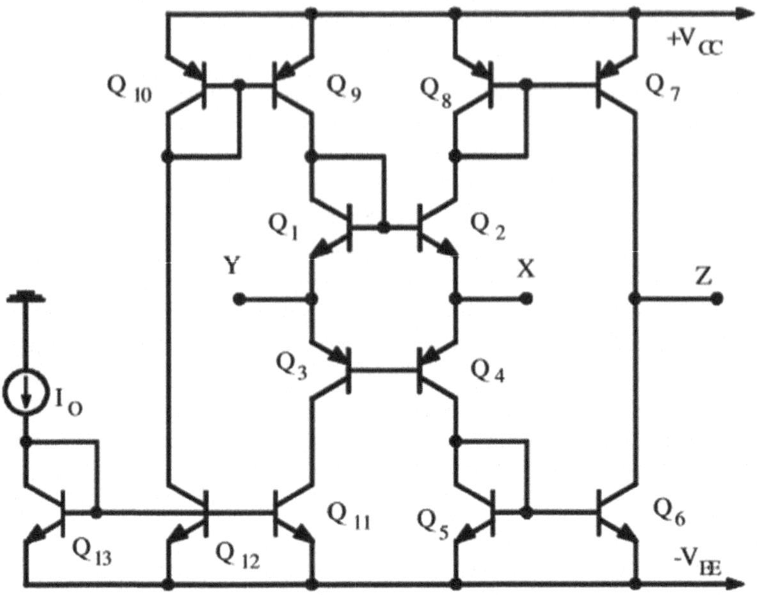

Fig. 3.22 Bipolar realization of CCCII. (Fabre et al. 1995b)

$$\begin{bmatrix} I_y \\ V_x \\ I_z \end{bmatrix} = \begin{bmatrix} 0 & 0 & 0 \\ 1 & R_x & 0 \\ 0 & \pm 1 & 0 \end{bmatrix} \begin{bmatrix} V_y \\ I_x \\ V_z \end{bmatrix} \tag{3.31}$$

where the positive sign denotes a positive current-controlled conveyor (CCCII+) and the negative sign denotes a negative current-controlled conveyor (CCCII−).

Bipolar realization circuit of CCCII1 is illustrated in Fig. 3.22 (Fabre et al. 1995b). For the circuit given in Fig. 3.22 the parasitic resistance R_x can be expressed as

$$R_x = \frac{V_{\mathrm{T}}}{2I_{\mathrm{o}}} \tag{3.32}$$

where V_{T} is the thermal voltage and I_{o} is the bias current of the CCCII as given in Fabre et al. (1995b). The three proposed circuits are shown in Fig. 3.23. The circuit of Fig. 3.23a comprises one CCCII, two capacitors, and one resistor and realizes band-pass filter at high output impedance. Routine analysis of this circuit yields

$$\frac{I_{BP}}{I_{in}} = \frac{\frac{1}{C_1 R_x} s}{s^2 + \frac{(C_1+C_2)}{R_1 C_1 C_2} s + \frac{1}{R_1 R_x C_1 C_2}} \tag{3.33}$$

and the parameters can be given as

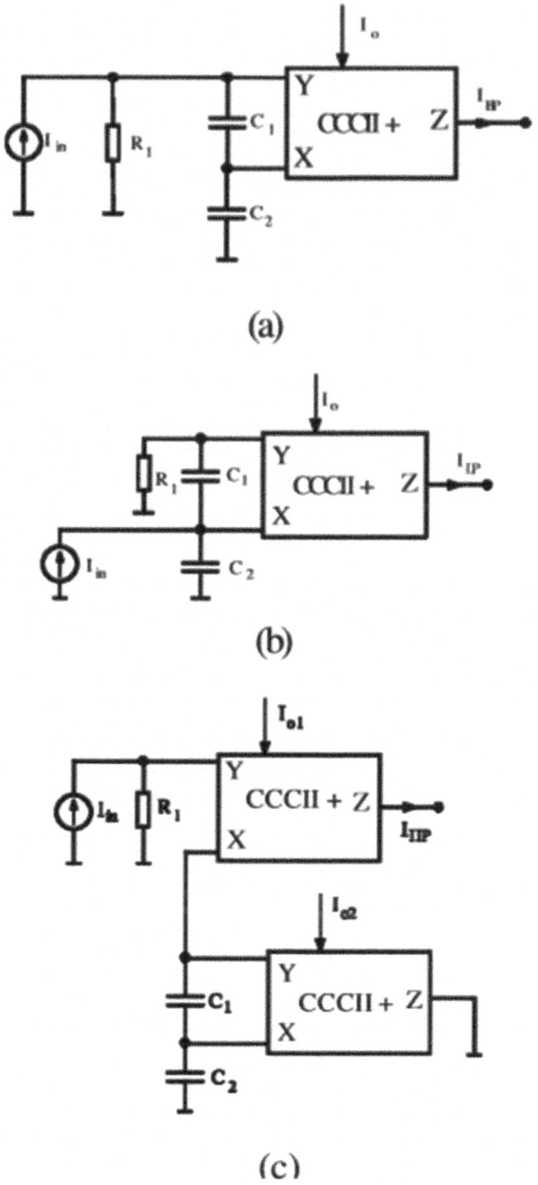

Fig. 3.23 Proposed current-mode current-controlled filters. (**a**) Band-pass filter, (**b**) low-pass filter, and (**c**) high-pass filter. (Minaei et al. 2001)

$$\omega_o = \sqrt{\frac{1}{R_1 R_x C_1 C_2}}$$
$$\frac{\omega_o}{Q} = \frac{(C_1 + C_2)}{R_1 C_1 C_2} \tag{3.34}$$

The third proposed circuit shown in Fig. 3.23c employs two CCCII+, two capacitors, and one resistor and produces high-pass response at high output impedance. The transfer function of this circuit is given as

$$\frac{I_{HP}}{I_{in}} = \frac{\frac{R_1}{R_{x1}} s^2}{s^2 + \frac{(C_1+C_2)}{R_{x1}C_1C_2} s + \frac{1}{R_{x1}R_{x2}C_1C_2}} \tag{3.35}$$

$$\omega_o = \sqrt{\frac{1}{R_{x1} R_{x2} C_1 C_2}}$$
$$\frac{\omega_o}{Q} = \frac{(C_1 + C_2)}{R_{x1} C_1 C_2} \tag{3.36}$$

Bipolar realization circuit of CCCII+ is illustrated in Fig. 3.22 (Fabre et al. 1995b). For the circuit given in Fig. 3.22 the parasitic resistance Rx can be expressed as $R_x = V_T/2I_o$ where V_T is the thermal voltage and I_o is the bias current of the CCCII as given in Fabre et al. (1995b). The three proposed circuits are shown in Fig. 3.23. The circuit of Fig. 3.23a comprises one CCCII+, two capacitors, and one resistor and realizes band-pass filter at high output impedance.

It can be seen that the parameter ω_o can be controlled electronically by adjusting the bias current I_{o2} without disturbing the parameter ω_o/Q. Sensitivity analysis shows that

$$S^{\omega_o}_{R_{x1}} = S^{\omega_o}_{R_{x2}} = S^{\omega_o}_{C_1} = S^{\omega_o}_{C_2} = -\frac{1}{2}, \quad S^{\omega_o}_{I_{o1}} = S^{\omega_o}_{I_{o2}} = \frac{1}{2}, \quad S^{\omega_o}_{V_T} = -1$$
$$S^{\omega_o/Q}_{C_1} = \frac{-C_2}{C_1 + C_2}, \quad S^{\omega_o/Q}_{C_2} = \frac{-C_1}{C_1 + C_2}, \quad S^{\omega_o/Q}_{R_{x1}} = -1, \quad S^{\omega_o/Q}_{R_{x2}} = 0$$
$$S^{\omega_o/Q}_{I_{ol}} = 1, \quad S^{\omega_o/Q}_{I_{o2}} = 0, \quad S^{\omega_o/Q}_{V_T} = -1$$

Electronically Tunable, Active-Only Floating Inductance Simulation

The applications and advantages in the realization of various filter transfer functions using current conveyors have received considerable attention (Wilson 1990). Moreover, because of the desire to produce solid-state filters without the use of physical coils, attention soon became focused on the simulation of floating inductor using active elements. A literature survey shows that a large number of circuit realizations for lossy and lossless floating inductors have been reported (Senani 1979, 1980a, b, 1987; Pal 1981a, b; Toumazou and Lidgey 1985; Higashimura and Fukui 1987, 1989; Hou et al. 1993; Chang et al. 1994; Kiranon and Pawarangkoon 1997). However, all of these circuits suffer from one or more of the following drawbacks:

1. The use of an excessive number of current conveyors (CCIIs) or operational amplifiers (OAs) (Pal 1981a, b; Toumazou and Lidgey 1985; Higashimura and Fukui 1987; Senani 1987; Hou et al. 1993; Kiranon and Pawarangkoon 1997)
2. Active or passive matching condition (Pal 1981a, b; Senani 1987; Chang et al. 1994)
3. Lack of electronic tunability (Senani 1979, 1980a, b, 1987; Pal 1981a, b; Toumazou and Lidgey 1985; Higashimura and Fukui 1987; Hou et al. 1993; Chang et al. 1994; Kiranon and Pawarangkoon 1997)

The recently introduced second-generation current-controlled conveyor (CCCII) (Fabre et al. 1996) allows current conveyor applications to be extended to the domain of electronically tunable functions. Although electronic tunability in inductance simulation circuits can be obtained using conventional operational transconductance amplifiers (OTAs) (Higashimura and Fukui 1989), comparing the transconductance (gm) of the OTA with the equivalent transconductance of the CCCII (1/Rx), it can be seen that for the same value of gm the biasing current of the CCCII will be four times less than the OTA (Fabre et al. 1996). Therefore, the use of CCCII instead of OTA leads to lower power consumption. Also, with very high values for the collector currents of the transistors, their maximum usable frequency will be reached sooner (Fabre et al. 1996). This consequently indicates that the frequency potential of circuits with CCCII will be much greater than that for OTA implementations. In this paper we present a new active-only inductance simulation circuit using only one dual-output CCCII and one operational amplifier (OA). The value of the simulated inductance can be controlled by adjusting the bias current of the CCCII over a wide range. The proposed circuit enjoys low active sensitivities and does not require any parameter matching condition. The use of only active elements in the circuit makes it suitable for use in integrated circuits (IC).

The proposed active-only floating inductance is shown in Fig. 3.24. It consists of a dual-output translinear bipolar second-generation current-controlled conveyor (CCCII) and an operational amplifier (OA). The dual-output translinear bipolar CCCII can be obtained by modifying the original circuit of the CCCII (Fabre et al. 1996) by adding additional cross-coupled current mirrors to obtain the required minus type output. Using the standard notations as in Fabre et al. (1996), the ideal dual-output CCCII can be characterized by

$$I_{z\pm} = \pm I_x, \quad I_y = 0, \quad V_x = I_x R_x + V_y,$$

where $R_x = V_T/2I_0$ is the input resistance at terminal X, V_T is the thermal voltage (26 mV at 27 °C), and Io is the bias current. The open-loop gain of a practical

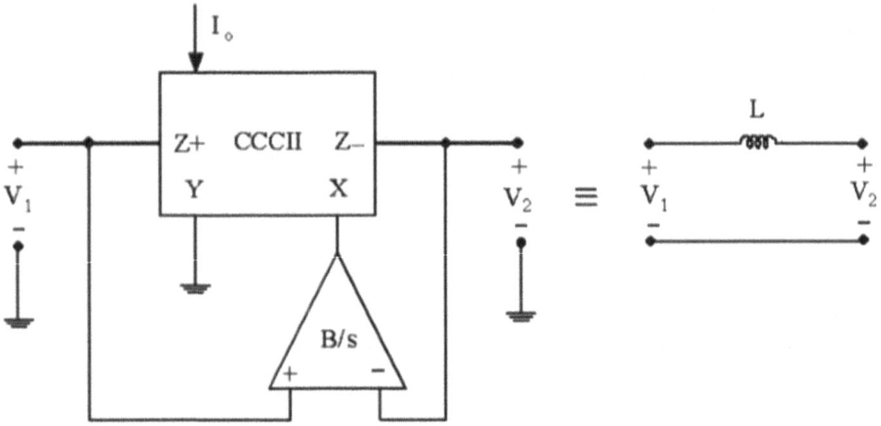

Fig. 3.24 Proposed active-only floating inductance simulator. (Mineai et al. 2003)

operational amplifier (OA) is represented by the well-known first pole roll-off characteristic (Abuelma'atti and Almaskati 1988) as

$$A(s) = \frac{A_o\omega_{p1}}{s + \omega_{p1}} = \frac{B}{s + \omega_{p1}} \tag{3.37}$$

In the frequency range Lp equation is reduced to

$$A(s) \cong \frac{B}{s} \tag{3.38}$$

This integrator model of OA is valid from a few kHz to a few hundred kHz range. Routine analysis of the circuit in Fig. 3.1 gives the short circuit admittance matrix as

$$\mathrm{Y} = \frac{B}{sR_x}\begin{bmatrix} 1 & -1 \\ -1 & 1 \end{bmatrix} \tag{3.39}$$

Therefore, the circuit realizes a lossless floating inductance given by

$$L = \frac{R_x}{B} = \frac{V_T}{2I_oB} \tag{3.40}$$

Equation (3.4) shows that the inductance value (L) can be tuned electronically by varying the bias current (I_0). Variation of the simulation inductance with the bias current of the CCCII is illustrated in Fig. 3.25. The plot of the frequency characteristic of the inductance simulation is given in Fig. 3.26.

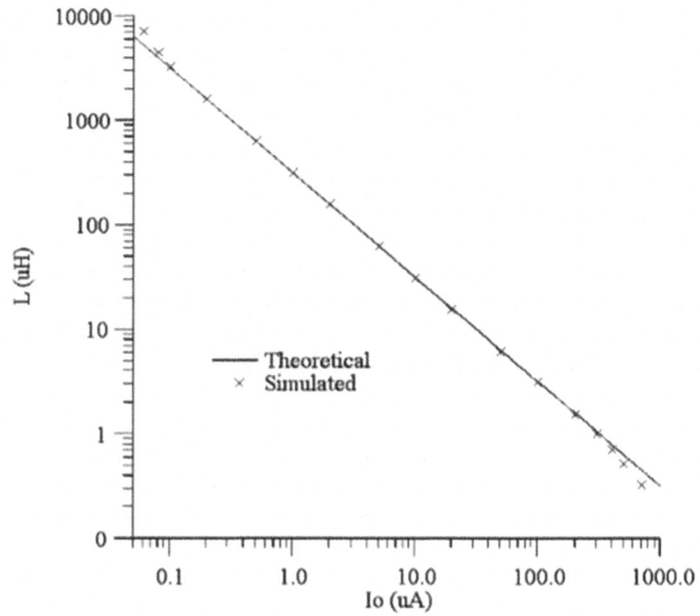

Fig. 3.25 Variation of the simulation inductance with the bias current of the CCCII. (Mineai et al. 2003)

Fig. 3.26 Plot of the frequency characteristic of the proposed inductance (Mineai et al. 2003)

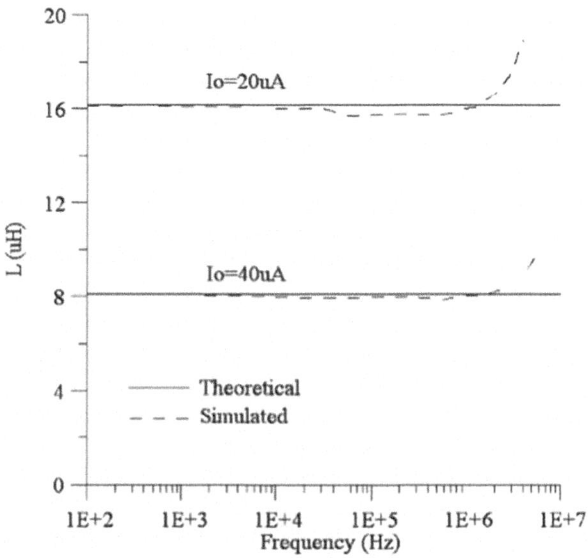

As an application the described circuit is used to construct a fourth-order elliptic LC low-pass filter shown in Fig. 3.27 with element values of $L_1 = 270$ μH, $L_2 = 380$ μH, $C_1 = 0.2$ μF, $C_2 = .399$ μF, $C_3 = 38$ μF, $R_s = R_L = 10$ (Filoramo et al. 2000). The filter specifications are as follows:

(a) 5 kHz bandwidth with lower than 0.1 dB in-band ripple
(b) Stop-band attenuation equal to 40 dB at 11 kHz
(c) Equal terminating resistance of 10

The corresponding biasing currents of the CCCIIs for simulation of inductors L_1 and L_2 are selected as $I_{o1} = 1.198$ mA and $I_{o2} = 0.842$ mA, respectively. The input dynamic range of the filter for output THD<1% is found to be 50 mV. The simulated frequency response of the filter is shown in Fig. 3.28.

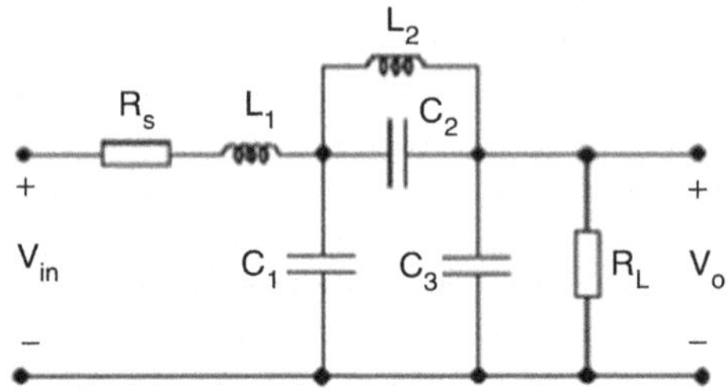

Fig. 3.27 Fourth-order elliptic low-pass filter. (Mineai et al. 2003)

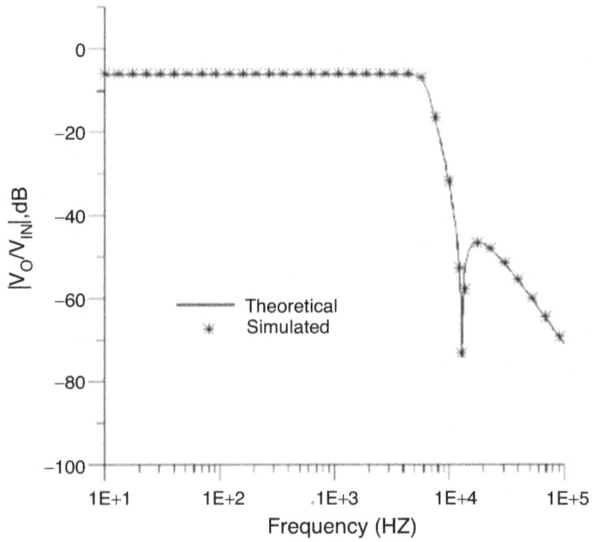

Fig. 3.28 Frequency response of the elliptic low-pass filter. (Mineai et al. 2003)

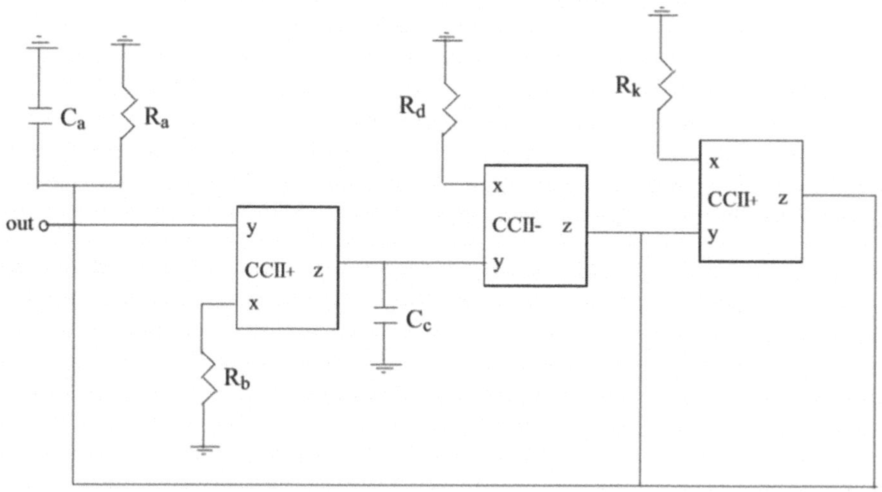

Fig. 3.29 CCII-based oscillator circuit. (Çam and Kuntman 1998)

A New CCII-Based Sinusoidal Oscillator Employing Grounded Resistors and Capacitors

In this section a new CCII-based double-element-controlled sinusoidal oscillator employing grounded resistors and capacitors is presented. The oscillator topology illustrated in Fig. 3.29 provides entirely independent control of oscillation condition and oscillation frequency and suitability for VLSI realization due to inclusion of only grounded passive elements. The circuit exhibits low active and passive sensitivities; voltage and current tracking error effects of the current conveyors can be eliminated easily with careful design. Theoretical results are confirmed with PSPICE simulations.

Oscillation condition and oscillation frequency are obtained as

$$R_a = R_k$$
$$\omega_0 = \sqrt{\frac{1}{R_b R_d C_a C_c}} \tag{3.41}$$

Oscillation condition and oscillation frequency are completely independent from each other.

Effect of CCII Non-idealities The dominant current conveyor non-idealities can be modeled as follows:

$$R_a = \frac{R_k}{\beta}$$
$$\omega_0 = \sqrt{\frac{\alpha\beta}{R_b R_d C_a C_c}} \tag{3.42}$$

where $\beta = 1-\varepsilon_v$ is the voltage gain, $\alpha = 1-\varepsilon_i$ is the current gain, and ε_i, ε_v denote current and voltage tracking errors, respectively.

Sensitivity Consideration The relative sensitivity of oscillation frequency with respect to resistors R_b, R_d, capacitors C_a, C_c, and tracking errors ε_i, ε_v can be obtained as

$$S_{R_b}^{\omega_0} = S_{R_d}^{\omega_0} = S_{C_c}^{\omega_0} = S_{C_a}^{\omega_0} = -0.5$$
$$S_{\alpha}^{\omega_0} = S_{\beta}^{\omega_0} = 0.5$$

It is clearly observed that the circuit has low passive and active sensitivity.

Simulations

To confirm theoretical results, the proposed oscillator circuit has been simulated using PSPICE program and the simulation result is given in Figs. 3.30 and 3.31. The circuit was constructed with the commercial current conveyor IC AD844 of analog device. Since AD844 is designed as CCII+, the CCII- was obtained by cascading two ICs where the X terminal of the second is driven from the Z terminal of the first AD844. The supply voltages were taken as $V_{DD} = 10$ V and $-V_{SS} = -10$ V.

Concluding Remarks

A new sinusoidal oscillator constructed with three CCIIs, two grounded capacitors, and four grounded resistors is introduced. Theoretical results are confirmed by PSPICE simulation. The most important advantage of the proposed topology is the possibility of entirely independent control of oscillation condition and oscillation frequency. Moreover, the oscillator is suitable for both monolithic IC and thin film fabrication due to inclusion of only grounded passive elements; as a result of this, it can be easily converted to voltage-controlled oscillator (VCO) using JFETs, MOS resistors, or OTAs as variable resistors instead of resistors R_b and R_d. The circuit shows low passive and active sensitivities, contains only CCIIs as active elements, and exhibits a large tuning range since double-resistance-controlled oscillation frequency is possible. With proper design current and voltage tracking errors can be eliminated easily.

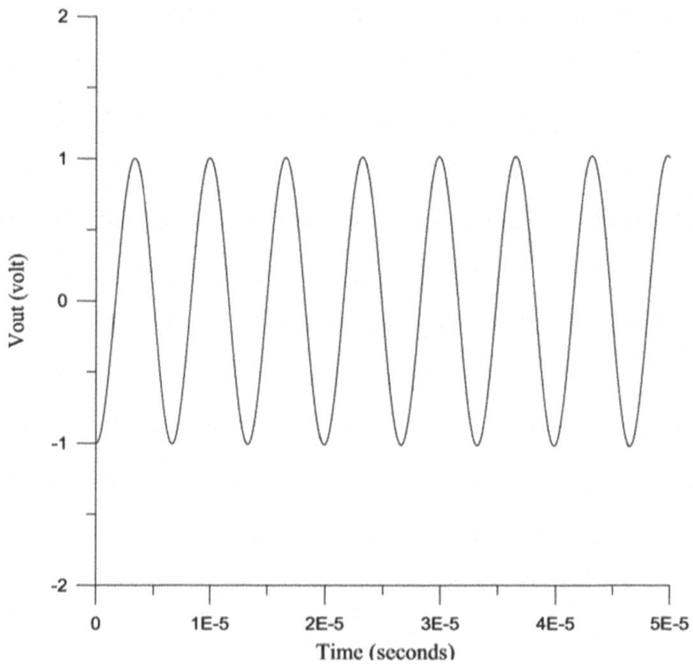

Fig. 3.30 PSPICE simulation result of the proposed circuit ($R_d = R_b = 1$ KΩ, $C_1 = C_2 = 1$ nF, $f = 143$ kHz). (Çam and Kuntman 1998)

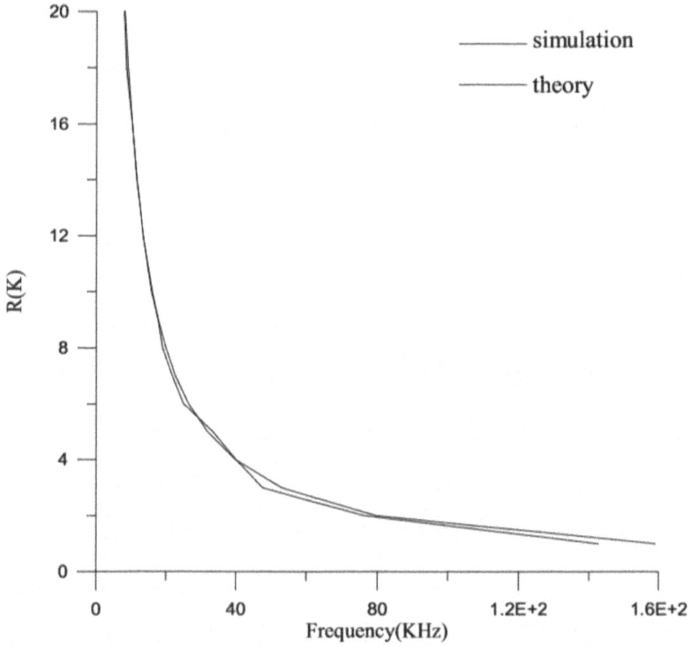

Fig. 3.31 Variation of the oscillation frequency with the $R_d = R_b = R$. (Çam and Kuntman 1998)

Differential Difference Current Conveyor (DDCC)

Several filter circuits are available in the literature using CCII as the active element, while none of them operates in the differential mode. This problem can be solved by using the recently proposed active elements, named differential difference current conveyor (DDCC) and differential voltage current conveyor (DVCC). Some application circuits based on these two new active elements have been tabulated. On the other hand, by introducing dual-output current conveyors (DO-CCIIs) in the filter design, new advantageous filter topologies could be realized.

The DO-DDCC, whose electrical symbol is shown in Fig. 3.32, is a six-terminal network with terminal characteristics described by

$$\begin{bmatrix} V_X \\ I_{Y1} \\ I_{Y2} \\ I_{Y3} \\ I_{Z1} \\ I_{Z2} \end{bmatrix} = \begin{bmatrix} 0 & 1 & -1 & 1 & 0 & 0 \\ 0 & 0 & 0 & 0 & 0 & 0 \\ 0 & 0 & 0 & 0 & 0 & 0 \\ 0 & 0 & 0 & 0 & 0 & 0 \\ 1 & 0 & 0 & 0 & 0 & 0 \\ -1 & 0 & 0 & 0 & 0 & 0 \end{bmatrix} \begin{bmatrix} I_X \\ V_{Y1} \\ V_{Y2} \\ V_{Y3} \\ V_{Z1} \\ V_{Z2} \end{bmatrix} \tag{3.43}$$

Figure 3.33 reflects a CMOS implementation example for DO-DDCC.

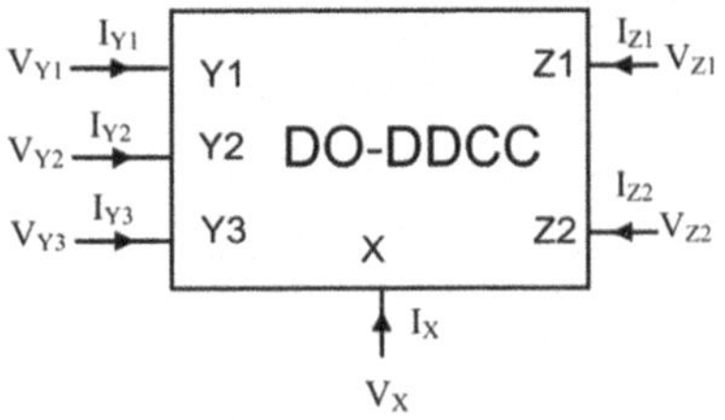

Fig. 3.32 Electrical symbol of DO-DDCC. (Ibrahim and Kuntman 2004)

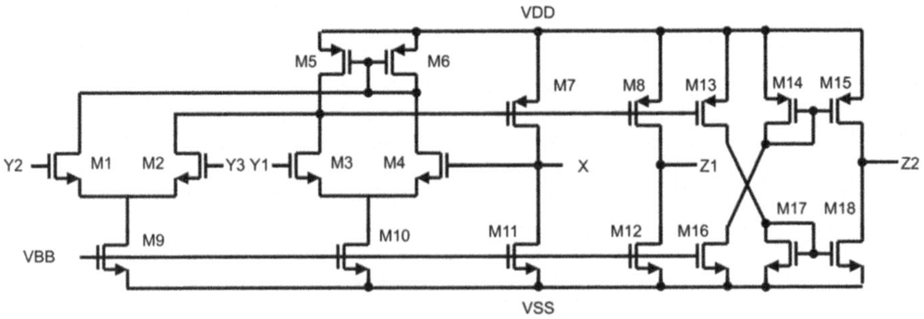

Fig. 3.33 A CMOS implementation for DO-DDCC. (Ibrahim and Kuntman 2004)

DVCC, Differential Voltage Current Conveyor

The DVCC was proposed first by Pal as a modified current conveyor and then developed and realized in CMOS technology by (Elwan and Soliman, 1996). It has the advantages of both the second-generation current conveyor (CCII) (such as large signal bandwidth, great linearity, wide dynamic range) and the differential difference amplifier (DDA) (such as high input impedance and arithmetic operation capability). This element is a versatile building block for applications demanding floating inputs. The electrical symbol of the DVCC is shown in Fig. 3.34 and its terminal relations are given by

$$\begin{bmatrix} V_X \\ I_{Y1} \\ I_{Y2} \\ I_{Z1} \\ I_{Z2} \end{bmatrix} = \begin{bmatrix} 0 & 1 & -1 & 0 & 0 \\ 0 & 0 & 0 & 0 & 0 \\ 0 & 0 & 0 & 0 & 0 \\ 1 & 0 & 0 & 0 & 0 \\ -1 & 0 & 0 & 0 & 0 \end{bmatrix} \begin{bmatrix} I_X \\ V_{Y1} \\ V_{Y2} \\ V_{Z1} \\ V_{Z2} \end{bmatrix} \tag{3.44}$$

Figure 3.35 shows a CMOS implementation example for the DVCC (Ibrahim et al. 2005a, b).

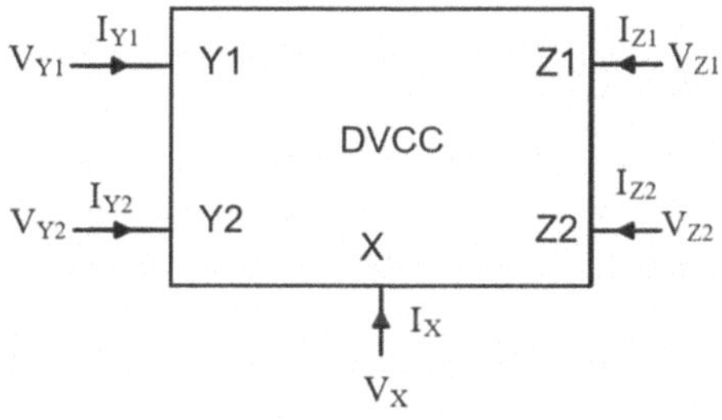

Fig. 3.34 Electrical symbol of the DVCC. (Ibrahim et al. 2005a, b)

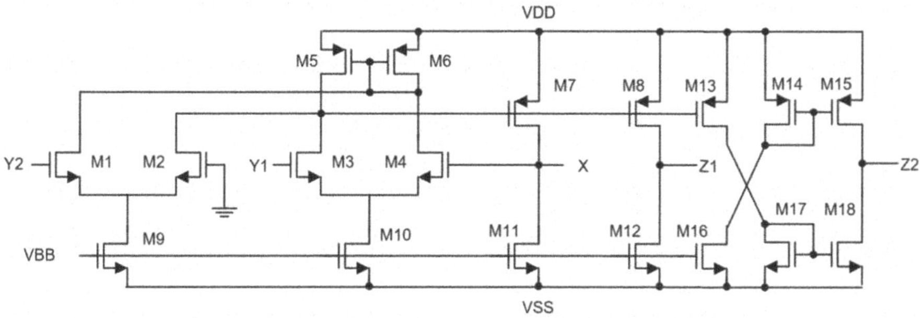

Fig. 3.35 The CMOS implementation of the DVCC. (Ibrahim et al. 2005a, b)

Design Example: A 22.5 MHz Current-Mode KHN Biquad Using Differential Voltage Current Conveyor and Grounded Passive Elements

Of the various methods of multifunction filter design, those based upon current conveyors (CC) or their variants have received recently more attention. On the other hand, the well-known Kerwin–Huelsman–Newcomb biquad, or KHN biquad, which is a filter circuit consists of two integrators and a summer offers several advantages such as low passive and active sensitivities, low component spread, and good stability behavior. Based on the active elements used in the integrator and summer circuits, several KHN biquads have been presented in literature.

In this part, we describe a CM KHN-biquad circuit, which employs three differential voltage current conveyors (DVCCs) as active elements together with two capacitors and four resistors. All the passive elements are grounded, which is important in integrated circuit implementation. The summer circuit of the KHN biquad can be implemented easily with a single DVCC due to the property of the DVCC of handling differential input voltages and providing two output currents of opposite signs. The three basic filter functions [high-pass (HP), band-pass (BP), and low-pass (LP)] are obtained simultaneously. In addition, the notch and all-pass functions can be realized by connecting proper output currents.

KHN Biquad

Using the terminal relation between the Y and X terminal and after the current to voltage (I → V) conversions, the CM KHN biquad can be implemented as shown in Fig. 3.36a. It can be observed that the first stage is the summer and the second and third stages are the integrator circuit stages. The node analysis of the circuit of Fig. 3.36a yields the following current transfer functions:

$$
\begin{aligned}
\frac{I_{\mathrm{HP}}}{I_{\mathrm{in}}} &= \frac{(R_1/R_2)s^2}{s^2 + (1/C_1R_2)s + (R_1/R_2R_3R_4C_1C_2)},\\
\frac{I_{\mathrm{BP}}}{I_{\mathrm{in}}} &= -\frac{(R_1/R_2R_3C_1)s}{s^2 + (1/C_1R_2)s + (R_1/R_2R_3R_4C_1C_2)},\\
\frac{I_{\mathrm{LP}}}{I_{\mathrm{in}}} &= \frac{(R_1/R_2R_3R_4C_1C_2)}{s^2 + (1/C_1R_2)s + (R_1/R_2R_3R_4C_1C_2)},
\end{aligned}
\tag{3.45}
$$

The natural angular frequency o and the quality factor Q of the filter can be expressed as

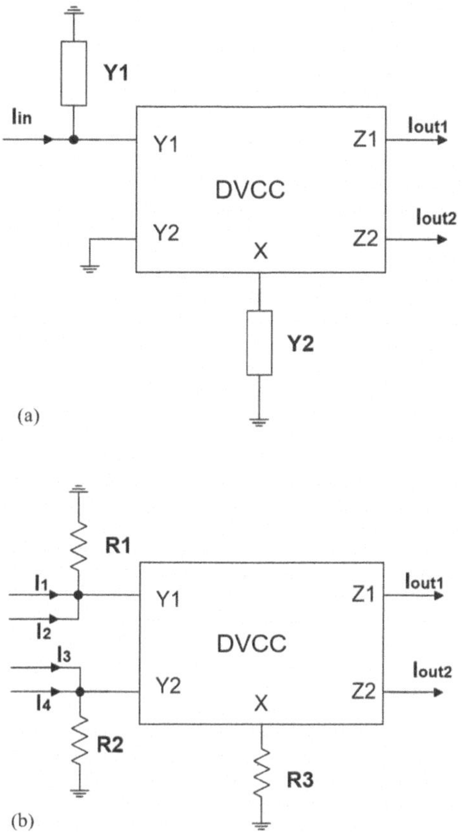

Fig. 3.36 (**a**) DVCC-based basic current processing block. (**b**) DVCC-based CM summer/substructure circuit. (Ibrahim et al. 2005a, b)

$$\omega_o = \sqrt{\frac{R_1}{R_2 R_3 R_4 C_1 C_2}}, \quad Q = \sqrt{\frac{R_1 R_2 C_1}{R_3 R_4 C_2}}. \tag{3.46}$$

It should be noted that ω_0 and Q are orthogonally adjustable. It means that o can be adjusted without disturbing the Q by changing R_2 and R_4 simultaneously while R_2/R_4 is constant. In addition, the three basic filter functions (HP, BP, and LP) are obtained simultaneously, all at high output impedance. Also, for $R_1 = R_2$, a notch function can be realized by connecting high-pass and low-pass outputs, i.e., by adding the corresponding currents without using additional active elements. Moreover, for $R_1 = R_2 = R_3$, an all-pass function can be obtained by connecting high-pass, low-pass, and band-pass outputs. The circuit has the advantage of employing grounded resistors, which enables using the simple realization of the electronic resistor. In the circuit shown in Fig. 3.36b the transistors are operating in saturation region. The matched transistors MR1 and MR2 are diode connected. A voltage V_{in} is applied to the central node of the circuit, developing a current Iin into the node. Using the square law characteristic, the drain currents in MR1 and MR2 can be expressed as

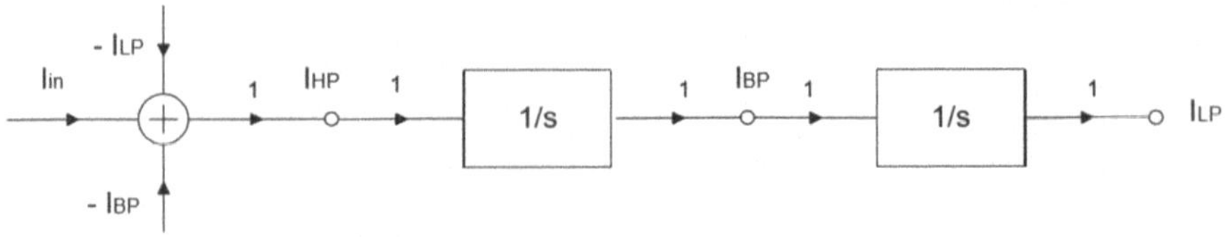

Fig. 3.37 The basic signal processing block of implementing CM KHN biquad. (Ibrahim et al. 2005a, b)

$$I_{\mathrm{DR1}} = \frac{\mu C_{\mathrm{OX}}}{2}\left(\frac{W}{L}\right)(V_{\mathrm{A}} - V_{\mathrm{in}} - V_{\mathrm{T}})^2 \tag{3.47}$$

$$I_{\mathrm{DR2}} = \frac{\mu C_{\mathrm{OX}}}{2}\left(\frac{W}{L}\right)(V_{\mathrm{in}} - V_{\mathrm{B}} - V_{\mathrm{T}})^2 \tag{3.48}$$

where μ is carrier mobility, C_{OX} is the gate capacitance per unit area, V_{T} is the threshold voltage, and W and L are the channel length and width, respectively. Selecting $V_{\mathrm{A}} = -V_{\mathrm{B}}$ and using Kirchhoff's current law (KCL), a simple algebraic manipulation can be given to the resistance as

$$R_{\mathrm{eq}} = \frac{V_{\mathrm{in}}}{I_{\mathrm{in}}} = \frac{L}{2\mu C_{\mathrm{OX}} W(V_{\mathrm{A}} - V_{\mathrm{T}})}. \tag{3.49}$$

The basic signal processing block of implementing CM KHN biquad is given in Fig. 3.37. The designed CM KHN biquad based on DVCCs and the realization of the electronic resistor are illustrated in Fig. 3.38.

In this work, a CM KHN-biquad circuit has been presented. The filter employs three DVCCs, two capacitors, and four resistors. All the passive elements are grounded, which is important in integrated circuit implementation. The filter provides the basic three filter functions (BP, HP, and LP) simultaneously. The notch and all-pass functions can be obtained by connecting appropriate output currents directly without using additional active elements. All the outputs are taken from high output impedances, which is important for cascade connections. Also the proposed filter exhibits low passive and active sensitivities. The filter is designed for operation at 22.5 MHz and tested using SPICE simulation program. The simulation results are given in Figs. 3.39, 3.40, and 3.41. Simulation results agree quite well with the theoretical analysis.

Third-Generation Current Conveyor, CCIII

Third-generation current conveyor concept is introduced by Fabre in 1995 (Fabre, 1995). The aim of this active element is to sense a current flowing through any element in a circuit and reflect it to a separate output terminal. The circuit symbol and definition equations are given below in Fig. 3.42. Furthermore, CMOS CCIII

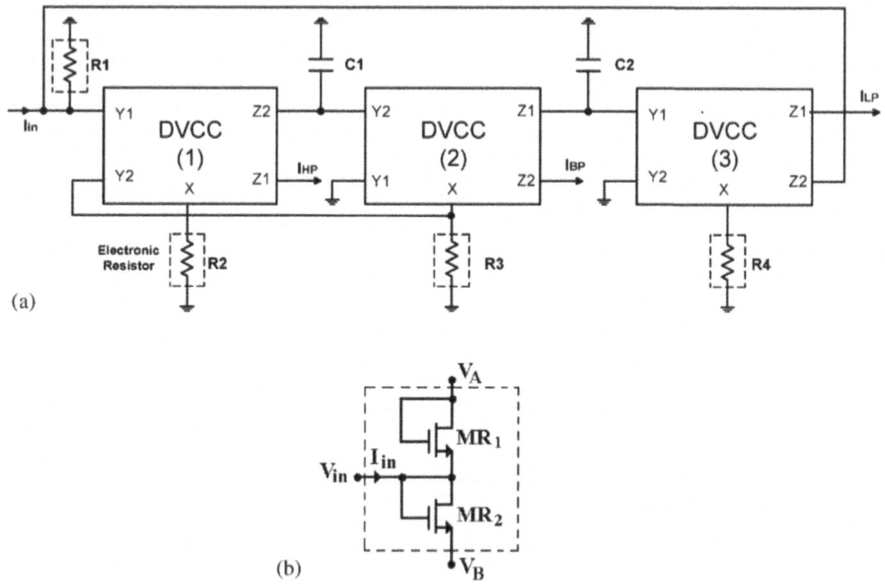

Fig. 3.38 (**a**) The proposed CM KHN biquad based on DVCCs. (**b**) Electronic resistor. (Ibrahim et al. 2005a, b)

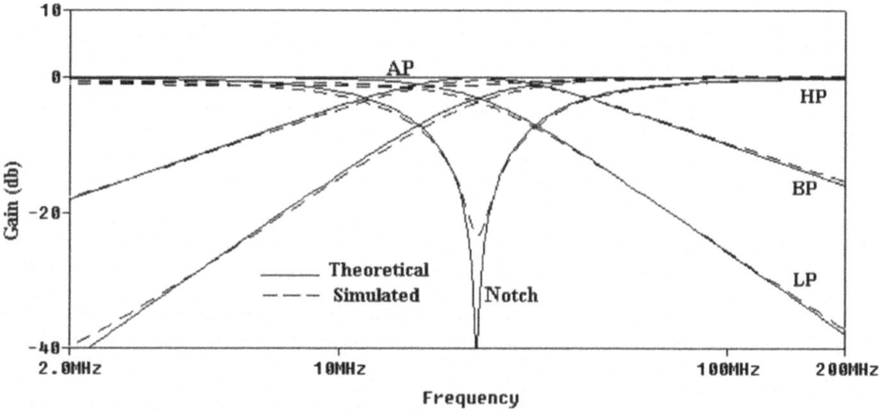

Fig. 3.39 The frequency response of the basic filter functions (BP, HP, LP, notch, and AP) for the proposed KHN biquad. (Ibrahim et al. 2005a, b)

realization and realization method with multi-output CCII are illustrated in Fig. 3.43. Note that voltages at Y and X terminals are equal while current through X terminal is negative of the current through Y terminal. As a result, if the designer inserts any floating element series to these terminals, the circuit acts as a current sensor, so that the current function is sensed and reflected to a separate output to use it anywhere in the circuit.

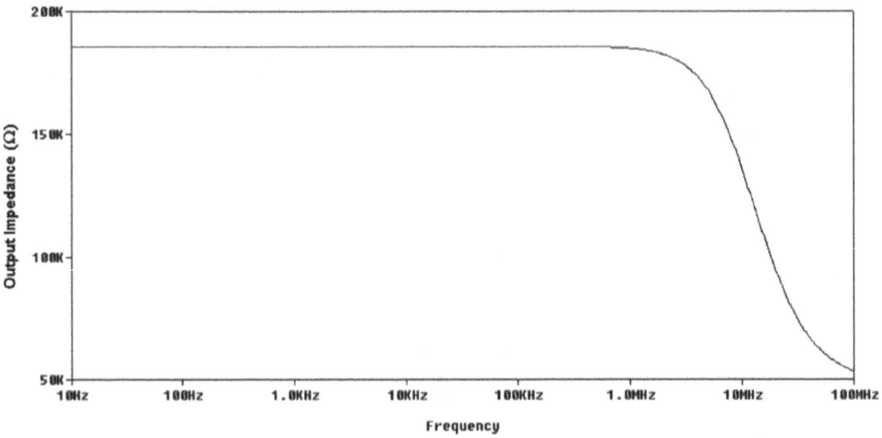

Fig. 3.40 The frequency response of the output impedance (BP, HP, and LP) for the proposed KHN biquad. (Ibrahim et al. 2005a, b)

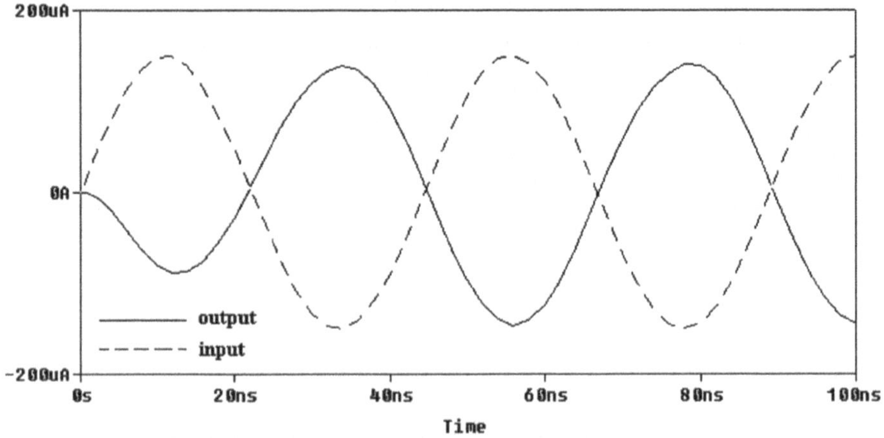

Fig. 3.41 The input and output waveforms of the BP response for a 22.5 MHz sinusoidal input current of 300 μA (peak to peak). (Ibrahim et al. 2005a, b)

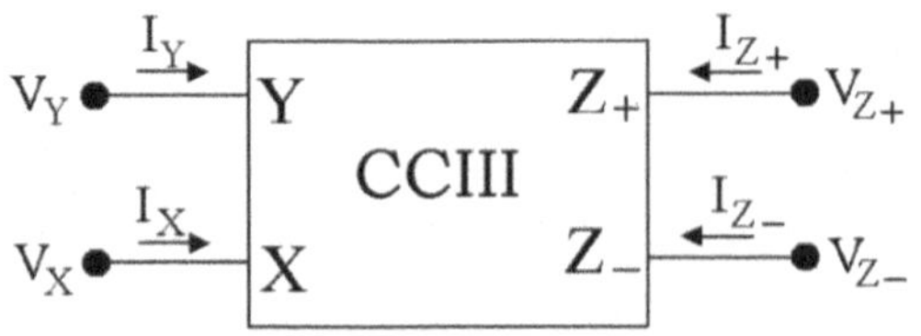

$$\begin{bmatrix} I_Y \\ V_X \\ I_{Z+} \\ I_{Z-} \end{bmatrix} = \begin{bmatrix} 0 & -1 & 0 & 0 \\ 1 & 0 & 0 & 0 \\ 0 & 1 & 0 & 0 \\ 0 & -1 & 0 & 0 \end{bmatrix} \begin{bmatrix} V_Y \\ I_X \\ V_{Z+} \\ V_{Z-} \end{bmatrix}$$

Fig. 3.42 Circuit symbol of CCIII and its definition equations. (Fabre et al. 1995a, b; Kuntman et al. 2000; Minaei et al. 2002)

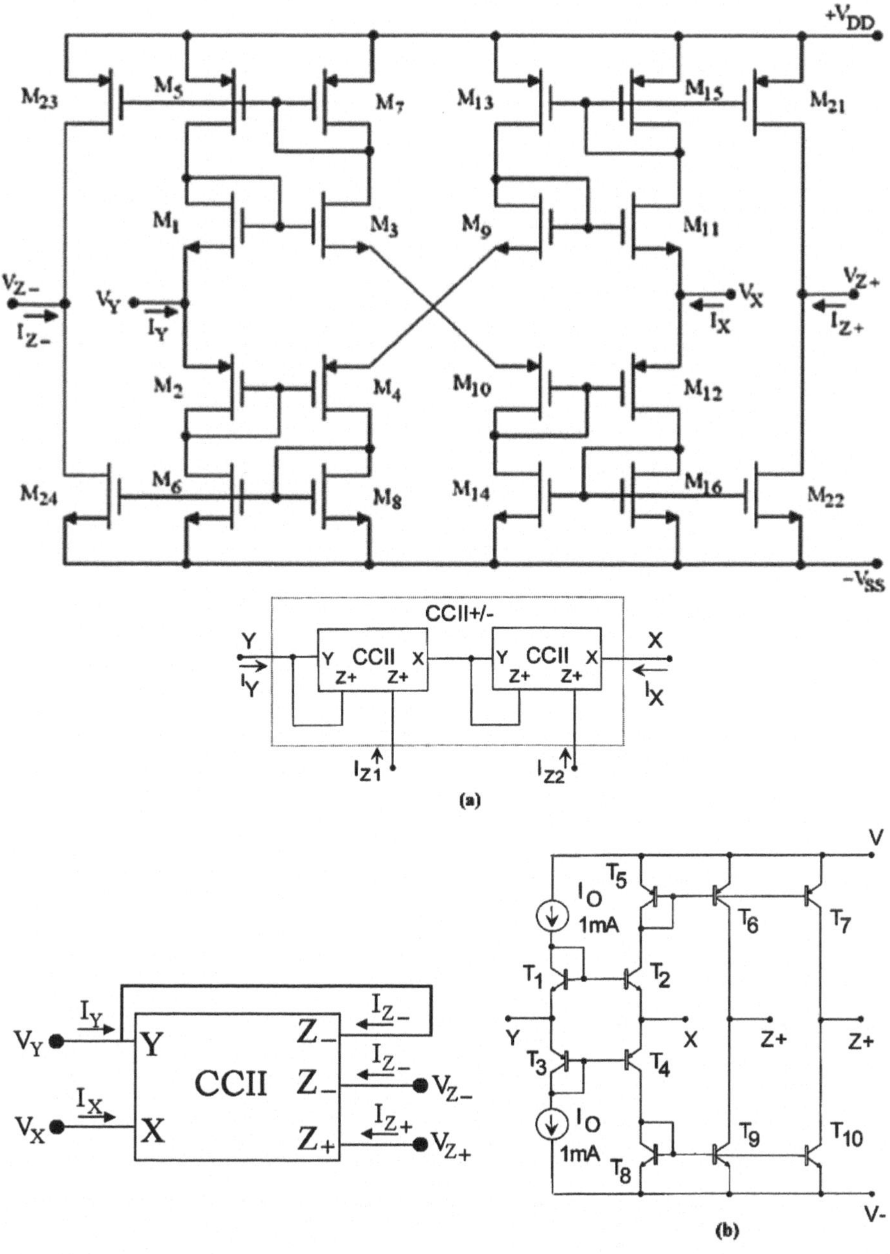

Fig. 3.43 CMOS and BJT CCIII realizations and realization method with multi-output CCII. (Fabre et al. 1995a, b; Kuntman et al. 2000; Minaei et al. 2002)

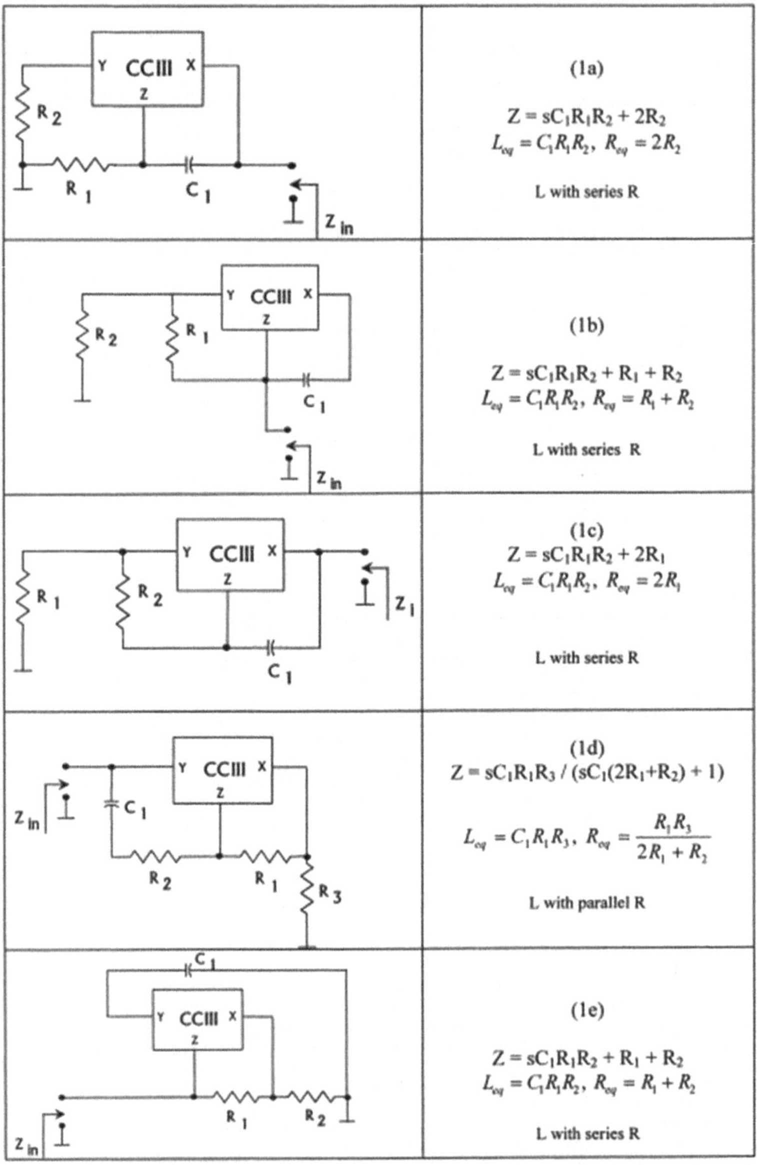

Fig. 3.44 Proposed five lossy grounded inductance simulator topologies. (Kuntman et al. 2000)

Design Example In the following five lossy inductor topologies employing one CCIII and different number of passive elements are presented. The proposed topologies require few passive components and no element matching conditions are necessary. Four different topologies for lossy series inductor and one topology for lossy parallel inductor are proposed. For the topologies the inductance L and the conductance G are independently adjustable.

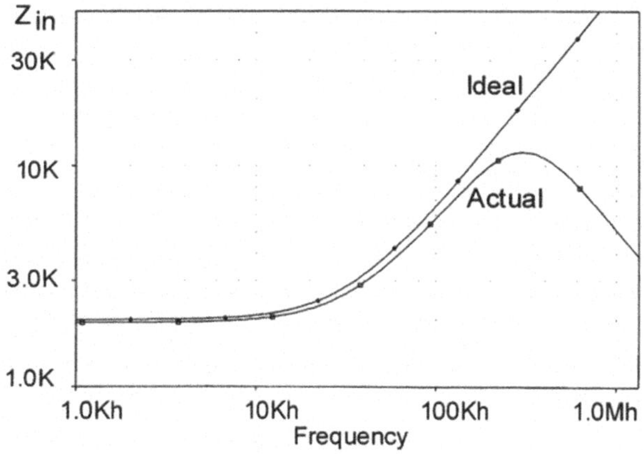

Fig. 3.45 The frequency response obtained from SPICE simulations for the series R–L simulator topology in Fig. 3.1c. (Kuntman et al. 2000)

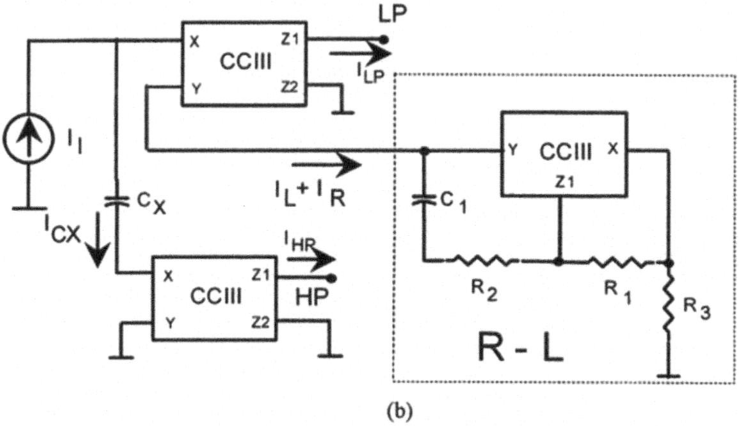

Fig. 3.46 An application of the proposed topologies: (**a**) parallel resonant circuit and (**b**) realization of parallel resonant circuit by using R–L simulator circuit illustrated in Fig. 3.1d and connecting a parallel capacitor CX to form a resonant circuit and to construct a current-mode multi-output filter employing only CCIIIs. (Kuntman et al. 2000)

Simulation Results, Discussion, and an Application Example: All of the circuits illustrated in Fig. 3.44 are tested with SPICE simulations. In this realization of the CCIII, a further z-output, namely, an inverting z-output, is available which provides the designer an additional possibility of obtaining high output impedance. The element values are chosen as $C_1 = 10$ nF, $R_1 = 1$ k, and $R_2 = 1$ k to obtain an inductance of $L_{\text{eq}} = 10$ mH and a series resistance of $R_{\text{eq}} = 2$ kOhm. The frequency response of the topology in Fig. 3.44c is chosen as an example to demonstrate the performance of the derived inductance simulators (Fig. 3.45). SPICE simulation results obtained for impedance against frequency characteristics are depicted in Fig. 3.46, with the simulation results obtained for the actual R–L circuit with the same values. The values obtained show that the circuit performs the inductance simulation well in the operation region of the CCIII. The actual circuit yields an impedance value of 1.94 kOhm at low frequencies which is ideally 2 kOhm. The deviation from the ideal characteristics at high frequencies is caused by the non-idealities of the actual CCIII. To illustrate an application of the proposed topologies, a current-mode multi-output filter employing only CCIIIs illustrated in Fig. 3.46 is designed and simulated. The basic cell is the parallel R–L simulator.

The transfer functions are given by

$$H_1(s) = \frac{I_R}{I_{\text{in}}} = \frac{\frac{G}{C}s}{s^2 + \frac{G}{C}s + \frac{1}{LC}}$$

$$H_2(s) = \frac{I_C}{I_{\text{in}}} = \frac{s^2}{s^2 + \frac{G}{C}s + \frac{1}{LC}}$$

$$H_3(s) = \frac{I_L}{I_{\text{in}}} = \frac{1}{CL}\frac{1}{s^2 + \frac{G}{C}s + \frac{1}{LC}} \tag{3.50}$$

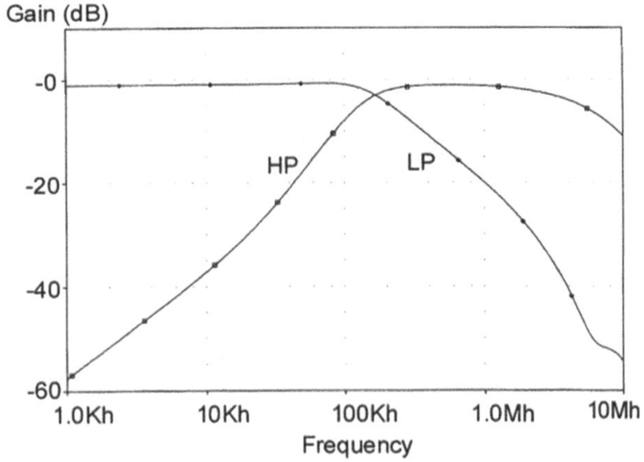

Fig. 3.47 The frequency response of the filter in Fig. 3.46. (Kuntman et al. 2000)

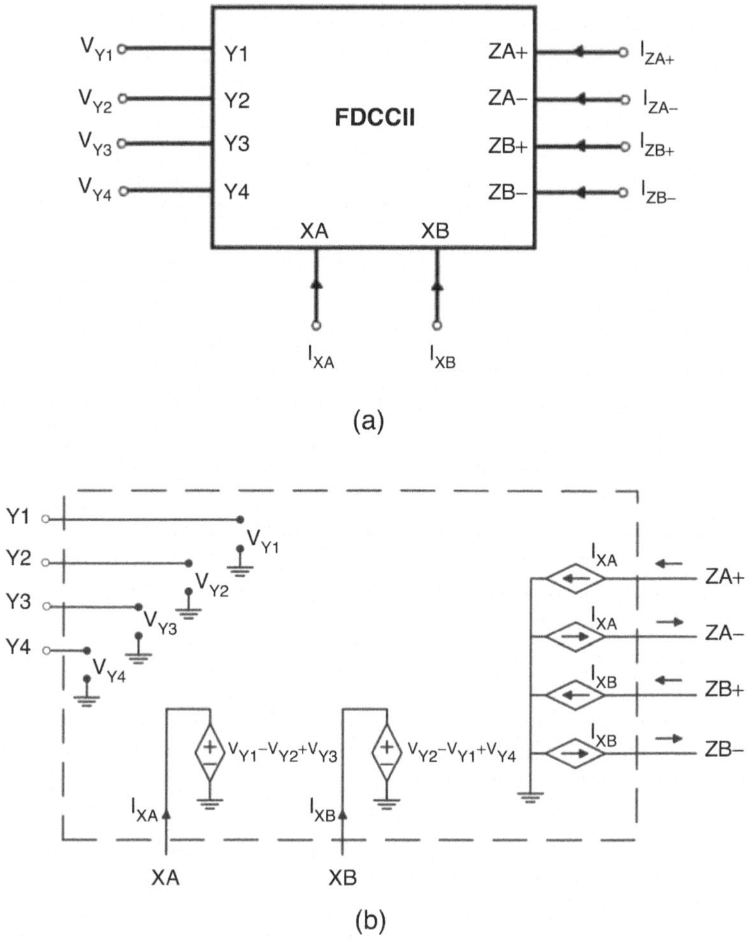

Fig. 3.48 (**a**) Block diagram and (**b**) equivalent circuit of FDCCII. (El-Adawy et al. 2000)

To pick up the capacitor and the parallel R–L circuit currents, two additional CCIIIs are introduced which provide the possibility of obtaining high output impedance for LP and HP filters. Note that the LP output yields actually the current of the parallel R–L circuit as the sum of IL 1 IR. However, by proper choice of element values LP character is obtained (Fig. 3.47).

Fully Differential Current Conveyor (FDCCII)

The fully differential current conveyor (FDCCII) is a versatile active element providing flexibility for analog designers. The block diagram of the FDCCII and its behavioral model which are shown in Fig. 3.48a, b, respectively, are defined by the following terminal equations:

$$\begin{bmatrix} V_{XA} \\ V_{XB} \\ I_{ZA+} \\ I_{ZA-} \\ I_{ZB+} \\ I_{ZB-} \end{bmatrix} = \begin{bmatrix} 1 & -1 & 1 & 0 & 0 & 0 \\ -1 & 1 & 0 & 1 & 0 & 0 \\ 0 & 0 & 0 & 0 & 1 & 0 \\ 0 & 0 & 0 & 0 & -1 & 0 \\ 0 & 0 & 0 & 0 & 0 & 1 \\ 0 & 0 & 0 & 0 & 0 & -1 \end{bmatrix} \begin{bmatrix} V_{Y1} \\ V_{Y2} \\ V_{Y3} \\ V_{Y4} \\ I_{XA} \\ I_{XB} \end{bmatrix}. \tag{3.51}$$

In this section, we describe a further implementation for a fully differential second-generation current conveyor (FDCCII) based on an improved active-feedback cascode current mirror (IAFCCM). The proposed FDCCII is compared with the cascode FDCCII (El-Adawy et al. 2000; Horng et al. 2006). It provides high output resistance at Z terminals. The circuit also exhibits excellent output–input gain accuracy both in voltage and current mode. The conventional CMOS realization of the cascode FDCCII is shown in Fig. 3.49 (El-Adawy et al. 2000).

Because of the good current transfer accuracy in cascode current mirrors, it is expected that the cascode FDCCII provides a suitable current and voltage transfer accuracy in applications. However, the output resistance of FDCCII must further be improved to enable easy cascading of FDCCII-based circuits in current-mode operation.

To increase the output resistance, a new FDCCII based on using IAFCCM (Zeki and Kuntman 1997) in the output stages of the conveyor is proposed. The proposed high-performance FDCCII is shown in Fig. 3.50. A major advantage of IAFCCM circuit is that the output conductance and the feedback capacitance are 100 times lower than the standard current mirror circuit (Zeki and Kuntman 1997).

It is shown that for an ideal FDCCII all the four Y terminals exhibit an infinite input resistance. The two ports X exhibit zero input resistance and the four ports Z exhibit an infinite output resistance.

Application Examples

Current-Mode Single-Input Four-Outputs (SIFO) Biquad Filter

Typical active filter examples are taken and investigated in the following: The first type proposed current-mode *single-input four-outputs (SIFO)* biquad filter as shown in Fig. 3.51. The proposed circuit which is composed of an FDCCII uses two capacitors and two resistors.

Note that the filter structure of Fig. 3.51 is a core structure where filter functions are available via currents flowing over grounded passive elements. It is possible to obtain BP1 and HP filter functions from a separate high output impedance by modifying the CMOS structure using Z-copy techniques for Z_{A-}, Z_B, and Z_{B-}

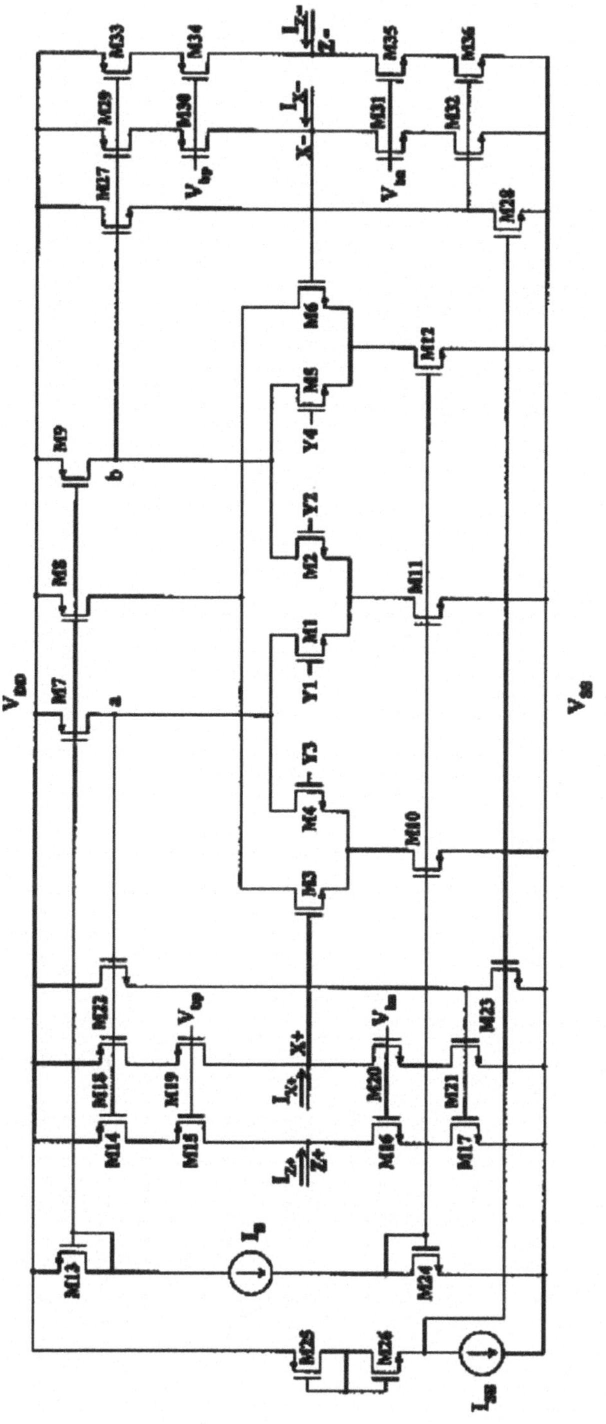

Fig. 3.49 Cascode multiple output fully differential second-generation current conveyor (FDCCII). (El-Adawy et al. 2000)

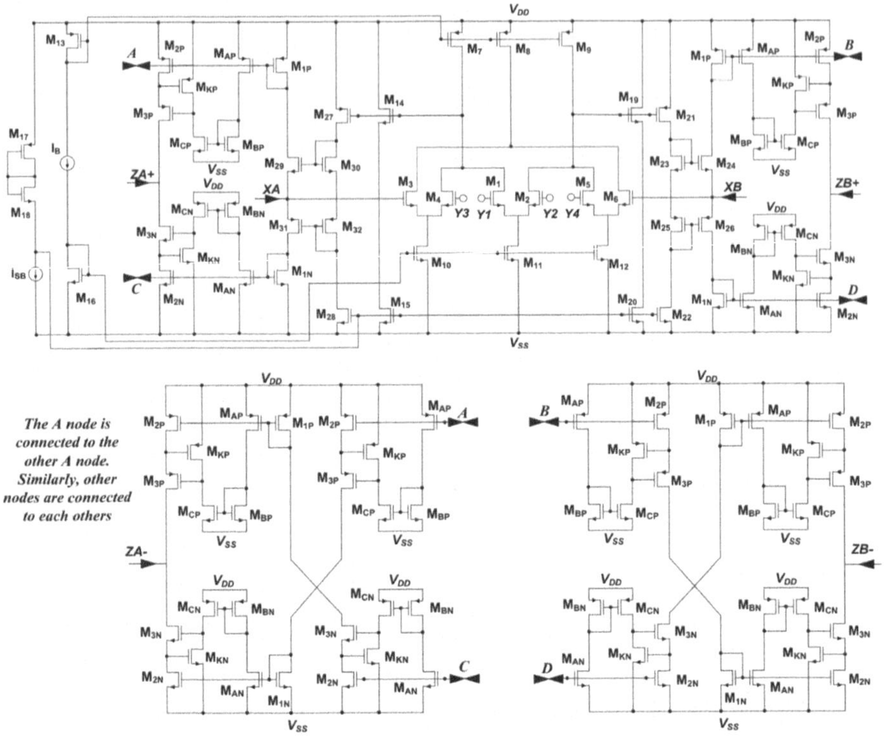

Fig. 3.50 CMOS FDCCII implementation. (Kacar et al. 2012)

Fig. 3.51 Proposed SIFO biquad filter employing FDCCII (Kacar et al. 2012)

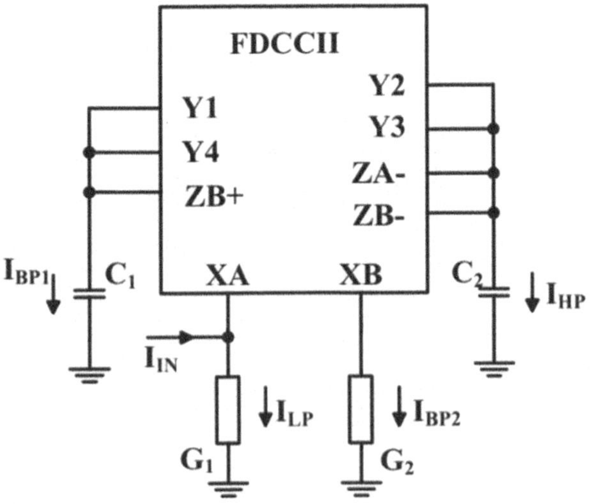

terminals. Fortunately, due to the behavior of the active elements, cascading of two filters of this type can be achieved simply by connecting the lower terminal of the related passive element directly to the input of the next filter stage instead of the ground. Due to internal grounded X_A and X_B terminals, used as input, the filters can be cascaded without any problem. In this case, no additional terminal is needed.

Current transfer functions of the circuit are given in Eq. (3.52). Routine analysis of this circuit yields the following current-mode filter transfer functions:

$$\begin{aligned}\frac{I_{LP}}{I_{IN}} &= \frac{G_1G_2}{s^2C_1C_2 + sC_1G_2 + G_1G_2},\\ \frac{I_{BP1}}{I_{IN}} &= \frac{C_1G_2s}{s^2C_1C_2 + sC_1G_2 + G_1G_2},\\ \frac{I_{BP2}}{I_{IN}} &= \frac{C_1G_2s}{s^2C_1C_2 + sC_1G_2 + G_1G_2},\\ \frac{I_{HP}}{I_{IN}} &= \frac{C_1C_2s^2}{s^2C_1C_2 + sC_1G_2 + G_1G_2}.\end{aligned} \tag{3.52}$$

Current-Mode MISO Biquad Filters

The second type of the proposed current-mode MISO biquad filters is shown in Fig. 3.50. Each of the proposed circuit, which is composed of an FDCCII, uses all grounded two capacitors and two resistors. In addition, the proposed circuits have high output impedance so these circuits enable easy cascadability.

Note that the circuits of Fig. 3.52 exhibit multiple-input and single-output configuration. As a result, similar to many other multiple-input filter structures available in the literature, multiple-output current followers/inverters are required to derive the input signals properly. The first circuit can be realized as multi-inputs single-output current-mode biquad filter and is shown in Fig. 3.52a. By routine circuit analysis, the current transfer function of the proposed FDCCII-based filter can be given by

$$I_{out} = \frac{I_3s^2C_1C_2 + s(I_4C_1G_2 - (I_1 + I_2)C_2G_2) + 2I_4G_1G_2}{s^2C_1C_2 + sC_1G_2 + 2G_1G_2}. \tag{3.53}$$

Depending on the current status of I_1, I_2, I_3, and I_4 in the numerator of Eq. (3.53), one of the following five filter functions is realized:

(i) LP: $I_2 = I_3 = 0$, $I_1 = I_4 = I_{IN}$.
(ii) BP: $I_2 = I_3 = I_4 = 0$, $I_1 = I_{IN}$ or $I_1 = I_3 = I_4 = 0$, $I_2 = I_{IN}$.
(iii) HP: $I_1 = I_2 = I_4 = 0$, $I_3 = I_{IN}$.

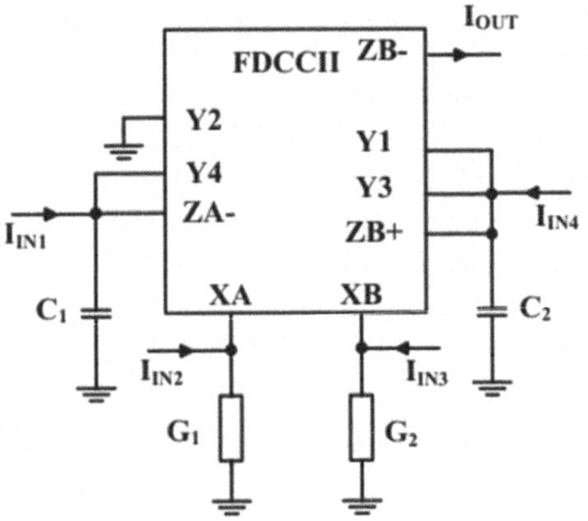

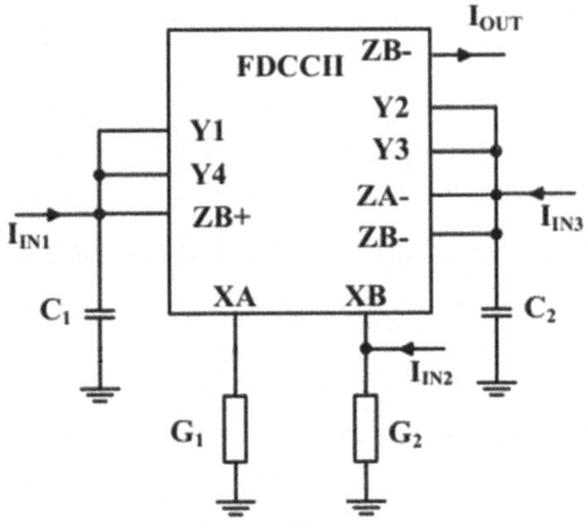

Fig. 3.52 Proposed MISO biquad filters employing FDCCII. (Kacar et al. 2012)

(iv) BS: $I_2 = 0$, $I_1 = I_3 = I_4 = I_{IN}$ and $C_1 = C_2$ or $I_1 = 0$, $I_2 = I_3 = I_4 = I_{IN}$ and $C_1 = C_2$.
(v) AP: $I_1 = I_2 = I_3 = I_4 = I_{IN}$ and $C_1 = C_2$.

From Eq. (3.53), the resonance angular frequency ω_{02} and the quality factor Q_2 are given by

$$\omega_{02} = \sqrt{\frac{2G_1G_2}{C_1C_2}}, \tag{3.54}$$

$$Q_2 = \sqrt{\frac{2G_1C_2}{G_2C_1}}. \tag{3.55}$$

Simulation Results

We simulated the FDCCII using the schematic implementation taken in Kacar et al. (2010). Because of the good current transfer accuracy in cascode current mirrors, it is expected that the cascode FDCCII provides a suitable current and voltage transfer accuracy in applications. However, the output resistance of FDCCII must further be improved to enable easy cascading of FDCCII-based circuits in current-mode operation. To increase the output resistance, a new FDCCII based on using IAFCCM (Zeki and Kuntman 1997) in the output stages of the conveyor is proposed. The high-performance FDCCII is shown in Fig. 3.50. A major advantage of IAFCCM circuit is that the output conductance and the feedback capacitance are 100 times slower than the standard current mirror circuit (Zeki and Kuntman 1997).

The supply voltages and biasing currents are given by $V_{DD} = -V_{SS} = 1.3$ V and $I_B = I_{SB} = 125$ μA, respectively. The dimensions of the MOS transistor for the FDCCII implementation are given in Table 3.4. The MOS transistors are simulated using TSMC CMOS 0.35 μm process model parameters.

For this simulation the passive component values were chosen as $R_1 = R_2 = 1$ kΩ and $C_1 = C_2 = 100$ pF, leading to a center frequency of f_0 =1.59 MHz and quality factor of $Q = 1$. Figure 3.53 shows the simulated results of low-pass, band-pass, and high-pass gain-frequency responses. It is noted that the results of circuit simulations agree quite well with the theory.

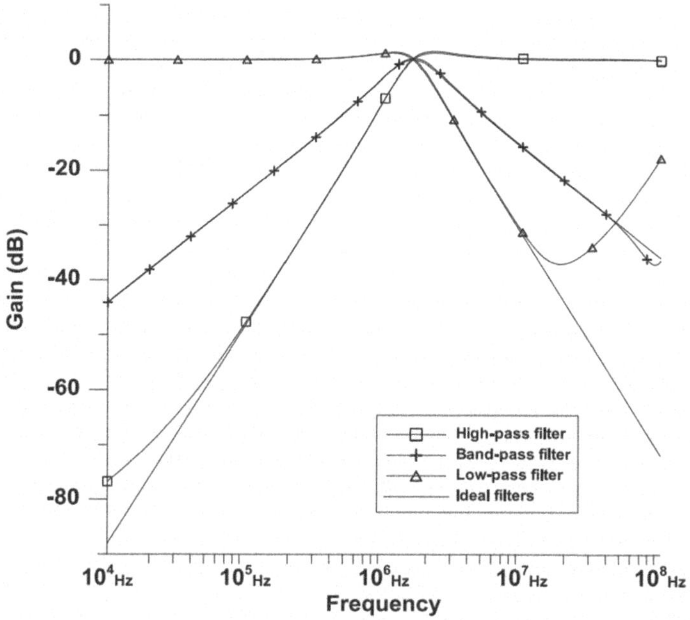

Fig. 3.53 Results of circuit simulations for basic filter responses of the proposed FDCCII-based CM SIFO biquad for $f_0 = 1.59$ MHz and $Q = 1$ (Kacar et al. 2012)

Concluding Remarks

In this section we describe new single-input four outputs (SIFO) current-mode multifunction filters using single FDCCII, two capacitors, and two resistors. The circuit has the following advantages:

(i) It can simultaneously realize low-pass, band-pass, and high-pass filter functions without changing the circuit topology and elements.
(ii) It uses minimum number of active and passive elements.
(iii) It has grounded resistors and grounded capacitors which can be easily fabricated for VLSI.
(iv) It does not require any parameter matching condition.
(v) It has low active and passive sensitivities. Other types of the proposed circuits are two new multi-inputs single-output (MISO) biquad filter circuits.

The salient features of the proposed circuits can now be summarized as follows:

(i) They use only a single active component and minimum passive components (two resistors and two capacitors).
(ii) They employ all grounded passive components and are convenient for IC integration.
(iii) They provide current output explicitly from a high output impedance node so they enable easy cascadability.
(iv) They realize the five standard filtering functions.
(v) They don't use inverting input current terminal for realizing the five standard filters. LTSPICE simulation results have been presented which confirm the workability of these circuits. The proposed CM biquad filters with these properties will be useful in many areas, for example, analog communication systems and in analog signal processing.

References

Abuelma'atti, M. T. and Almaskatı, R. H. (1988). Research Note: Two novel voltage (current) controlled integrable oscillators. International journal of electronics, 64(4), 611–615.

Abuelma'atti, M. T. and Tasadduq, N. A. (1998a). A novel single-input multiple-output current-mode current-controlled universal filter. Microelectronics Journal, 29(11), 901–905.

Abuelma'Atti, M. T. and Tasadduq, N. A. (1998b). New current-mode current-controlled filters using the current-controlled conveyor. International Journal of Electronics, 85(4), 483–488.

Abuelma'atti, M. T., Khan, M. H. and Al-Zaher, H. A. (1998). Simulation of active-only floating inductance. Frequenz, 52(7–8), 161–164.

Acar, C. (1996). Nth-order lowpass voltage transfer function synthesis using CCII+ s: signal-flow graph approach. Electronics letters, 32(3), 159–160.

Acar, C. and Kuntman, H. (1996). Limitations on input signal level in current-mode active-RC filters using CCIIs. Electronics Letters, 32(16), 1461–1462.

Acar, C. and Kuntman, H. (1999). Limitations on input signal level in voltage-mode active-RC filters using current conveyors. Microelectronics Journal, 30(1), 69–76.

Acar, C. and Ozoguz, S. (1996). High-order voltage transfer function synthesis using CCII+ based unity gain current amplifiers. Electronics letters, 32(22), 2030–2031.

Acar, C. and Özoğuz, S. (2000). Nth-order current transfer function synthesis using current differencing buffered amplifier: signal-flow graph approach. Microelectronics Journal, 31(1), 49–53.

Alami, M. and Fabre, A. (1991). Insensitive current-mode bandpass filter implemented from two current conveyors. Electronics letters, 27(11), 897–899.

Aronhime, P., Nelson, D. and Adams, C. (1990). Applications of a first-generation current conveyor in current-mode circuits. Electronics Letters, 26(18), 1456–1457.

Chang, C. M. (1991). Current mode allpass/notch and bandpass filter using single CCII. Electronics Letters, 27(20), 1812–1813.

Chang, C. M. (1993a). Current-mode lowpass, bandpass and highpass biquads using two CCIIs. Electronics Letters, 29(23), 2020–2021.

Chang, C. M. (1993b). Novel universal current-mode filter with single input and three outputs using only five current conveyors. Electronics Letters, 29(23), 2005–2007.

Chang, C. M. and Chen, P. C. (1991). Realization of current-mode transfer function using second-generation current conveyors. International Journal of Electronics, 71(5), 809–815.

Chang, C. M., Chien, C. C. and Wang, H. Y. (1993). Universal active current filters using single second-generation current conveyor. Electronics Letters, 29(13), 1159–1160.

Chang, C. M., Wang, H. Y. and Chien, C. C. (1994). Realization of series impedance functions using one CCII+. International journal of electronics, 76(1), 83–85.

Cicekoglu, O. and Kuntman, H. H. (1998). Circuit design techniques: AD-HOC methods or systematic generation methods. In Proceedings of the Tenth International Conference on Microelectronics (Cat. No. 98EX186) (pp. 187–190). IEEE.

Çam, U. and Kuntman, H. (1998). A new CCII-based sinusoidal oscillator providing fully independent control of oscillation condition and frequency. Microelectronics Journal, 29(11), 913–919.

Çam, U., Çiçekoğlu, O. and Kuntman, H. (1999). A new four terminal floating nullor based single-input three-output current-mode multifunction filter. Microelectronics Journal, 30(2), 115–118.

Çiçekoğlu, O., Özcan, S. and Kuntman, H. (1999). Insensitive multifunction filter implemented with current conveyors and only grounded passive elements. Frequenz, 53(7–8), 158–160.

Çiçekoğlu, O., Tarim, N. and Kuntman, H. (2002). Wide dynamic range high output impedance current-mode multifunction filters with dual-output current conveyors. AEU-International Journal of Electronics and Communications, 56(1), 55–60.

El-Adawy, A. A., Soliman, A. M. and Elwan, H. O. (2000). A novel fully differential current conveyor and applications for analog VLSI. IEEE Transactions on Circuits and Systems II: Analog and Digital Signal Processing, 47(4), 306–313.

Elwan, H. O. and Soliman, A. M. (1996). A novel CMOS current conveyor realization with an electronically tunable current mode filter suitable for VLSI. IEEE Transactions on Circuits and Systems II: Analog and Digital Signal Processing, 43(9), 663–670.

Fabre, A. (1995). Third-generation current conveyor: a new helpful active element. Electronics Letters, 31(5), 338–339.

Fabre, A. and Alami, M. (1995). Universal current mode biquad implemented from two second generation current conveyors. IEEE Transactions on Circuits and Systems I: Fundamental Theory and Applications, 42(7), 383–385.

Fabre, A., Saaid, O. and Barthelemy, H. (1995a). On the frequency limitations of the circuits based on second generation current conveyors. Analog Integrated Circuits and Signal Processing, 7(2), 113–129.

Fabre, A., Saaid, O., Wiest, F. and Boucheron, C. (1995b). Current controlled bandpass filter based on translinear conveyors. Electronics Letters, 31(20), 1727–1728.

Fabre, A. and Alami, M. (1997). On the Macromodelling Approach of First Generation Current Conveyors. In ECCTD'97 (p. 1). NC.

Fabre, A. and Mimeche, N. (1994). Class A/AB second-generation current conveyor with controlled current gain. Electronics Letters, 30(16), 1267–1269.

Fabre, A., Martin, F. and Hanafi, M. (1990). Current mode allpass/notch and bandpass filters with reduced sensitivities. Electronics Letters, 26(18), 1495–1496.

Fabre, A., Saaid, O., Wiest, F. and Boucheron, C. (1996). High frequency applications based on a new current controlled conveyor. IEEE Transactions on Circuits and Systems I: Fundamental Theory and Applications, 43(2), 82–91.

Fabre, A., Saaid, O., Wiest, F. and Boucheron, C. (1997). Low power current-mode second-order bandpass IF filter. IEEE Transactions on Circuits and Systems II: Analog and Digital Signal Processing, 44(6), 436–446.

Ferri, G. and Guerrini, N. C. (2003). Low-voltage low-power CMOS current conveyors. Springer Science & Business Media.

Filoramo, P., Giustolisi, G., Palmisano, G. and Palumbo, G. (2000). Approach to the design of low-voltage SC filters. IEE Proceedings-Circuits, Devices and Systems, 147(3), 196–200.

Gunes, E. O. and Anday, F. (1995). Realisation of nth-order voltage transfer function using CCII+. Electronics Letters, 31(13), 1022–1023.

Higashimura, M. and Fukui, Y. (1989). Brief communication Simulation of lossless floating inductance using two current conveyors and an operational transconductance amplifier. International Journal of Electronics, 66(4), 633–638.

Higashimura, M. and Fukui, Y. (1987). Novel method for realising lossless floating immittance using current conveyors. Electronics letters, 23(10), 498–499.

Higashimura, M. and Fukui, Y. (1990). Realization of current mode all-pass networks using a current conveyor. IEEE transactions on circuits and systems, 37(5), 660–661.

Hou, C. L., Chen, R. D., Wu, Y. P. and Hu, P. C. (1993). Realization of grounded and floating immittance function simulators using current conveyors. International journal of electronics, 74(6), 917–923.

Ibrahim, M. A. and Kuntman, H. (2004). A novel high CMRR high input impedance differential voltage-mode KHN-biquad employing DO-DDCCs. AEU-International Journal of Electronics and Communications, 58(6), 429–433.

Ibrahim, M. A., Kuntman, H., Ozcan, S., Suvak, O. and Cicekoglu, O. (2004). New first-order inverting-type second-generation current conveyor-based all-pass sections including canonical forms. Electrical Engineering, 86(5), 299–301.

Ibrahim, M. A., Kuntman, H. and Cicekoglu, O. (2003). First-order all-pass filter canonical in the number of resistors and capacitors employing a single DDCC. Circuits, Systems and Signal Processing, 22(5), 525–536.

Ibrahim, M. A., Kuntman, H. and Cicekoglu, O. (2005a). Single DDCC biquads with high input impedance and minimum number of passive elements. Analog Integrated Circuits and Signal Processing, 43(1), 71–79.

Ibrahim, M. A., Minaei, S. and Kuntman, H. (2005b). A 22.5 MHz current-mode KHN-biquad using differential voltage current conveyor and grounded passive elements. AEU-International Journal of Electronics and Communications, 59(5), 311–318.

Ikeda, K. and Tomita, Y. (1994). Realization of current-mode biquadratic filter using CC IIs with current followers. Electronics and Communications in Japan (Part II: Electronics), 77(1), 99–107.

İbrahim, M. A., Minaei, S. and Kuntman, H. (2006). DVCC based differential-mode all-pass and notch filters with high CMRR. International Journal of Electronics, 93(04), 231–240.

Kacar, F. and Kuntman, H. (2008). A new dual-X CMOS second generation current conveyor (DXCCII). In 2008 15th IEEE International Conference on Electronics, Circuits and Systems (pp. 1–4). IEEE.

Kacar, F., Yesil, A. and Kuntman, H. (2012). Current-mode biquad filters employing single FDCCII. Radioengineering, 21(4), 1269–1278.

Kaçar, F., Metin, B., Kuntman, H. and Cicekoglu, O. (2010). A new high-performance CMOS fully differential second-generation current conveyor with application example of biquad filter realisation. International Journal of Electronics, 97(5), 499–510.

Khan, I. A. and Zaidi, M. H. (2000). Multifunctional translinear-C current-mode filter. International Journal of Electronics, 87(9), 1047–1051.

Kiranon, W. and Pawarangkoon, P. (1997). Floating inductance simulation based on current conveyors. Electronics letters, 33(21), 1748–1749.

Kuntman, H. (1994). Simple and accurate nonlinear OTA macromodel for simulation of CMOS OTA-C active filters. International journal of electronics, 77(6), 993–1006.

Kuntman, H. and Özpinar, A. (1998). On the realization of DO-OTA-C oscillators. Microelectronics Journal, 29(12), 991–997.

Kuntman, H., Çiçekoğlu, O. and Özoğuz, S. (2002). A modified third generation current conveyor, its characterization and applications. Frequenz, 56(1–2), 47–54.

Kuntman, H., Gülsoy, M. and Çiçekoğlu, O. (2000). Actively simulated grounded lossy inductors using third generation current conveyors. Microelectronics Journal, 31(4), 245–250.

Liu, S. I., Tsao, H. W. and Wu, J. (1990a). Cascadable current-mode single CCII biquads. Electronics Letters, 26(24), 2005–2006.

Liu, S. I., Tsao, H. W., Wu, J. and Lin, T. K. (1990b). MOSFET capacitor filters using unity gain CMOS current conveyors. Electronics Letters, 26(18), 1430–1431.

Minaei, S., Cicekoglu, O., Kuntman, H. and Türköz, S. (2003). Electronically tunable, active only floating inductance simulation. International Journal of Electronics, 89(12), 905–912.

Minaei, S., Çiçekoglu, O., Kuntman, H. and Türköz, S. (2001). High output impedance current-mode lowpass, bandpass and highpass filters using current controlled conveyors. International Journal of Electronics, 88(8), 915–922.

Minaei, S., Yildiz, M., Kuntman, H. and Turkoz, S. (2002). High performance CMOS realization of the third generation current conveyor (CCIII). In The 2002 45th Midwest Symposium on Circuits and Systems, 2002. MWSCAS-2002. (Vol. 1, pp. I–307). IEEE.

Minaei, S., Sayin, O. K. and Kuntman, H. (2006a). A new CMOS electronically tunable current conveyor and its application to current-mode filters. IEEE Transactions on Circuits and Systems I: Regular Papers, 53(7), 1448–1457.

Minaei, S., Sayin, O. K. and Kuntman, H. (2006b). Nth-order current transfer function synthesis using a high-performance electronically tunable current conveyor. In 2006 IEEE Mediterranean Electrotechnical Conference (pp. 15–18). IEEE.

Pal, K. (1981a). New inductance and capacitor floatation schemes using current conveyors. Electronics letters, 17(21), 807–808.

Pal, K. (1981b). Novel floating inductance using current conveyors. Electronics letters, 17(18), 638–638.

Papazoglou, C. A. and Karybakas, C. A. (1997). Noninteracting electronically tunable CCII-based current-mode biquadratic filters. IEE Proceedings-Circuits, Devices and Systems, 144(3), 178–184.

Roberts, G. W. and Sedra, A. S. (1989). All current-mode frequency selective circuits. Electronics Letters, 25(12), 759–761.

Sayın, O.K. (2004). "CMOS ECCII ile yüksek dereceden ayarlanabilir aktif süzgeç tasarımı" M.Sc. Thesis, Istanbul Technical University, Institute of Science and Technology.

Sedra, A. S. and Smith, K. (1970). A second-generation current conveyor and its applications. IEEE Transactions on circuit theory, 17(1), 132–134.

Sedra, A. S., Roberts, G. W. and Gohh, F. (1990). The current conveyor: history, progress and new results. IEE Proceedings G-Circuits, Devices and Systems, 137(2), 78–87.

Senani, R. (1979). Novel active RC circuit for floating-inductor simulation. Electronics Letters, 15(21), 679–680.

Senani, R. (1980a). New tunable synthetic floating inductors. Electronics letters, 16(10), 382–383.

Senani, R. (1980b). Novel circuit implementation of current conveyors using an OA and an OTA. Electronics Letters, 1(16), 2–3.

Surakampontorn, W. and Kumwachara, K. (1992). CMOS-based electronically tunable current conveyor. Electronics letters, 28(14), 1316–1317.

Surakampontorn, W. and Thitimajshima, P. (1988). Integrable electronically tunable current conveyors. In IEE Proceedings G (Electronic Circuits and Systems) (Vol. 135, No. 2, pp. 71–77). IET Digital Library.

Tarim, N., Yenen, B. and Kuntman, H. (1996). Simple and accurate nonlinear current conveyor macromodel. In Proceedings of 8th Mediterranean Electrotechnical Conference on Industrial Applications in Power Systems, Computer Science and Telecommunications (MELECON 96) (Vol. 1, pp. 447–450). IEEE.

Tek, H. and Anday, F. (1989). Voltage transfer function synthesis using current conveyors. Electronics Letters, 25(23), 1552–1553.

Toumazou, C. and Lidgey, F. J. (1985). Floating-impedance convertors using current conveyors. Electronics Letters, 21(15), 640–642.

Toumazou, C., Lidgey, F. J. and Haigh, D. (Eds.). (1990). Analogue IC design: the current-mode approach (Vol. 2). Presbyterian Publishing Corp.

Wilson, B. (1990). Recent developments in current conveyors and current-mode circuits. IEE Proceedings G (Circuits, Devices and Systems), 137(2), 63–77.

Chapter 4
The Operational Transresistance Amplifier (OTRA)

Basic Concept

The operational transresistance amplifier (OTRA) is one of the most important building blocks of analog integrated circuits and systems. The growing demand for mobile communications improved the researches for high-frequency applications. In the newly designed circuit topologies for high frequencies, the current-mode approach is preferred rather than the traditional voltage-mode structures. The current-mode structures have higher bandwidths. Operational transresistance amplifier (OTRA), which is commercially available under the name of Norton amplifier, has attracted attention due to its advantages in the current-mode circuit design (Salama and Soliman 1999; Chen et al. 1995; Kuntman 2010; Kuntman 2011; Kuntman and Uygur 2012). Low input and output impedances and device gain which is bandwidth independent are the main advantageous properties of the OTRA. However, the commercial realizations are unable to provide a true virtual ground at the input terminals and they allow the input current to flow in one direction only. In order to remove these disadvantages of the OTRA, several works were performed by the researchers and several topologies suitable for VLSI realization are proposed in the literature.

But these solutions are complex structures and do not operate properly in low power supplies like 1.2 V in the submicron technologies. In today's technology, circuits using power supplies as 1 V and fabricated in the CMOS 65 nm technology could be designed. CMOS 45 nm technology products were available in the year 2007 with a power supply of 0.6 V. So for the future design concept the main interest is designing circuitries with low power supplies. This demand leads to designing a high-performance differential OTRA for the current-mode analog system design, by using the STMicroelectronics CMOS 0.13 m technology for 1.2 V power supply.

Both the input and output terminals are characterized by low impedance. The input terminals are virtually grounded, leading to circuits that are insensitive to stray

H. Kuntman, D. Özenli, *Trends in Circuit Design for Analog Signal Processing*,
Analog Circuits and Signal Processing,
https://doi.org/10.1007/978-3-030-96836-6_4

capacitance as reported in the literature. For ideal operation, the transresistance gain (Rm) approaches infinity forcing the input currents to be equal. Thus, the OTRA must be used in a negative feedback configuration in a way that is similar to conventional op-amps. OTRA has similar transmission properties to the current feedback op-amp, but with two low input impedances and two low output impedances for the differential OTRA.

Since the input terminals of OTRA are at ground potential, most effects of parasitic capacitances and resistances disappear. Thus, it is possible to obtain very accurate and cascadable transfer functions by using this device in a negative feedback loop. There are many applications of immittance simulators such as oscillator design, filter design, and cancellation of parasitic elements. Many immittance simulators employing various active components such as current conveyor, current feedback op-amp, and four-terminal floating nullor are available in the literature. From these circuits some are related with simulation of series immittance function. Moreover, several specific circuits for the simulation of R–L and C–D immittances have been reported. A recent work presents universal series and parallel immittance simulator topologies using two FTFNs (Cam et al. 2000). General immittance simulator circuits that enable simulation of all possible form of inductors such as lossy and lossless types are proposed in the literature. However, they are using three (CCII+)s. Numerous floating immittance simulator topologies are also available in the literature. In this section we present new grounded parallel immittance simulator topologies employing operational transresistance amplifier to provide further possibilities to the designers in the realization of analog signal processing circuits.

Single-Output OTRA

The circuit symbol of the OTRA is illustrated in Fig. 4.1. The port relations of an OTRA can be characterized by the following matrix equation:

$$\begin{bmatrix} V_{\mathrm{p}} \\ V_{\mathrm{n}} \\ V_{\mathrm{z}} \end{bmatrix} = \begin{bmatrix} 0 & 0 & 0 \\ 0 & 0 & 0 \\ R_{\mathrm{m}} & -R_{\mathrm{m}} & 0 \end{bmatrix} \begin{bmatrix} I_{\mathrm{p}} \\ I_{\mathrm{n}} \\ I_{\mathrm{z}} \end{bmatrix} \tag{4.1}$$

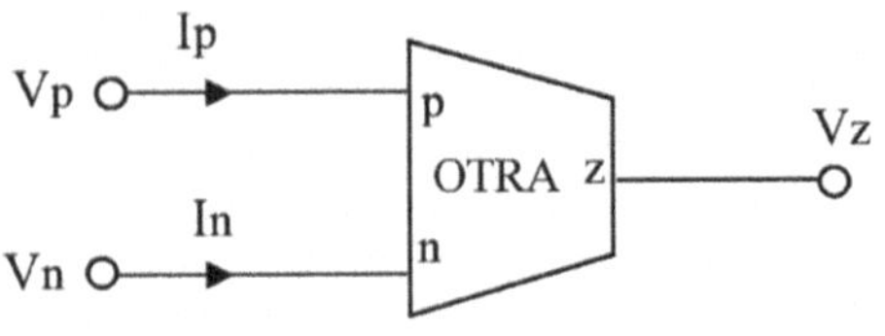

Fig. 4.1 Circuit symbol of the OTRA. (Salama and Soliman 1999; Chen et al. 1995)

For the OTRA both the input terminals and the output terminals have low impedance, thereby eliminating response limitations incurred by capacitive time constants. The input terminals are internally grounded leading to circuits that are insensitive to stray capacitances. For ideal operation, the transresistance Rm approaches infinity forcing the input currents to be equal. Thus, the OTRA is used in a feedback configurations in a way that is similar to the op-amp.

The growing demand for mobile communications has led to high level of chip integration and directed research toward the field of high-frequency applications. In the newly designed circuit topologies for high frequencies, the current-mode approach is preferred rather than the traditional voltage-mode structures. OTRA (operational transresistance amplifier), which is commercially available under the name of Norton amplifier, has attracted attention due to its advantages in the current-mode circuit design. Low input and output impedances and a bandwidth independent of the device gain can be considered the main advantageous properties of the OTRA. These commercial realizations don't provide a true virtual ground at the input terminals and they allow the input current to flow in one direction only. In order to remove these disadvantages of the OTRA, some topologies are proposed in the literature. But these solutions are both complex structures and do not operate properly at low power supplies like 1.2 V if they are realized with submicron technologies. A CMOS realization example of the OTRA is illustrated in Fig. 4.2.

In today's technology, circuits which use power supplies as 1 V and fabricated in the CMOS 0.09 μm technology can be designed, and the process improvement works on the CMOS 65 nm technology with a power supply of 0.9 V are still going on. Also CMOS 45 nm technology is available with a power supply of 0.6 V. So for the future design concept the main interest is designing circuitries with low power supplies. This demand leads to designing a high-performance CMOS differential OTRA for the current-mode analog system design. For these reasons, using the STMicroelectronics CMOS 0.13 μm technology, a differential OTRA is designed for

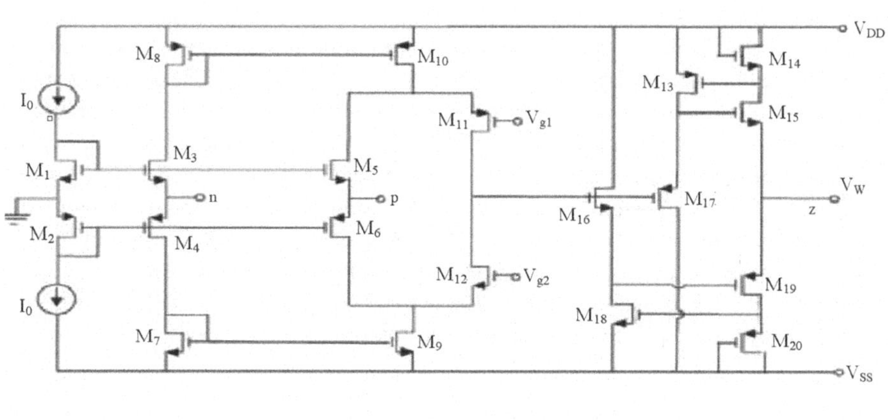

Fig. 4.2 A CMOS implementation of the OTRA. (Cam et al. 2003)

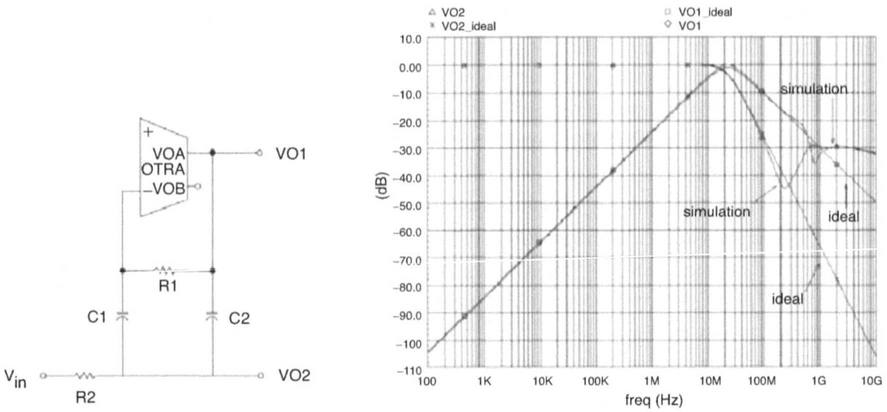

Fig. 4.3 The band-pass and low-pass filters with OTRA and their frequency responses. (Duruk and Kuntman 2005)

1.2 V power supply. This new CMOS differential OTRA topology is characterized by the CADENCE simulation tool and the characteristic results showing its high performance are given. A filter design example is given in Fig. 4.3.

Low-pass and band-pass filters with single CMOS differential OTRA structures are tested with simulations to verify the theoretical results.

Design Example: Two OTRA-Based Grounded Immittance Simulator Topologies

The proposed circuits are shown in Fig. 4.4. Routine analysis of these circuits yields the input impedances, equivalent inductances, and equivalent resistances as given in Table 4.1. From Table 4.1 one can observe that the proposed circuits can simulate various combinations of parallel lossy inductances. Note that RC/CR transformation makes active simulation of parallel C–D immittances possible, which is an important research topic in active network synthesis. From comparison of the proposed immittance simulator topologies in Table 4.1, it is concluded that the first circuit can simulate (+L) with (+R). Thus, it can be used to implement a current-mode filter. The second circuit can simulate (−L) with (R). This is especially important if the parasitics have to be cancelled. The third circuit can simulate pure (+L).

Note that RC/CR transformation makes active simulation of parallel C–D immittances possible, which is an important research topic in active network synthesis.

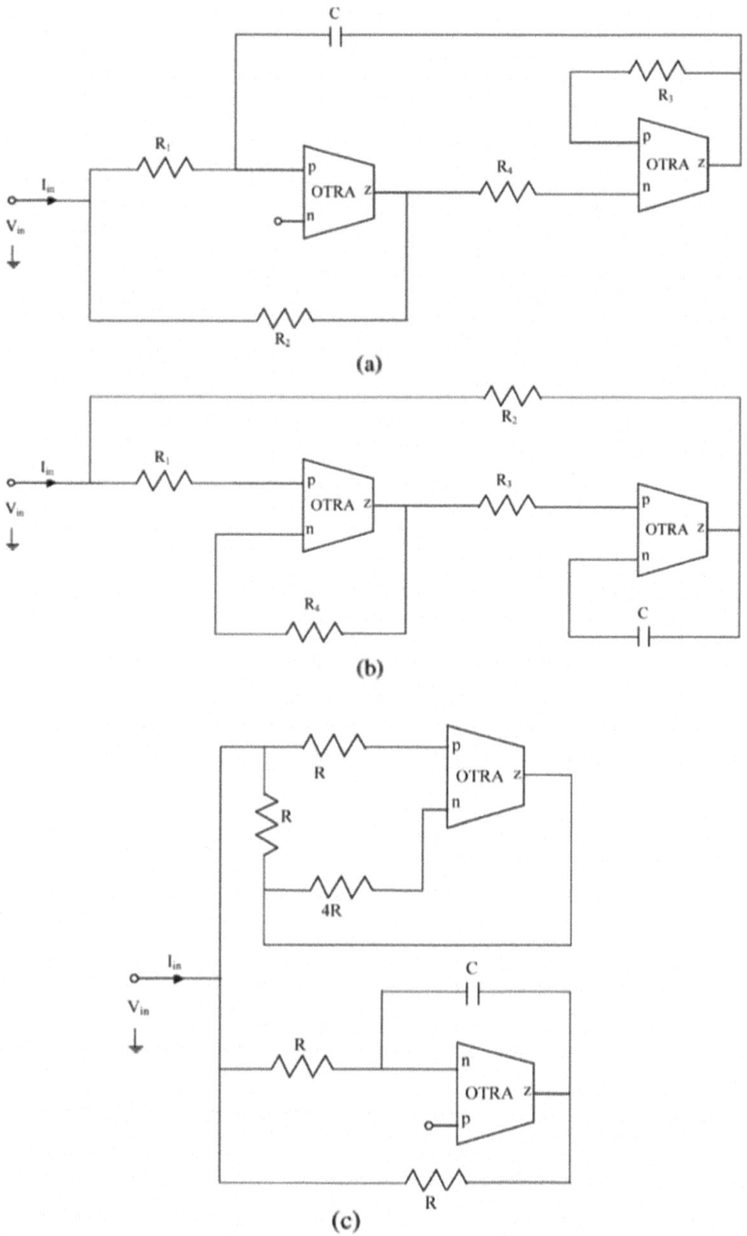

Fig. 4.4 The proposed immittance simulator topologies. (Cam et al. 2003)

Table 4.1 Various types of inductors realized with the presented circuits

Figure	L_{eq}	G_{eq}	Type
4.4a	$\frac{CG_4}{G_1G_2G_3}$	G_1+G_2	L parallel with R
4.4b	$-\frac{CG_4}{G_1G_2G_3}$	G_1+G_2	−L parallel with R
4.4c	$\frac{C}{G^2}$	–	Pure L

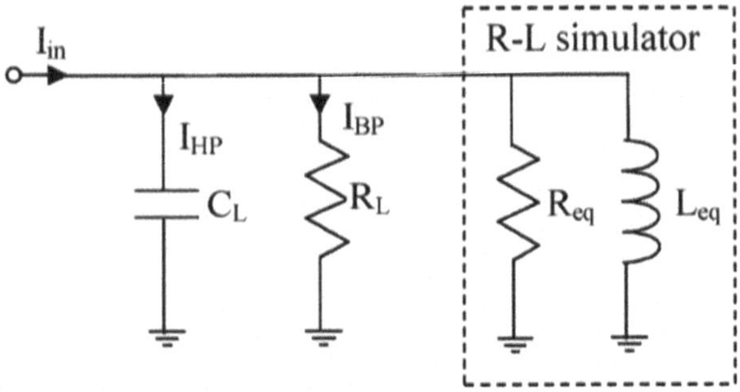

Fig. 4.5 A current-mode multi-output filter realized with the proposed parallel R–L simulator. (Cam et al. 2004)

Application Example and Simulation Results

A current-mode dual-function filter was designed and simulated to illustrate an application of the proposed topologies. The basic cell is a parallel R–L simulator with a parallel capacitor CL and RL to form the resonant circuit shown in Fig. 4.5. In this figure the actively simulated R–L circuit in Fig. 4.5a replaces the parallel R–L circuit.

The transfer functions are given by

$$I_{\mathrm{HP}}/I_{\mathrm{in}} = s^2/\Delta. \tag{4.2}$$

$$I_{\mathrm{BP}}/I_{\mathrm{in}} = (G_L/C_L)s/\Delta \tag{4.3}$$

$$\Delta = s^2 + \left(G_{\mathrm{eq}} + G_L\right)s/C_L + 1/L_{\mathrm{eq}}C_L \tag{4.4}$$

Capacitor or the resistor current can easily be sensed by additional current buffers and high output impedances:

The element values of the realized filter are chosen as follows: $C_{\mathrm{L}} = 10$ nF, $R_{\mathrm{L}} = 1$ kΩ, $R_1 = 2$ kΩ, $R_2 = 2$ kΩ, $R_3 = 1$ kΩ, $R_4 = 1$ kΩ, and $C = 1$ nF; thus, an inductor with $L_{\mathrm{eq}} = 4$ mH with a parallel resistance $R_{\mathrm{eq}} = 1$ kΩ is obtained which results in a pole frequency $f_{\mathrm{p}} = 25164$ Hz and a quality factor $Q_{\mathrm{p}} = 0.79$.

The designed filter circuit is simulated with PSPICE program using two AD844-based realizations of OTRA shown in Fig. 4.6. Supply voltages are taken as $V_{\mathrm{DD}} = 10$ V and $V_{\mathrm{SS}} = -10$ V. The simulation result of the filter response is given in Fig. 4.7, which is in good agreement with the predicted theory. The deviations in the frequency response of the filter from theoretical values are caused by the non-idealities of the OTRA. There are many applications that require a narrow

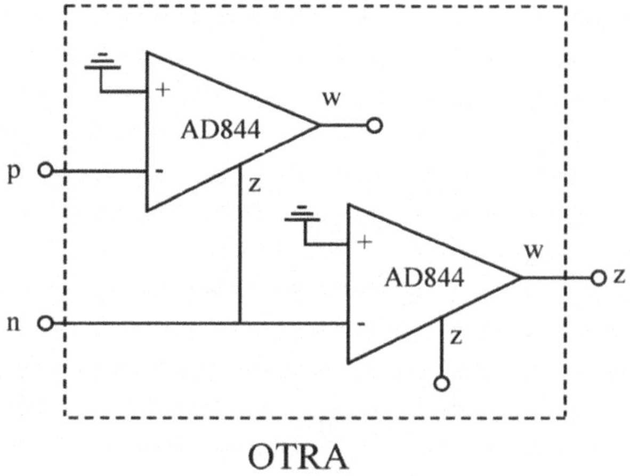

Fig. 4.6 Realization of an OTRA with two AD844 CCII+. (Cam et al. 2004)

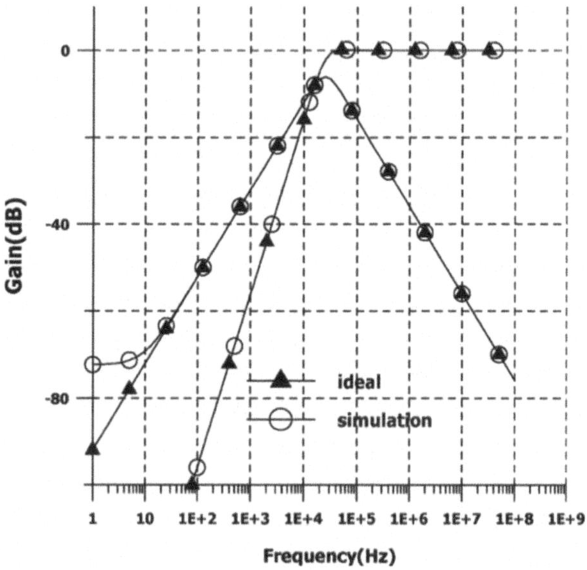

Fig. 4.7 The frequency response of the current-mode filter in Fig. 4.5. (Cam et al. 2004)

band band-pass-tuned amplifiers such as video signal processing, TV receivers, and wireless communications stages. A fourth-order band-pass filter was implemented which is shown in Fig. 4.8. Two AD 844 CCII ICs of analog devices were used as a voltage-to-current converter and a current buffer. It is possible to adjust the center frequency of individual stages as stagger or synchronized tuning. In this study synchronized tuning of stages is chosen as $f_c = 129.9$ kHz. The required passive element values in Fig. 4.8 are $R_{L1} = 1$ kΩ, $R_{L2} = 1$ kΩ, $L_{eq} = 1$ mH, $R_{eq} = 0.5$ kΩ, and $C_L = 1.5$ nF. The simulated R_{eq} and L_{eq} can easily be realized by using the circuit in Fig. 4.2a with $R_1 = 1$ kΩ, $R_2 = 1$ kΩ, $R_3 = 1$ kΩ, $R_4 = 1$ kΩ, and $C = 1$ nF. The frequency response of the fourth-order band-pass filter is shown in Fig. 4.9.

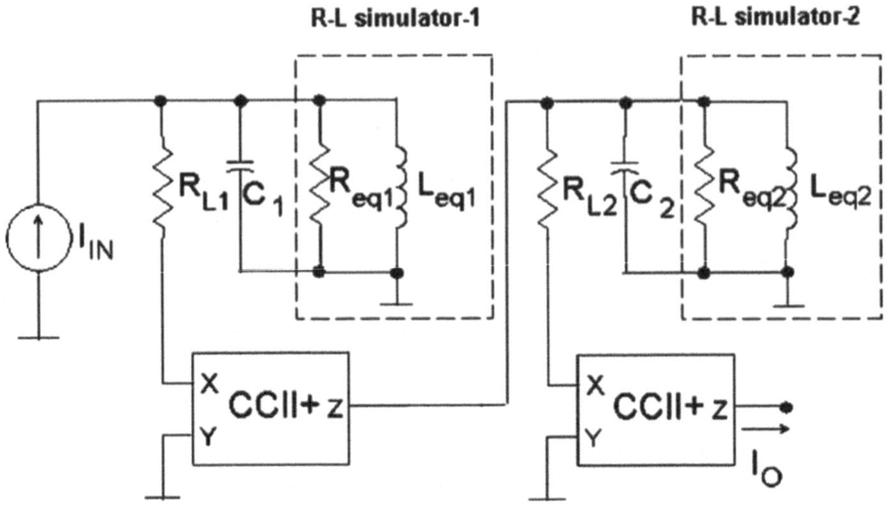

Fig. 4.8 A fourth-order band-pass filter realized with two R–L simulators. (Cam et al. 2004)

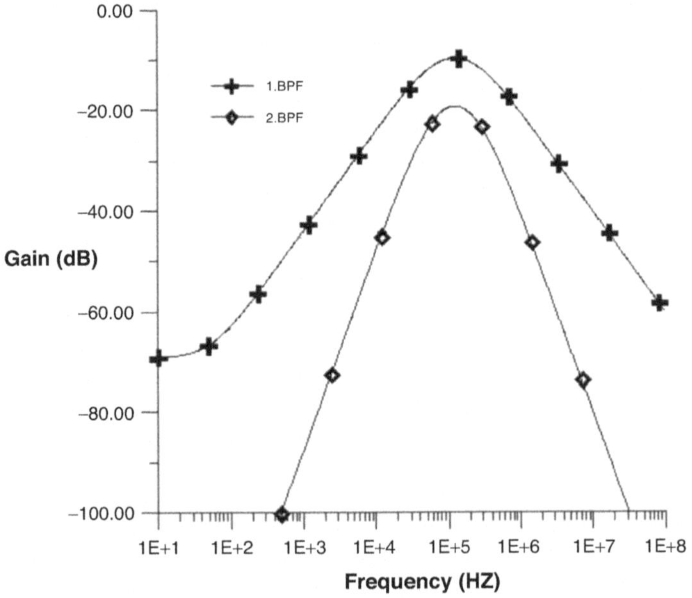

Fig. 4.9 PSPICE simulation results of frequency response of the fourth-order band-pass filter. (Cam et al. 2004)

In this part, three circuits for actively simulation of parallel immittances are described. All circuits employ only two OTRAs as the active element. A fourth-order band-pass filter is realized to illustrate the practical use of the proposed topologies as an application example. PSPICE simulation results are given to verify the theory. It is expected that the proposed circuits will be useful in the design of analog signal processing applications such as filters and oscillators and will provide therefore further possibilities to the designer in the realization of analog circuits.

Low-Voltage CMOS Differential OTRA for Submicron Technologies

The differential OTRA is a five-terminal analog building block with a describing matrix in the form

$$\begin{bmatrix} V_1 \\ V_2 \\ V_{\mathrm{OA}} \\ V_{\mathrm{OB}} \end{bmatrix} = \begin{bmatrix} 0 & 0 & 0 & 0 \\ 0 & 0 & 0 & 0 \\ R_{\mathrm{m}} & -R_{\mathrm{m}} & 0 & 0 \\ -R_{\mathrm{m}} & R_{\mathrm{m}} & 0 & 0 \end{bmatrix} \begin{bmatrix} I_1 \\ I_2 \\ I_{\mathrm{OA}} \\ I_{\mathrm{OB}} \end{bmatrix} \tag{4.5}$$

Circuit symbol of the differential OTRA is illustrated in Fig. 4.10.

The proposed CMOS differential OTRA is illustrated in Fig. 4.11. This circuitry also includes the four power-down transistors, which creates the enabled power supplies named as "bes" and "top." OTRA is active when the "EN" input is at the V_{SS} voltage level. When the "EN" input is at the V_{DD} voltage level, all the system is in the power-down mode and the output "V_{OA}" and "V_{OB}" behaves as a high output impedance. Mbescap and Mtopcap are MOS capacitances, which filter the power supply to ground, and avoid nodes "bes" and "top" from being floating nodes in the power-down mode. The basic cell consists of six transistors, which form a low-voltage current mirror topology (M_1–M_6) as illustrated in Fig. 4.12. The input currents are directly applied to the I_1 and I_2 nodes. These inputs are connected to the source of the PMOS and the drain of the NMOS in the input. "c" and "d" node voltages are changing due to the input current differences. So in this stage input current difference is converted to differential voltage ("c" and "d" nodes). Also a resistor network (R_3–R_7) is connected to the I_1 and I_2 inputs as illustrated in Fig. 4.11. The large resistors R_3–R_6 which have the same resistor value (12 k) are

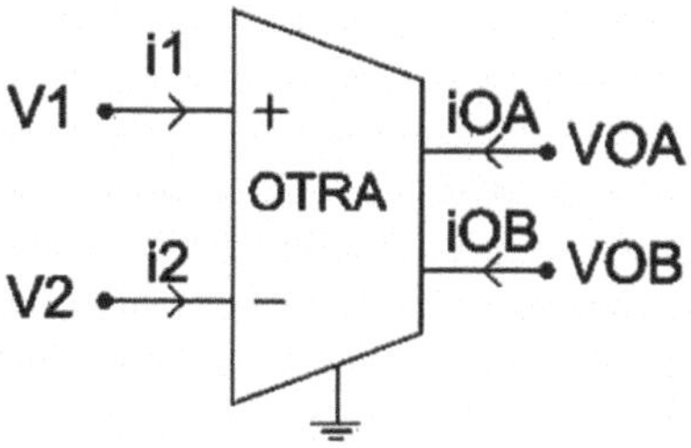

Fig. 4.10 Circuit symbol of the differential OTRA. (Duruk et al. 2007)

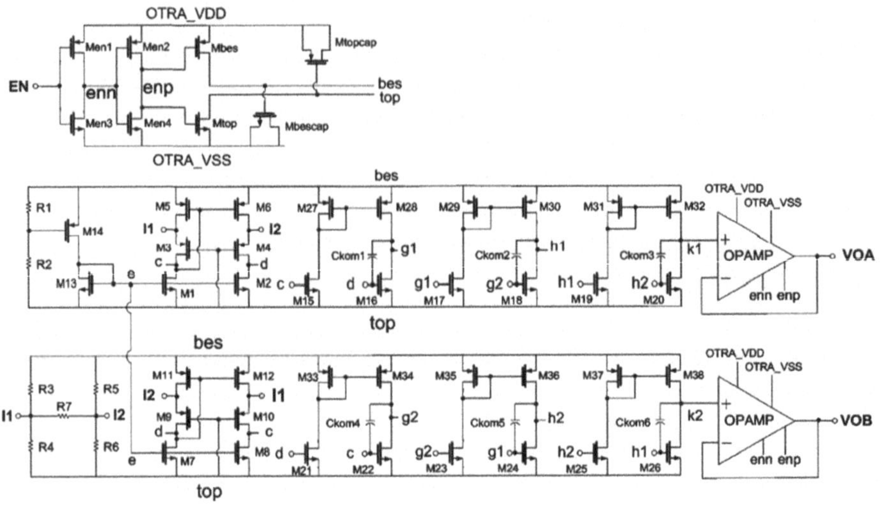

Fig. 4.11 Designed differential CMOS OTRA topology. (Duruk et al. 2007)

Fig. 4.12 Basic cell of the OTRA input stage. (Duruk et al. 2007)

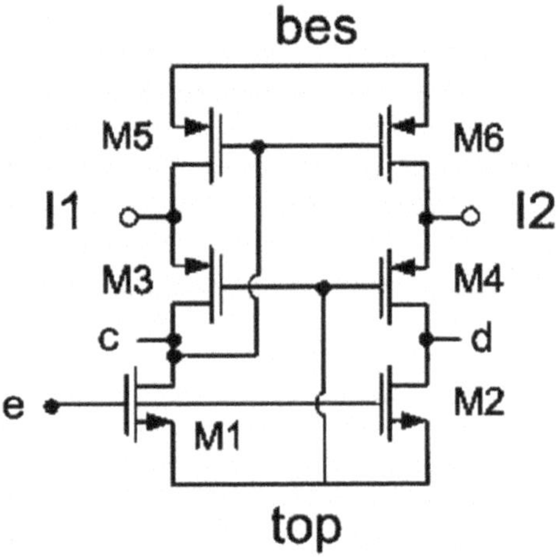

generating the biasing voltage for I_1 and I_2, which would be ideally in the ground potential (0 V). R_7 resistor is to force the input resistance of the OTRA in low values. If one basic cell is used, the OTRA input will not be symmetrical. For the given basic cell, the input I_1 is formed by the use of PMOS (M_3) and pseudo-diode connected PMOS (M_5) transistors, but at the input I_2 there is no pseudo-diode connection. So "c" and "d" nodes cannot form any symmetrical signals. For that reason, a second basic cell is placed into the design, by replacing the input pins, meaning that input I_2 is applied to the pseudo-diode connected input part of the basic cell and input I_1 is applied to the other input as illustrated in Fig. 4.11 (M_7–M_{12}).

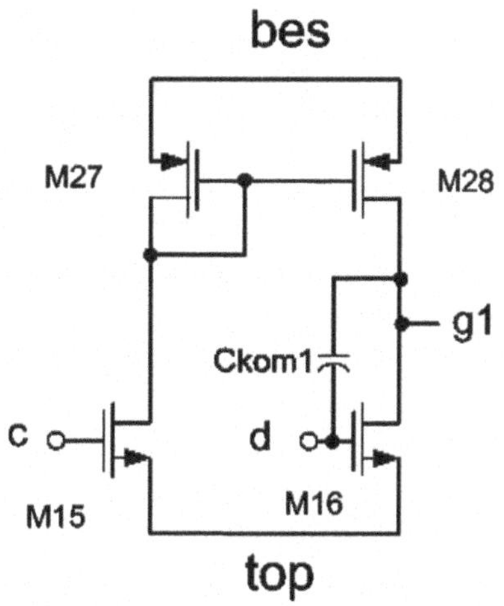

Fig. 4.13 Basic cell of the OTRA gain stage. (Duruk et al. 2007)

The basic cell of the gain part is composed of four transistors, namely, M_{15}, M_{16}, M_{27}, and M_{28}, as illustrated in Fig. 4.11. This differential amplifier is not a classical amplifier with a fixed current source biasing the NMOS transistors M_{15} and M_{16}. This amplifier can work independent of any fixed current (Fig. 4.13).

This cell is converting the two input differential signals to one single-ended signal. In order to have a symmetrical output in g1 and g2 nodes, two basic cells are used within the same idea in the input part of the OTRA. For the second basic cell, which consists of the transistors M21, M22, M33, and M34 in Fig. 4.11, the input pins reversed. So both I_1 and I_2 inputs are directly connected to one diode-connected PMOS (M27 and M33) and one not-diode-connected PMOS (M34 and M28). The first gain outputs, which are "g1" and "g2" in names, are connected to the second same structured gain stage like the first gain stage. The outputs of the second gain stage, which are "h1" and "h2" in names, are connected to the third, same structured gain stage as the first gain stage. The outputs of the third gain stage, which are "k1" and "k2" in names, are the non-buffered dual outputs of differential OTRA. The non-buffered dual outputs are buffered with the unity gain configuration by using op-amp, which is shown in Fig. 4.14. The V_{OA} is the same functional output as the V_O in the classical single-output OTRA, and V_{OB} is the dual output of V_{OA}, which just differs from the V_{OA} output in the phase level.

Characterization of the CMOS differential OTRA: The proposed CMOS differential OTRA is simulated in the CADENCE simulation program, with the STMicroelectronics CMOS 0.13 μm technology Spectre models by using the transistor and element values listed in Table 4.2 for OTRA and Table 4.3 for op-amp.

Power supplies are chosen as ±0.6 V, and the operating temperature is taken as 27 °C. The load capacitances used in the simulations are 5 pF. The typical DC simulation results are shown in Fig. 4.15. According to these results the maximum output voltages are reached with at least 100 μA of input difference current. The maximum output voltage in the DC analysis is 0.531 mV, and the minimum output

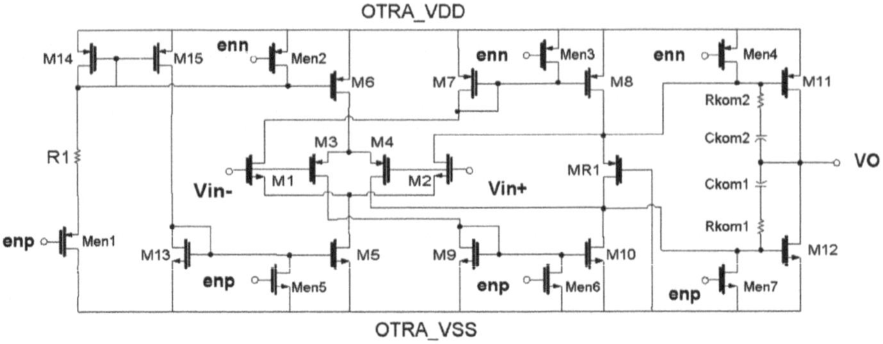

Fig. 4.14 Op-amp used in the OTRA as a unity gain output buffer. (Duruk et al. 2007)

Table 4.2 Transistor dimensions (W/L) in the proposed CMOS differential OTRA

Transistor name	W/L
M1, M2, M7, M8	64 μm/0.3 μm
M3. M4. M9. M10	200 μm/0.15 μm
M5, M6, M11, M12	20 μm/0.15 μm
M14	12.24 μm/0.3 μm
M13	5 μm/0.3 μm
M15–M26	1 μm/0.25 μm
M27–M38	3.34 μm/0.25 μm
Men1, Men2	28.8 μm/0.13 μm
Mcn3. Men4	13.6 μm/0.13 μm
Mbes	561 μm/0.15 μm
Mtop	140 μm/0.15 μm
Mbescap, Mtopcap	64.84 μm/5.28 μm
Other devices	Value
Ckoml-Ckom6	5 fF
R1	6 kΩ
R2–R6	12 Ω
R7	30 Ω

voltage is −0.531 mV with 100 μA input current. Another typical DC test simulation result is shown in Fig. 4.16. According to these results, it is shown that this OTRA can work within the limits of −2 and 2 mA. If one of the input nodes is biased with more than 2 mA, it is also possible to work. But the 2 mA input current limit will be more than enough for a classical operation. The transresistance gain is shown in Fig. 4.17. The transresistance gain is 38 660 V/A. The −3 dB point of the transresistance gain is at 29.2 MHz. Also the unity gain bandwidth is 2.51 GHz. This bandwidth region gives the opportunity of designing high bandwidth filters, inductance simulators, and other topologies. The performance characteristics of the CMOS differential OTRA are shown in Table 4.4.

Table 4.3 Transistor dimensions (W/L) and other device values in the proposed op-amp

Transistor name	W/L
M1. M2	8.5 μm/0.3 μm
M3. M4	41 μm/0.3 μm
M5	11.2 μm/0.4 μm
M6	32.64 μm/0.4 μm
M7, M8	8.7 μm/0.3 μm
M9. M10	3.57 μm/0.3 μm
M11	175 μm/0.3 μm
M12	70 μm/0.3 μm
M13	3.4 μm/0.4 μm
M14, M15	8.16 μm/0.4 μm
MR 1	1 μm/10 μm
Men 1	43.2 μm/0.13 μm
Men2–Men7	1 μm/0.13 μm
Other devices	Value
Ckoml. Ckom2	400 fF
Rkoml. Rkom2	1400 Ω
R 1	6 kΩ

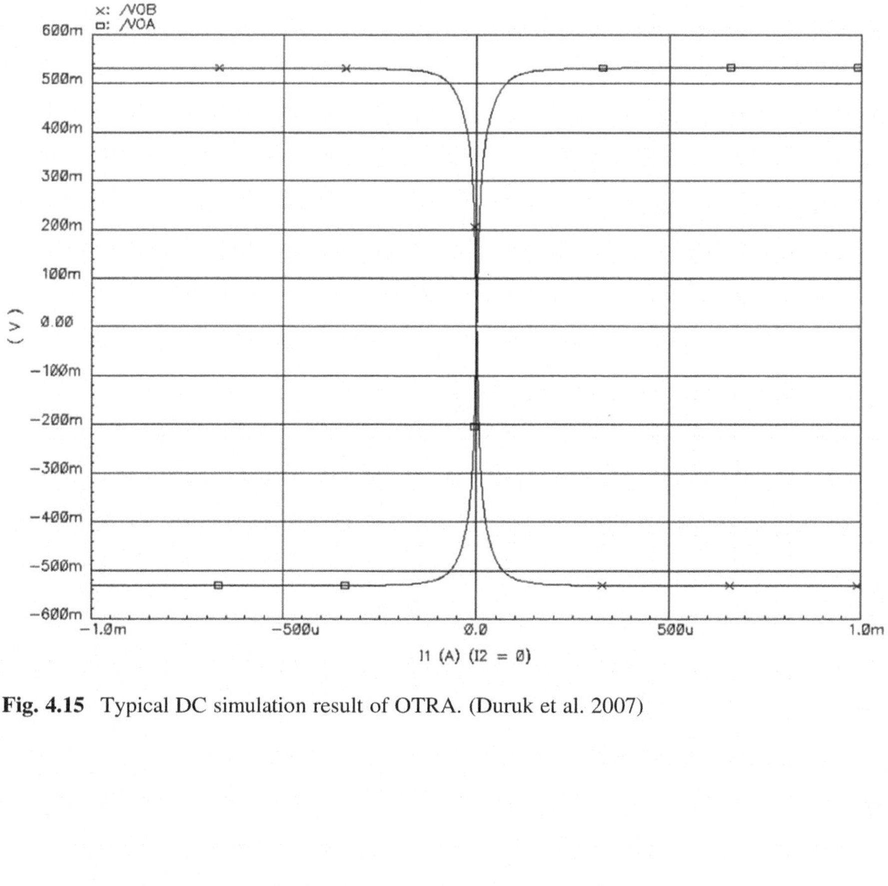

Fig. 4.15 Typical DC simulation result of OTRA. (Duruk et al. 2007)

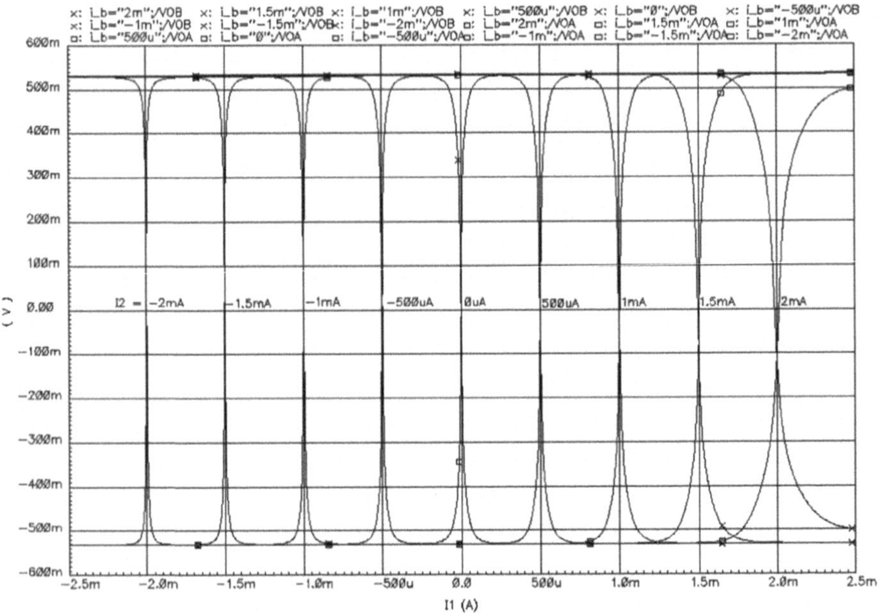

Fig. 4.16 Typical DC simulation result of OTRA with different I2 input currents between −2 and 2. (Duruk et al. 2007)

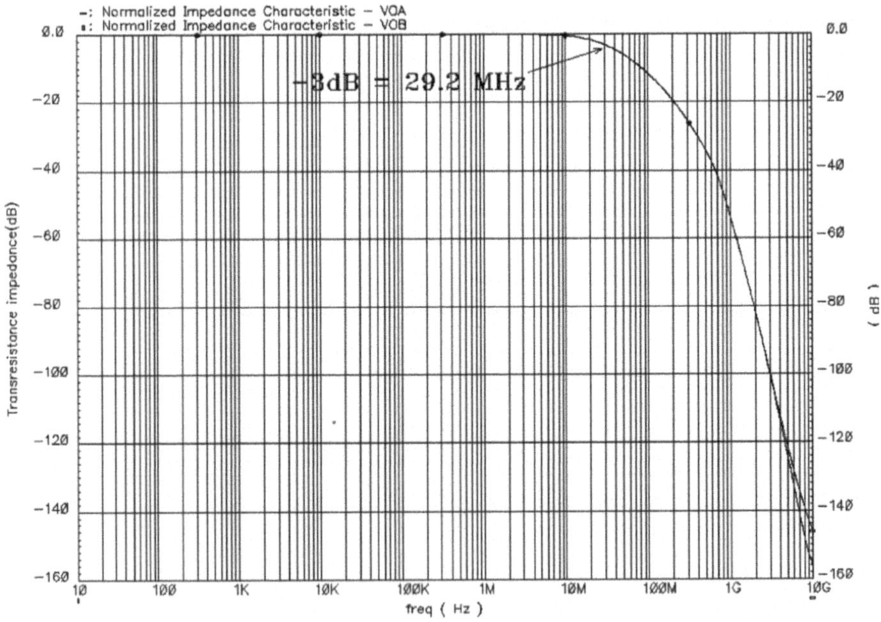

Fig. 4.17 Typical AC simulation result, normalized transresistance gain of the OTRA. (Duruk et al. 2007)

Table 4.4 Performance of the proposed CMOS differential OTRA

Power supply	±0.6 V
Maximum output voltage	0.531 V
Minimum output voltage	−0.531 V
Input resistance	$R\_I_1 = R\_I_2 = 51.8\ \Omega$
Output resistance	1.45 Ω
Input offset current	2 nA
Transresistance gain (Rm) (DC)	38 660 V/A
Transresistance gain bandwidth (−3dB)	29.2 MHz
Transresistance unity gain bandwidth	2.51 GHz
Leakage current	6.51 μA
Power consumption	9.9 mA, 11.88 mW
Expected silicon area	134 μm × 72 μm

Universal MOS-C Filter Design Example and Simulation Results

The performance of the CMOS differential OTRA proposed is demonstrated on a design example of universal MOS-C filter structure. This filter is based on the MOS structure illustrated in Fig. 4.18. In this structure, when two NMOS transistors operate in linear region, the nonlinearity effects are cancelled since they have the same drain and source voltages. In this operating mode, the following function appears:

$$I_1 - I_2 = G(V_i - V_2), \tag{4.6}$$

where

$$\begin{aligned} G &= K_{\mathrm{n}}(V_{\mathrm{a}} - V_{\mathrm{b}}), \\ K_{\mathrm{n}} &= \mu_{\mathrm{n}} C_{\mathrm{ox}} \frac{W}{L}. \end{aligned} \tag{4.7}$$

The universal MOS-C filter topology illustrated in Fig. 4.19 is similar in principle to the presented work in the literature (Salama and Soliman 1999). The original filter topology has one input. The transistors and capacitors have to be removed for the desired filter characteristics. For obtaining a real universal filter the topology is extended to three inputs. The low-pass (LP), band-pass (BP), high-pass (HP), notch (BR), and all-pass (AP) filter characteristics are obtained by selecting the proper input pins without removing any elements from the filter. The transfer functions of the filters are given by the following equations:

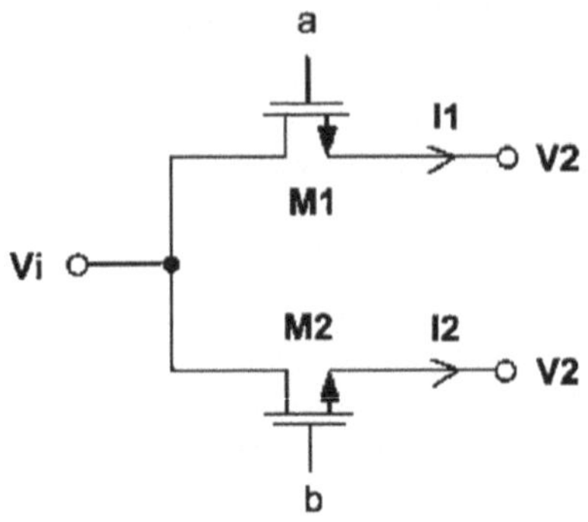

Fig. 4.18 MOS structure ($V2 \rightarrow 0$). (Duruk et al. 2007)

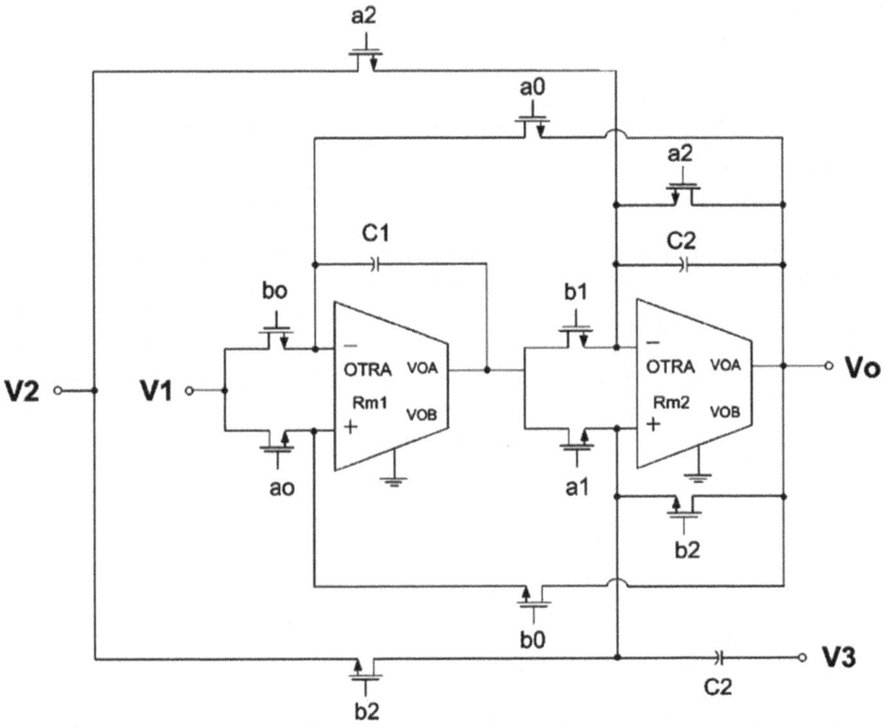

Fig. 4.19 Universal MOS-C filter topology with CMOS OTRA. (Duruk et al. 2007)

$$\begin{aligned} &V_2 = 0, V_3 = 0 \rightarrow \text{Lowpass}, \\ &\frac{V_0}{V_1} = \frac{G_1 G_0}{s^2 C_1 C_2 + s C_1 G_2 + G_1 G_0}, \\ &V_1 = 0, V_3 = 0 \rightarrow \text{Bandpass}, \\ &\frac{V_0}{V_2} = \frac{-s C_1 G_2}{s^2 C_1 C_2 + s C_1 G_2 + G_1 G_0}, \end{aligned} \tag{4.8}$$

$$\begin{aligned} &V_1 = 0, V_2 = 0 \rightarrow \text{Highpass}, \\ &\frac{V_0}{V_3} = \frac{s^2 C_1 C_2}{s^2 C_1 C_2 + s C_1 G_2 + G_1 G_0}, \\ &V_1 = V_3 = V_i, V_2 = 0 \rightarrow \text{Band reject (Notch)} \end{aligned} \tag{4.9}$$

$$\begin{aligned} &\frac{V_0}{V_i} = \frac{s^2 C_1 C_2 + G_1 G_0}{s^2 C_1 C_2 + s C_1 G_2 + G_1 G_0}, \\ &V_1 = V_2 = V_3 = V_i, \rightarrow \text{Allpass}, \\ &\frac{V_0}{V_i} = \frac{s^2 C_1 C_2 - s C_1 G_2 + G_1 G_0}{s^2 C_1 C_2 + s C_1 G_2 + G_1 G_0}. \end{aligned} \tag{4.10}$$

The cutoff frequency for the second-order filters is given by

$$f_{(-3\text{ dB})} = \frac{1}{2\pi} \sqrt{\frac{G_1 G_0}{C_1 C_2}}. \tag{4.11}$$

The quality factor of this filter is

$$Q = \frac{1}{G_2} \sqrt{\frac{G_0 G_1 C_2}{C_1}}. \tag{4.12}$$

The sensitivities of the universal filter are given as

$$\begin{aligned} &S_{G_0}^{\omega_0} = S_{G_1}^{\omega_0} = -S_{C_1}^{\omega_0} = -S_{C_2}^{\omega_0} = \frac{1}{2}, \quad S_{G_2}^{\omega_0} = 0, \\ &S_{G_0}^{Q} = S_{G_1}^{Q} = S_{C_2}^{Q} = -S_{C_1}^{Q} = \frac{1}{2}, \quad S_{G_2}^{Q} = -1. \end{aligned} \tag{4.13}$$

The universal MOS-C OTRA filter frequency response (BP, HP, LP, notch, AP) is shown in Fig. 4.20. The dependence of the output harmonic distortion of low-pass filter on input voltage amplitude with 1 MHz signal is illustrated in Fig. 4.21.

The CMOS OTRA MOS-C universal filter is simulated using CADENCE simulation program, and the STMicroelectronics CMOS 0.13 μm technology Spectre models. The transistor dimensions and element values for CMOS differential OTRA and CMOS op-amp are listed in Tables 4.2 and 4.3. Power supplies are chosen as ±0.6 V, and the operating temperature is taken as 27 °C. The receiver of the global system for mobile (GSM) cellular telephone comprises two IF stages. The frequency

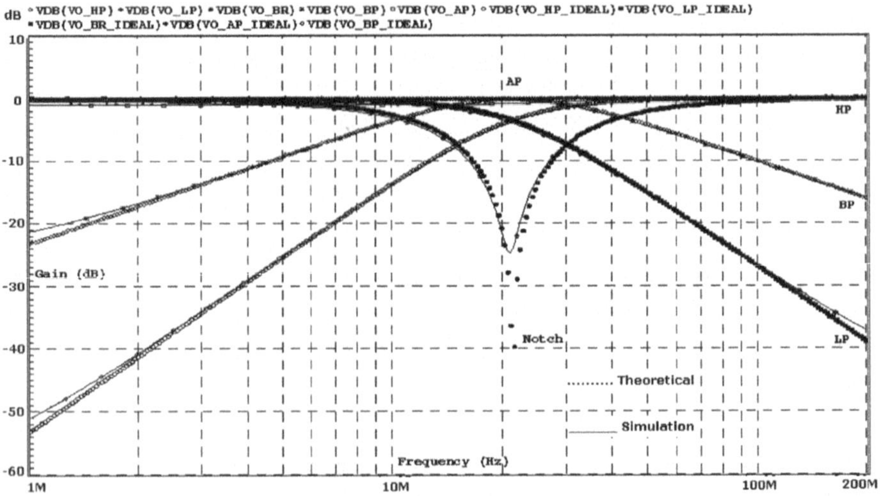

Fig. 4.20 Universal MOS-C OTRA filter frequency response (BP, HP, LP, notch, AP). (Duruk et al. 2007)

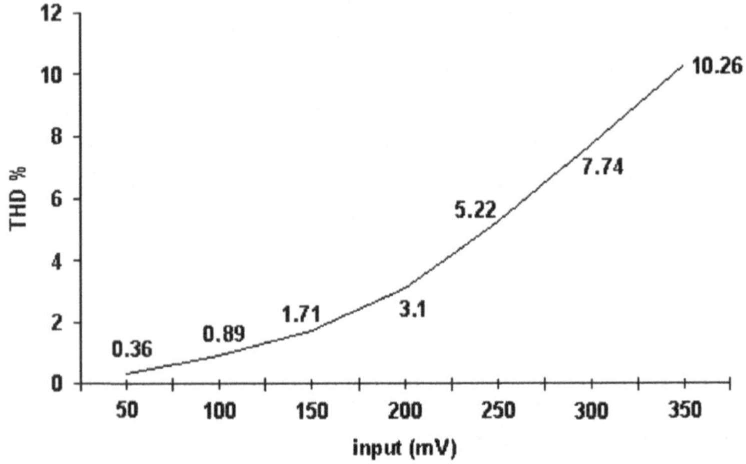

Fig. 4.21 Dependence of the output harmonic distortion of low-pass filter on input voltage amplitude with 1 MHz signal. (Duruk et al. 2007)

of the second stage is about 20 MHz. At present time these filters are obtained using separate surface acoustic wave (SAW) filter. These SAW filters exhibit losses, which are more than 7 dB. Also they are expensive and generally require additional matching components to be interfaced perfectly with their driving and loading circuit. Therefore, the design of active filter operation in IF2 stages instead of SAW filters appears very attractive. Because of this application the universal MOS-C OTRA filter is targeted to have the 22.5 MHz cutoff frequency.

References

Analog Devices Inc. Data Sheet, AD844 Current Feedback OpAmp.

Chen, J. J., Tsao, H. W., Liu, S. I. and Chiu, W. Parasitic-capacitance-insensitive current-mode filters using operational transresistance amplifiers. IEE Proceedings-Circuits, Devices and Systems, 142(3), 186–192, 1995.

Çam, U., Cicekoglu, O. and Kuntman, H. Universal series and parallel immittance simulators using four terminal floating nullors. Analog Integrated Circuits and Signal Processing, 25(1), 59–66, 2000.

Çam, U., Kaçar, F., Cicekoglu, O., Kuntman, H., and Kuntman, A. Novel two OTRA-based grounded immitance simulator topologies. Analog Integrated Circuits and Signal Processing, 39(2), 169–175, 2004.

Çam, U., Kaçar, F., Cicekoglu, O., Kuntman, H., and Kuntman, A., "Novel Grounded Parallel Immittance Simulator Topologies Employing Single OTRA", AEU-International Journal of Electronics and Communications, 57, No. 4, 287–290, 2003.

Duruk, A. and Kuntman, H. A new CMOS differential OTRA design for the low voltage power supplies in the sub-micron technologies. Turkish Journal of Electrical Engineering & Computer Sciences, 13(1), 23–38, 2005.

Duruk, A., Güneş, E. O. And Kuntman, H. A new low voltage CMOS differential OTRA for sub-micron technologies. AEU-International Journal of Electronics and Communications, 61(5), 291–299, 2007.

Kuntman, H. H. New advances and possibilities in active circuit design. In Proc. 10th International Conference on Development and Application Systems (pp. 9–18), 2010.

Kuntman, H. New trends in circuit design for analog signal processing. In 2011 7th International Conference on Electrical and Electronics Engineering (ELECO) (pp. II–26). IEEE, 2011.

Kuntman, H. H. and Uygur, A. New possibilities and trends in circuit design for analog signal processing. In 2012 International Conference on Applied Electronics (pp. 1–9). IEEE, 2012.

Salama, K. N., and Soliman, A. M. CMOS operational transresistance amplifier for analog signal processing. Microelectronics Journal, 30(3), 235–245, 1999.

Chapter 5
FTFN: Four-Terminal Floating Nullor

Basic Concept

Current-mode circuits have received significant attention due to their particular advantages compared with voltage-mode circuits. They offer the designer several salient features such as inherently wide bandwidth, greater linearity, wider dynamic range, simple circuitry, and low power consumption (Toumazou et al. 1990). Analog designers use mostly second-generation current conveyors (CCIIs) rather than operational amplifiers (op-amp) as the active element in analog integrated circuit (IC) design because of their advantages and simultaneously voltage copying and current copying properties (Toumazou et al. 1990). Recently, attention has turned to the use of the four-terminal floating nullor (FTFN) as current-mode active element since it has been shown that the FTFN is a more flexible, versatile, and stable active element than an op-amp and a CCII (Senani 1987; Higashimura 1991a, b; Liu 1995; Abuelma'Atti and Al-Zaher 1999; Cam et al. 1999). Generally, the FTFN has been implemented either with the supply current sensing method (SCSM) using an op-amp and current mirrors or with the commercial current feedback op-amp ICs AD844 of analog devices (Higashimura 1991a, b; Liu 1995; Abuelma'Atti and Al Zaher 1999; Cam et al. 1999). The main limitation of op-amp and cross-coupled current mirrors of FTFN type is that feedback can be established only between the voltage output, while the current at the novel terminal cannot be established by feedback (Cabeza et al. 1994). The performance of the FTFN is limited also by critical op-amp specifications. For example, if an FTFN is used to implement a constant bandwidth variable gain voltage amplifier circuit, the frequency responses are constrained by the finite gain-bandwidth product of the op-amp. When CCII-based implementation is used, the number of active element increases to realize a specified transfer function. On the other hand, in modern mixed analog/digital IC design the preferred technology is CMOS owing to its low power consumption and high integration density consideration. Therefore, there is a special interest for

H. Kuntman, D. Özenli, *Trends in Circuit Design for Analog Signal Processing*, Analog Circuits and Signal Processing,
https://doi.org/10.1007/978-3-030-96836-6_5

CMOS implementation of the FTFN (Laopoulos et al. 1995). It is the aim of this study to yield simple and high-performance CMOS FTFN topologies suitable for VLSI technologies. In this study, we present a new CMOS FTFN circuit based on common-source and common-gate amplifiers and floating current source.

It was demonstrated recently that the FTFN is a more general and flexible building block compared to the active elements mentioned above. This has led to a growing interest in the design of amplifiers, gyrators, inductance simulators, oscillators, and current-mode filters which use FTFN as the active element. It is more suitable to exploit FTFN as an active element in current-mode circuit design since it has been shown that an FTFN is the most flexible and versatile building block in active network synthesis. The symbol of the FTFN and traditional implementations of the FTFN are illustrated in Figs. 5.1 and 5.2, respectively.

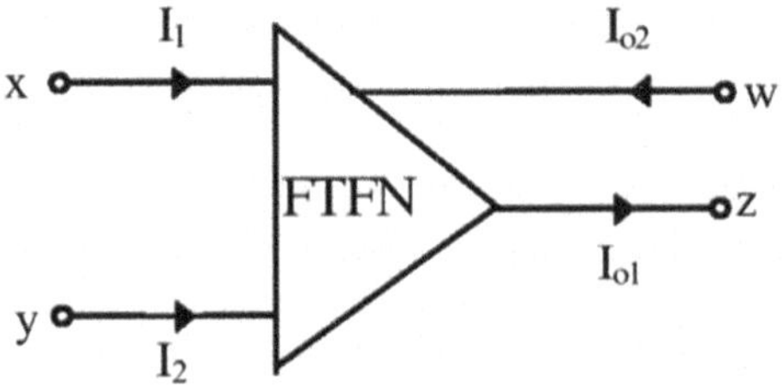

Fig. 5.1 Symbol of the FTFN. (Higashimura 1991a, b; Liu 1995; Abuelma'Atti and Al Zaher 1999; Cam et al. 1999)

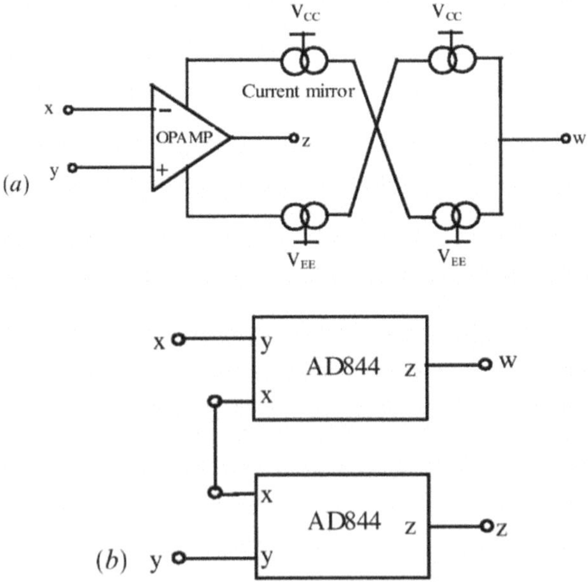

Fig. 5.2 Traditional implementations of the FTFN. (**a**) Op-amp and cross-coupled current mirror-type FTFN. (**b**) Realization of an FTFN with commercial AD844 ICs. (Higashimura 1991a, b; Liu 1995; Liu 1997; Liu and Liao 1996; Liu and Lee 1996; Abuelma'Atti and Al Zaher 1999; Cam et al. 1999)

The port relations of an FTFN as shown in Fig. 5.1 can be characterized as

$$I_1 = I_2 = 0 \quad I_{o1} = I_{o2} \quad V_x = V_y \tag{5.1}$$

The FTFN is ideally a transconductance amplifier featuring infinite gain and two output currents. The basic equation describing its operation is

$$I_w = I_z = G_m \cdot \left(V_x - V_y\right) \tag{5.2}$$

An FTFN is actually equivalent to an ideal nullor or is called operational floating amplifier, as shown in Fig. 5.3.

For a finite open-loop transconductance gain G_m, the difference between the two differential voltages increases as G_m decreases. Therefore, the open-loop transconductance gain should be as large as possible to achieve high-performance operation.

Typical examples of CMOS FTFN realization are shown in Fig. 5.4. It is also possible to realize FTFN-based nonlinear circuits to replace op-amp-based nonlinear structures. A good example from chaotic communication is the following inductorless realization of Chua's circuit using an FTFN-based nonlinear resistor and an FTFN-based inductance simulator.

The circuit schematic of a further new FTFN is shown in Fig. 5.5. It consists of a common-source and common-gate amplifiers (Gray et al. 2001) and a floating current source (Arbel and Goldminz 1992). Transistors M_1–M_{11} perform the common-source and common-gate amplification which provide large voltage gain over a wide frequency range. This voltage is then converted into two balanced output currents by the floating current source, formed from the M_{12}–M_{17} transistors, which requires no transistor matching constraint unlike the current mirror structure.

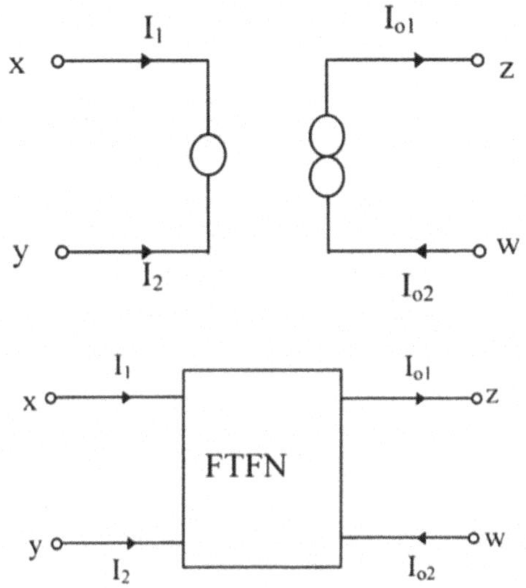

Fig. 5.3 Nullor model and symbol of the FTFN. (Cam et al. 1999)

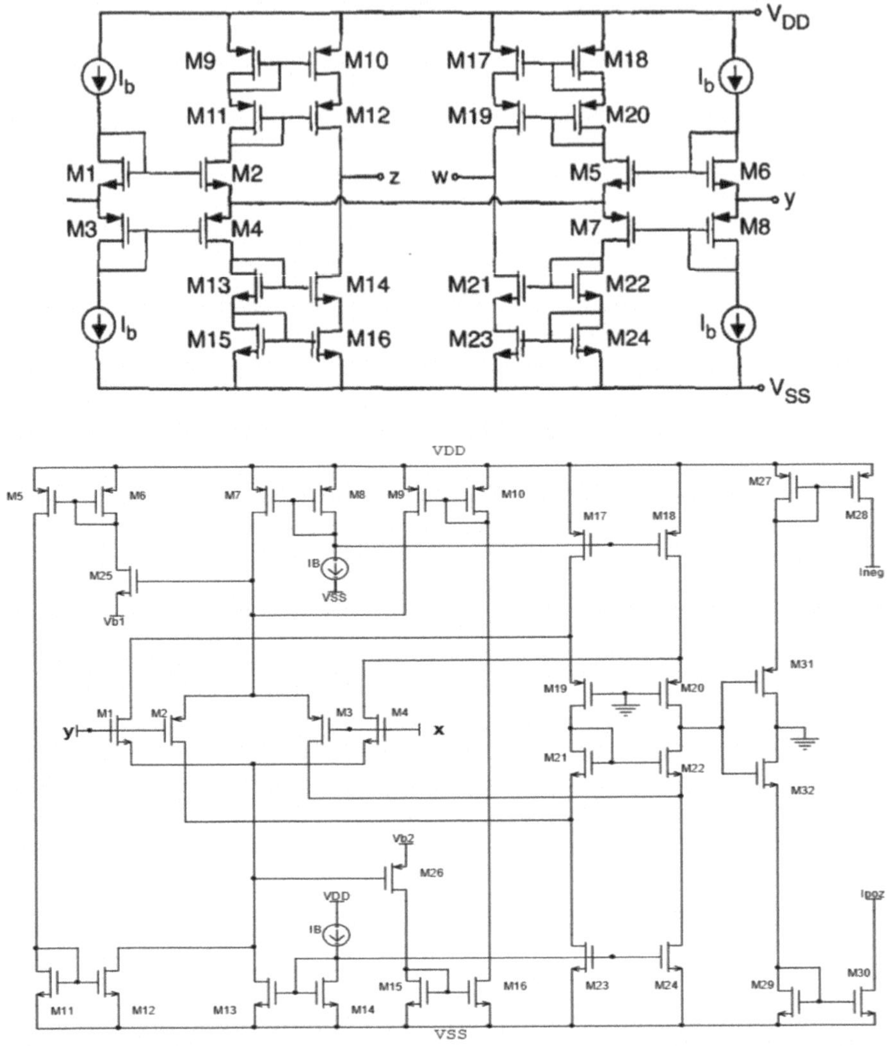

Fig. 5.4 Two realization examples of CMOS FTFN realization. (Kılıç et al. 2002, 2004)

Essentially, this stage is just two matched CMOS inverters biased by two current sources and operates as an analog transconductor stage. This current output stage is also used in some applications such as for the design of balanced output CMOS OTA design.

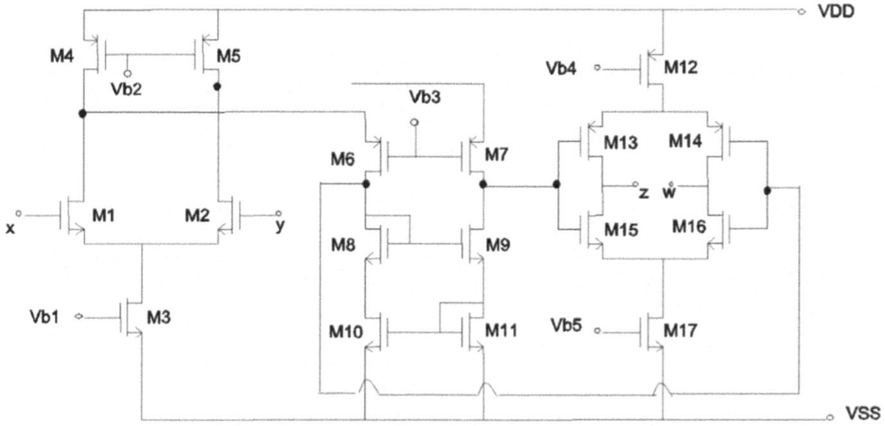

Fig. 5.5 Circuit schematic diagram of the further FTFN. (Çam et al. 2000; Çam and Kuntman 2000)

The overall magnitude of gain, magnitude of transconductance, and output resistance of the described FTFN can be given as

$$K_v = \frac{V_z}{V_x} = g_{m1}\frac{1}{\frac{g_{o2}}{g_{m9,}r_{09}} + \frac{g_{o2}+g_{os}}{g_{m7}r_{o7}}}\left(\frac{g_{m13}g_{m14}}{g_{m13}+g_{m14}} + \frac{g_{m15}g_{m16}}{g_{m15}+g_{m16}}\right)\frac{1}{g_{ol3}+g_{ol5}} \tag{5.3}$$

$$G_m = \frac{I_z}{V_x} = g_{m1}\frac{1}{\frac{g_{o2}}{g_{m9}r_{o9}} + \frac{g_{o2}+g_{o5}}{g_{m7}r_{o7}}}\left(\frac{g_{m13}g_{m14}}{g_{m13}+g_{m14}} + \frac{g_{m15}g_{m16}}{g_{m15}+g_{m16}}\right) \tag{5.4}$$

$$R_{oz} = \frac{1}{g_{o13}+g_{ol5}} \qquad R_{ow} = \frac{1}{g_{o14}+g_{ol6}} \tag{5.5}$$

Simulation Results

The designed CMOS FTFN circuit is simulated with a PSPICE simulation program with TUBITAK YITAL CMOS process.

The simulated DC response of the proposed circuits is shown in Fig. 5.6a. A linear current following operation can be seen over a wide dynamic current range in Fig. 5.6b. The open-loop voltage gain of the circuit given in Fig. 5.7a shows that the −3dB cutoff frequency is nearly 1.2 MHz, whereas the unity gain bandwidth is close to 250 MHz. The open-loop transconductance of the circuit given in Fig. 5.7b shows also that the transconductance of the proposed circuit is greater than that of other circuits in the literature.

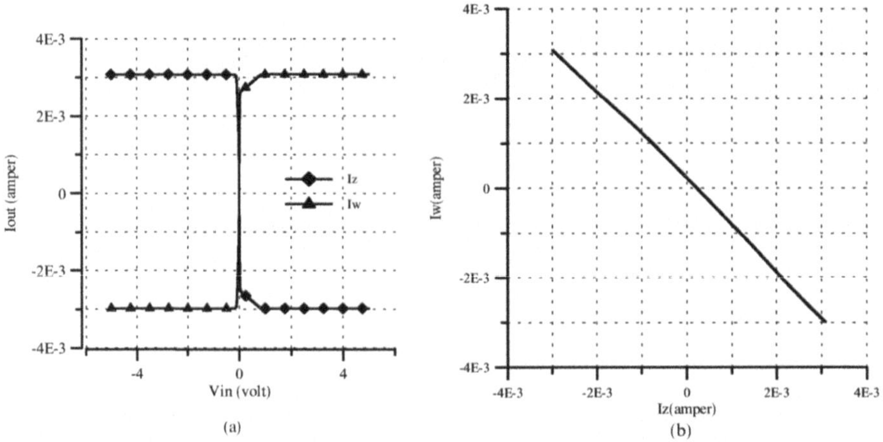

Fig. 5.6 (**a**) Simulated DC response of the proposed circuit. (**b**) Current following characteristic from node Z to node W. (Çam and Kuntman 2000)

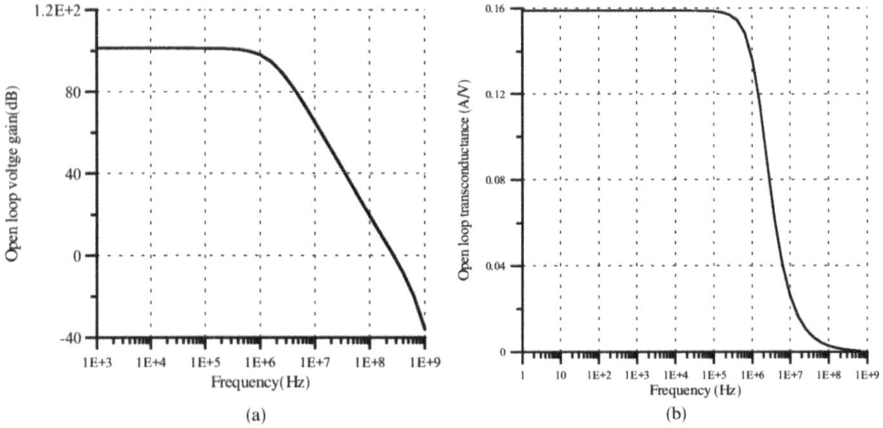

Fig. 5.7 (**a**) Simulated open-loop voltage gain of the proposed circuit. (**b**) Simulated open-loop transconductance of the proposed circuit. (Çam and Kuntman 2000)

The realization of a current-mode band-pass filter shown in Fig. 5.8a is chosen as an illustrative example of the proposed circuits (Liu 1995). The transfer function of the current-mode band-pass filter is

$$\frac{I_{\text{out}}}{I_{\text{in}}} = -\frac{s/3C_1R_2}{s^2 + s\left(\frac{1}{C_1R_1} + \frac{1}{3C_4R_2} - \frac{1}{3C_1R_2}\right) + \frac{1}{3C_1C_4R_1R_2}} \tag{5.6}$$

where the natural frequency and quality factor of the filter are given by

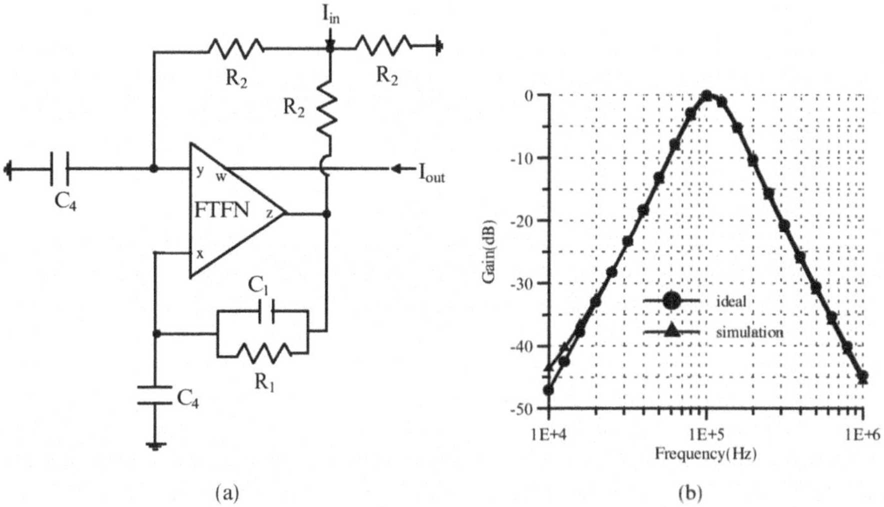

Fig. 5.8 (**a**) Current-mode band-pass filter. (**b**) PSPICE simulation result using the FTFN in Fig. 5.2 and ideal model, $C_1 = C_4 = 0.5$ nF; $R_1 = 3$ kΩ; $R_2 = 1$ kΩ. (Çam and Kuntman 2000)

$$\begin{aligned} Q &= \frac{\sqrt{3C_1C_4R_1R_2}}{3C_4R_2 + C_1R_1 - C_4R_1} \\ \omega &= \frac{1}{\sqrt{3C_1C_4R_1R_2}} \end{aligned} \tag{5.7}$$

All simulations were performed using a 1 kΩ load. The simulation results are found to agree well with the theoretical predictions.

In this section, a further CMOS realization of an FTFN has been described. The proposed circuit combines the advantages of a cascade of common-source and common-gate amplifiers and floating current source. The simulation results exhibit that the presented circuit design offers a practical alternative solution to the CMOS FTFN integration problem. The proposed circuit provides high performance in terms of voltage and current tracking errors, dynamic range, and frequency response compared with the CMOS FTFN in the literature. The feasibility of the circuit has also been tested on a current-mode band-pass filter.

Universal Series and Parallel Immittance Simulators Using Four-Terminal Floating Nullors

The current-mode analog circuits employing active elements such as OTAs, current conveyors, DOOTAs, DO-CCIIs, and FTFNs play an important role in the IC design, since these active elements exhibit greater linearity, wider bandwidth, and wider dynamic range over the voltage-mode counterparts. The four-terminal floating nullor (FTFN) has been also receiving considerable attention recently as it has been shown that an FTFN is a very flexible and versatile building block in active network synthesis. This leads to a growing interest in the design of amplifiers, oscillators, filters, and floating immittance simulators which use FTFN as an active element. Active simulation of grounded immittance functions has a wide range of applications in oscillator design, active filters, and cancellation of parasitic elements. Simulation of grounded immittance using op-amp and current conveyor is well known. There are several publications on the realization of grounded immittance functions using current conveyors. However, only few papers related with simulation of grounded immittance function using the FTFN exist in the literature. Most of the presented op-amp and CCII-based topologies presented up to now realize some of possible inductor±resistor±capacitor combinations using the same configuration. Wang and Lee proposed an impedance simulator which can realize only series and parallel $(R)\pm(L)$ and $(\pm R)\pm(\pm L)$ or $(C)\pm(D)$ and $(\pm C)\pm(\pm D)$ using different single FTFN and passive component configurations. In this section a parallel and a series immittance simulator are presented. The presented circuits are universal and able to simulate a $(+G)$ $(+L)$ $(+C)$ for which all values are independently adjustable.

The series immittance simulator employs two FTFNs and seven passive components, and the parallel immittance simulator uses two FTFNs and nine passive components. These configurations represent general forms. Reduction in the number of active and passive elements is possible for special cases. The simulator circuits do not require any component matching constraint, except for the realization of pure inductance. The performance of the proposed universal circuits is demonstrated on sinusoidal oscillator and second-order band-pass filter obtained from the parallel immittance simulator by PSPICE simulations (Fig. 5.9).

Routine analysis of the proposed series immittance simulator shown in Fig. 5.10 yields the input impedance expressed as

$$Z_{\text{in}} = \frac{1}{y_1} + \frac{1}{y_2} + \frac{y_4}{y_3 y_7} - \frac{y_4}{y_3 y_5} - \frac{y_4}{y_3 y_6} \tag{5.8}$$

If the admittances were chosen as $y_1 = G_1$, $y_2 = sC_2$, $y_3 = G_3$, $y_4 = sC_4 + G_4$, $y_5 = G_5$, $y_6 = sC_6$, $y_7 = G_7$, the input impedance becomes

$$\begin{aligned} Z_{\text{in}} = & \frac{1}{G_1} + \frac{G_4}{G_3 G_7} - \frac{G_4}{G_3 G_5} - \frac{C_4}{G_3 C_6} \\ & + s\left(\frac{C_4}{G_3 G_7} - \frac{C_4}{G_3 G_5}\right) + \frac{1}{s}\left(\frac{1}{C_2} - \frac{G_4}{G_3 C_6}\right) \end{aligned} \tag{5.9}$$

which represents a series connection of C_{eq}, L_{eq}, and G_{eq} which are given by

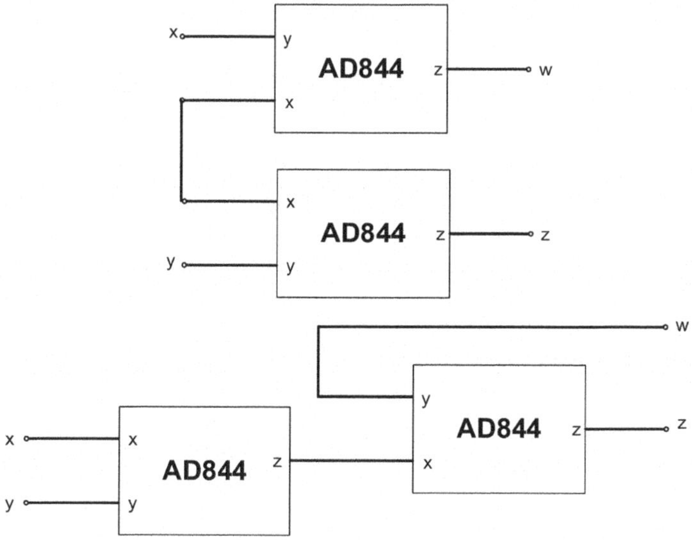

Fig. 5.9 Realization of the FTFN with commercially available CFOA-IC, AD844 from analog devices: (**a**) FTFN+ and (**b**) FTFN−. (Çam et al. 2000)

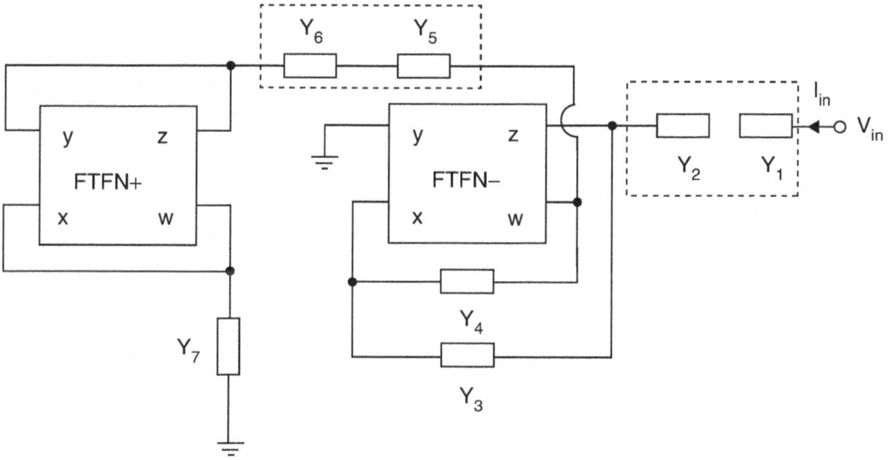

Fig. 5.10 Universal series immittance simulator. (Çam et al. 2000)

$$C_{eq} = \frac{C_2C_6G_3}{G_3C_6 - G_4C_2}$$
$$L_{eq} = \frac{C_4}{G_3G_7} - \frac{C_4}{G_3G_5} \qquad (5.10)$$
$$G_{eq} = \frac{1}{\frac{1}{G_1} + \frac{G_4}{G_3G_7} - \frac{G_4}{G_3G_5} - \frac{C_4}{G_3C_6}}$$

It can be observed from the equation that the circuit simulates series $(\pm G)$, $(\pm L)$, and $(\pm C)$ immittance by selecting passive component values appropriately.

The proposed parallel immittance simulator is shown in Fig. 5.11. Routine analysis yields the input impedance as

$$Y_{in} = \frac{i_{in}}{v_{in}} = y_1 + y_2 + \frac{y_1y_2}{y_3} - \frac{y_4y_5}{y_6} + y_7 \qquad (5.11)$$

If we choose $y_1 = G_1$, $y_2 = G_2$, $y_3 = sC_3$, $y_4 = sC_4 + G_4$, $y_5 = sC_5 + G_5$, $y_6 = sC_6$, $y_7 = sC_7$, then the input impedance becomes

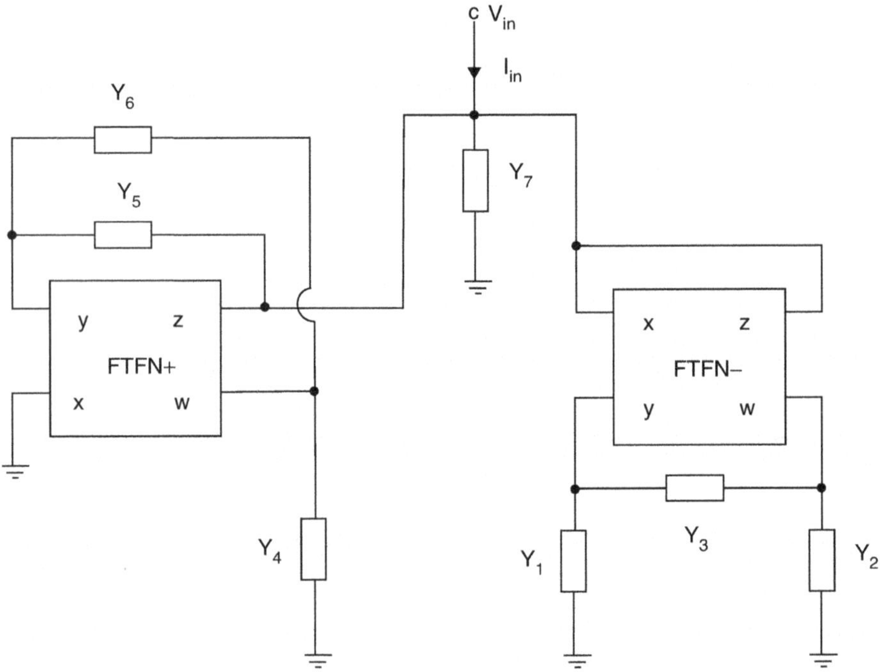

Fig. 5.11 Universal parallel immittance simulator. (Çam et al. 2000)

$$
\begin{aligned}
Y_{\text{in}} = \quad & G_1 + G_2 - \frac{G_4C_5 + C_4G_5}{C_6} + s\left(C_7 - \frac{C_4C_5}{C_6}\right) \\
+ \quad & \frac{1}{s}\left(\frac{G_1G_2}{C_3} - \frac{G_4G_5}{C_6}\right)
\end{aligned} \tag{5.12}
$$

which represents a parallel connection of C_{eq}, L_{eq}, and G_{eq} which are given by

$$
\begin{aligned}
C_{\text{eq}} &= C_7 - \frac{C_4}{C_6}C_5 \\
L_{\text{eq}} &= \frac{C_3C_6}{C_6G_1G_2 - C_3G_4G_5} \\
G_{\text{eq}} &= G_1 + G_2 - \frac{C_5}{C_6}G_4 - \frac{C_4}{C_6}G_5
\end{aligned} \tag{5.13}
$$

It can be observed from the equation that the circuit simulates parallel ($\pm G$), ($\pm L$), and ($\pm C$) immittance by selecting passive component values appropriately.

The performance of the proposed universal parallel immittance simulator is demonstrated on chosen typical application examples by SPICE simulation. The first chosen application example is the realization of a sinusoidal oscillator. Parallel connection of L and C with ($+G$) produces growing sinusoidal oscillations where oscillation frequency can be given by

$$
f_{\text{osc}} = \frac{1}{2\pi\sqrt{LC}} \tag{5.14}
$$

The sinusoidal oscillator circuit obtained from the proposed parallel immittance simulator is shown in Fig. 5.11. If resistance and capacitance values are chosen as:

$R_1 = 1$ kΩ, $R_2 = 1$ kΩ, $C_3 = 0.5$ nF, $R_4 = 5$ kΩ, $C_4 = 1$ nF, $R_5 = 1$ kΩ, $C_5 = 1$ nF, $C_6 = 0.75$ nF, $C_7 = 2$ nF resulting in $L_{\text{eq}} = 0.57$ mH, $C_{\text{eq}} = 0.67$ nF which gives an oscillation frequency of $f_{OSC} = 258$ kHz theoretically.

The sinusoidal oscillator circuit obtained from the proposed parallel immittance simulator to test the functionality of the circuit is illustrated in Fig. 5.12. In recent works it was shown that the FTFNs can be constructed with commercially available CFOAIC AD844 of analog devices. The realization of the FTFN with AD844 is illustrated in Fig. 5.9. Using AD844 macromodels for the FTFN SPICE simulations are performed. The supply voltages are chosen as $V_{DD} = 10$V and $V_{SS} = -10$V. The time domain response of the oscillator obtained from SPICE simulations is given in Fig. 5.13.

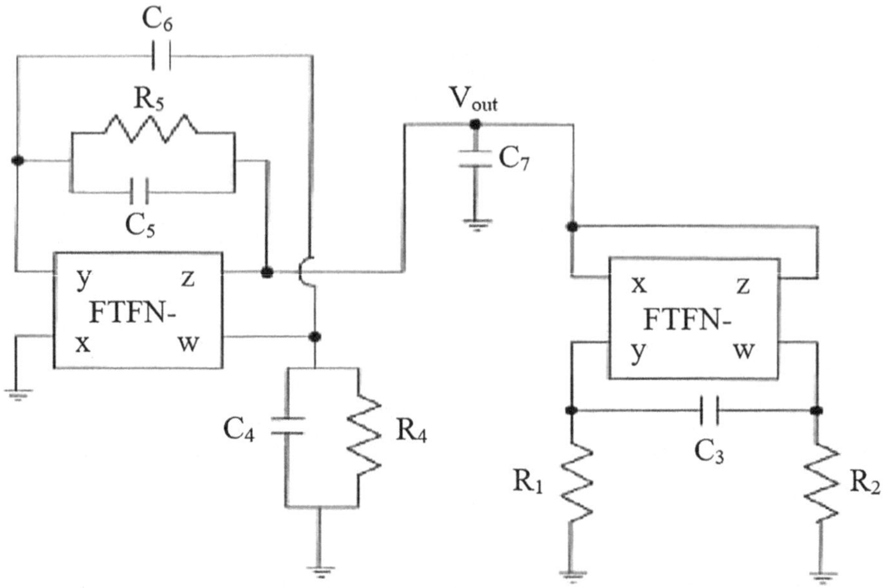

Fig. 5.12 The sinusoidal oscillator circuit obtained from the proposed parallel immittance simulator to test the functionality of the circuit. (Çam et al. 2000)

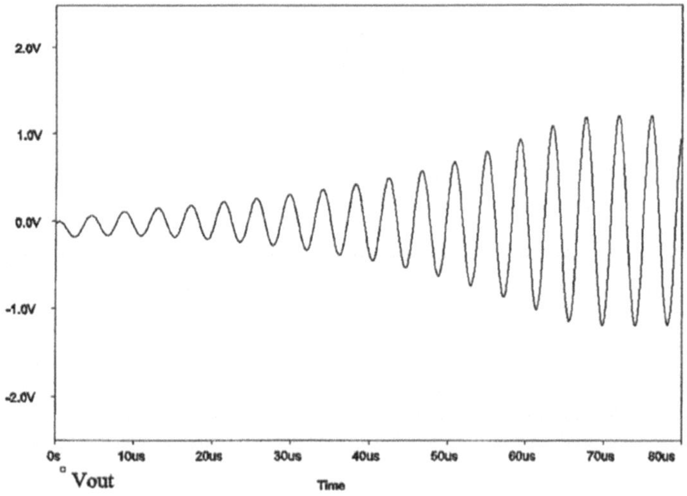

Fig. 5.13 PSPICE simulation results of the oscillator using the FTFN in Fig. 5.12 for growing oscillations ($R_1 = 1$ kΩ, $R_2 = 1$ kΩ, $C_3 = 0.5$ nF, $R_4 = 5$ kΩ, $C_4 = 1$ nF, $R_5 = 1$ kΩ, $C_5 = 1$ nF, $C_6 = 0.75$ nF, $C_7 = 2$ nF). (Çam et al. 2000)

Concluding Remarks

In this section, new universal series and parallel immittance simulators are introduced. The series immittance simulator employs two FTFNs, five resistors, and three capacitors, and parallel immittance simulator uses two FTFNs, five capacitors, and four resistors. Strong reduction in the number of active and passive elements is possible for special cases. The proposed simulators are able to realize series and parallel (+*G*), (+*L*), (+*C*) for which all values are independently adjustable. No component matching constraints are imposed for the simulators, except for the realization of pure inductance. The performance of the proposed universal immittance simulators is demonstrated on two chosen typical application examples by PSPICE simulations.

A New Four-Terminal Floating Nullor-Based Single-Input Three-Output Current-Mode Multifunction Filter

Circuit Description

The proposed circuit contains four FTFNs and six passive components, four of them grounded. Considering the proposed configuration in Fig. 5.14, routine analysis yields the following transfer functions:

$$\begin{aligned}\frac{I_{\mathrm{LP}}}{I_{\mathrm{in}}}&=\frac{G_1G_2}{s^2C_3C_5+s(C_3+C_5)G_1+G_1G_2}\\ \frac{I_{\mathrm{HP}}}{I_{\mathrm{in}}}&=\frac{-s^2C_5C_6}{s^2C_3C_5+s(C_3+C_5)G_1+G_1G_2}\\ \frac{I_{\mathrm{BP}}}{I_{\mathrm{in}}}&=\frac{-sC_5G_4}{s^2C_3C_5+s(C_3+C_5)G_1+G_1G_2}\end{aligned}\tag{5.15}$$

where the pole natural frequency and pole quality factor of the proposed filter are expressed as

$$\begin{aligned}\omega_0&=\sqrt{\frac{G_1G_2}{C_3C_5}},\\ \frac{\omega_0}{Q}&=\frac{G_1(C_3+C_5)}{C_3C_5},\\ Q&=\sqrt{\frac{G_2}{G_1}}\frac{\sqrt{C_3C_5}}{C_3+C_5}\end{aligned}\tag{5.16}$$

Thus, second-order current-mode low-pass, high-pass, and band-pass filter characteristics can be simultaneously realized without changing the circuit configuration.

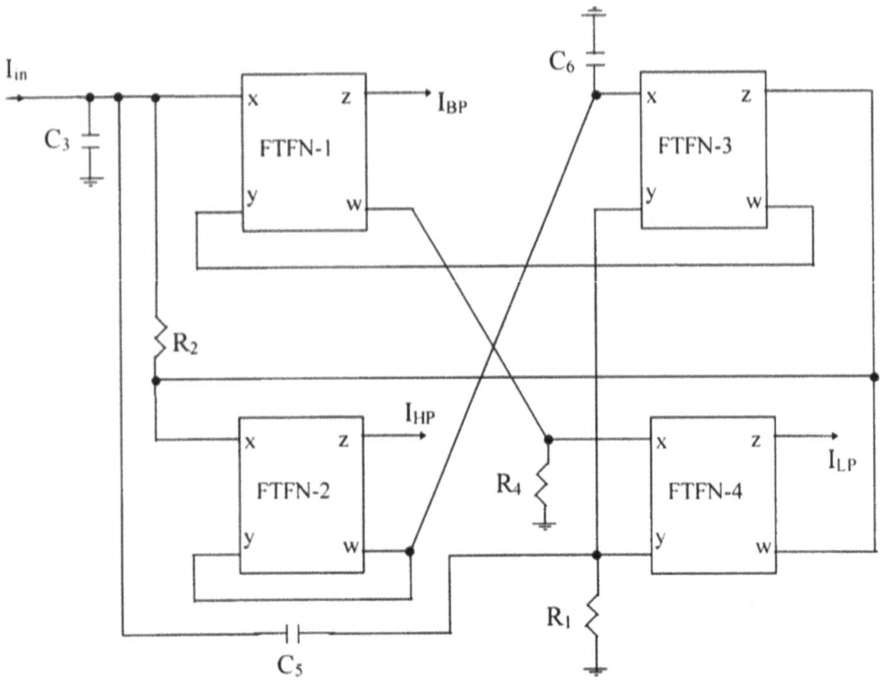

Fig. 5.14 The proposed current-mode multifunction filter. (Çam et al. 1999)

Furthermore, it should be noted that v_0 can be adjusted independently from ω_0/Q by changing the value of R_2. Moreover, to realize only band-pass and low-pass filter functions with the same circuit configuration, C_6 can be eliminated. Sensitivity analysis of the proposed filter with respect to passive elements yields

$$\begin{aligned} S_{R_1}^{\omega_0} &= S_{R_2}^{\omega_0} = S_{C_3}^{\omega_0} = S_{C_5}^{\omega_0} = -S_{R_1}^{Q} = S_{R_2}^{Q} = -\frac{1}{2} \\ S_{C_3}^{Q} &= S_{C_5}^{Q} = \frac{C_5 - C_3}{2(C_3 + C_5)} \end{aligned} \tag{5.17}$$

To verify the theoretical analysis the proposed circuit has been simulated using PSPICE program and simulation results are given in Fig. 5.15. The circuit was constructed with the commercial current feedback op-amp IC AD844 of analog devices. PSPICE simulations were performed using AD844 macromodel of analog devices. The passive elements of the filter were chosen as $R_1 = 20$ kOhm, $R_4 =$ 10 kOhm, $C_5 = C_6 = 1$ nF, and $\omega_0/(2Q_p) = 11{,}254$. The PSPICE simulation results given in Fig. 5.15 for the low-pass, high-pass, and band-pass filter characteristics verify the theoretical analysis.

In this section, we investigated a single-input three-outputs (SITO) current-mode multifunction filter using four FTFNs, three capacitors, and three resistors. The circuit has the following advantages:

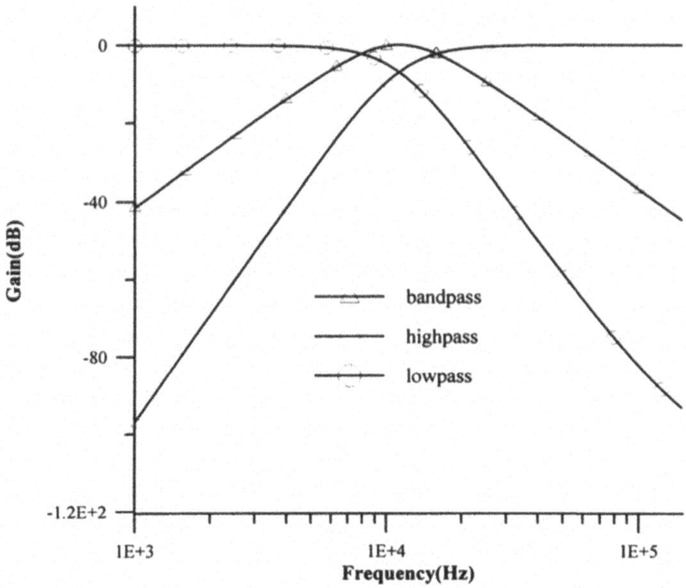

Fig. 5.15 PSPICE simulation results of the proposed multifunction filter with $R_1 = 20$ kOhm, $R_2 = R_4 = 10$ kOhm, $C_3 = C_5 = C_6 = 1$ nF. (Çam et al. 1999)

It can simultaneously realize low-pass, band-pass, and high-pass filter functions without changing the circuit topology and elements. It does not require any parameter matching condition. It has low passive sensitivities. ω_0, Q, and ω_0/Q of the filter are insensitive to current tracking errors of the FTFNs and ω_0 of the filter is also insensitive to voltage tracking errors of the FTFNs. It has high output impedances which can enable easy cascadability. It has the possibility of independent adjustment of ω_0 without disturbing ω_0/Q.

Theoretical analysis is confirmed by PSPICE simulations and simulation results show very close agreement with the theory.

Realization of Inductorless Chua's Circuit Using FTFN-Based Nonlinear Resistor and Inductance Simulator

FTFN-based Chua's circuit were performed by some researchers (Kılıç et al. 2002, 2004). An inductorless realization of the autonomous Chua circuit using the FTFN is presented (Kılıç et al. 2004). In this realization, a new version of the autonomous Chua circuit has been considered using FTFN-based active circuit topologies: both the nonlinear resistor, namely, Chua's diode, and inductor element. The presented modifications provide an alternative solution to the integration problem of the chaotic circuits using CMOS VLSI technologies. The performance of the proposed inductorless Chua circuit is demonstrated with PSPICE simulations.

Since Chua's circuit is an extremely simple system, and yet it exhibits a rich variety of bifurcations and chaos, it has been studied extensively and accepted as a paradigm for studying with chaos and chaotic signals. Several realizations of Chua's circuit have been proposed in the literature. The methodologies used in these realizations can be divided into two basic categories. In the first approach, a variety of circuit topologies have been considered for realizing the nonlinear resistor, NR in Chua's circuit. Of the circuit topologies proposed for nonlinear resistor, Kennedy's op-amp-based nonlinear resistor, which is formed by connecting two voltage-controlled negative impedance converters in parallel, has been accepted as a standard. This realization uses two conventional voltage op-amps (VOA) and six resistors. The main idea in the second approach related to the implementation of Chua's circuit is its inductorless realization. For this purpose, various circuit topologies providing a synthetic inductor, i.e., by inductance simulation, are used instead of the inductor in Chua's circuit. The inductance simulators used in such realizations are generally based on circuit elements such as op-amps and OTAs (Torres and Aguirre 2000). Recently, a CFOA-based synthetic inductor (Senani and Gupta 1998) and an FTFN-based inductance simulator (Kılıç et al. 2002) have been proposed to realize Chua's circuit and other chaotic circuits without coils. In addition to using an inductance simulator for an inductorless Chua circuit, another approach has been developed based on replacing the LC resonator of Chua's circuit with an RC configuration. In this study, we also present a new inductorless realization of Chua's circuit using an FTFN-based nonlinear resistor and an FTFN-based inductance simulator.

Circuit Description of FTFN-Based Chua's Circuit

Chua's circuit, shown in Fig. 5.16a, contains three energy storage elements, a linear resistor, and a nonlinear resistor NR, namely, Chua's diode. Chua's circuit is defined by the following state equations:

$$\begin{aligned} L\frac{\mathrm{d}i_L}{\mathrm{dt}} &= -V_{C2} - i_L \cdot R_S \\ C_2\frac{\mathrm{d}V_{C2}}{\mathrm{dt}} &= i_L - \frac{1}{R}(V_{C2} - V_R) \\ C_1\frac{\mathrm{d}V_R}{\mathrm{dt}} &= \frac{1}{R}(V_{C2} - V_R) - f(V_R) \end{aligned} \tag{5.18}$$

where $f(V_R)$ is a piecewise-linear function, shown graphically in Fig. 5.16b, defined by

$$i_R = f(V_R) = G_bV_R + \frac{1}{2}\cdot(G_a - G_b) \times (|V_R + B_P| - |V_R - B_P|) \tag{5.19}$$

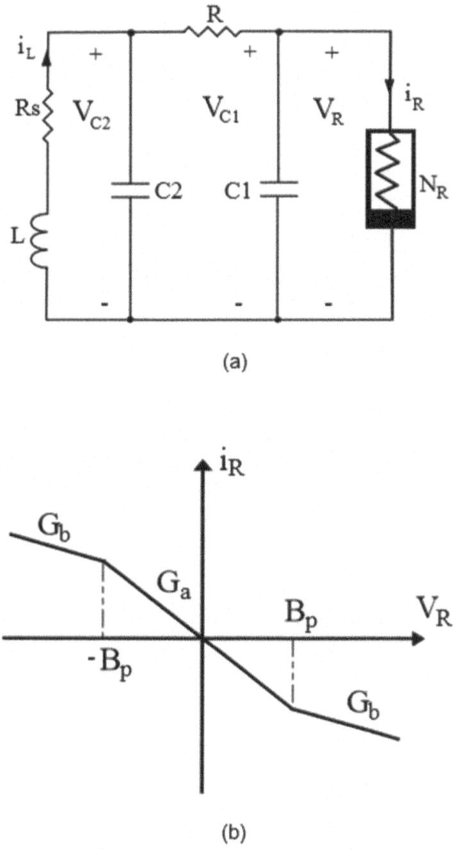

Fig. 5.16 (**a**) Chua's circuit. (**b**) The three-segment piecewise linear characteristic of the nonlinear resistor in Chua's circuit. (Kılıç et al. 2004)

where G_a and G_b are the slopes in the inner and outer regions, respectively, and $\pm B_p$ denote the breakpoints.

FTFN-based version of Chua's diode: A realization of Chua's diode, proposed by Kennedy, which is designed by connecting two voltage-controlled negative impedance converters in parallel, has been accepted as the standard for discrete implementation. The op-amp-based nonlinear resistor structure with its simulated DC characteristic is shown in Fig. 5.17. As shown in Fig. 5.17a, Kennedy's realization uses two op-amps, operating in both their linear and nonlinear region, and six resistors. The slopes and breakpoints are chosen as $G_a \cong -0.756$ mA/V, $G_b \cong -0.409$ mA/V, and $\pm B_p = \pm 1$ V with the circuit parameters in Fig. 5.17a. Since it has a simple and easily configurable circuit structure, most of the experimental studies with Chua's circuit in the literature have been performed using Kennedy's implementation (Fig. 5.18).

FTFN-based inductance simulator: The FTFN-based floating inductance simulator is shown in Fig. 5.19. Routine analysis yields equivalent inductance between terminals 1 and 2 as

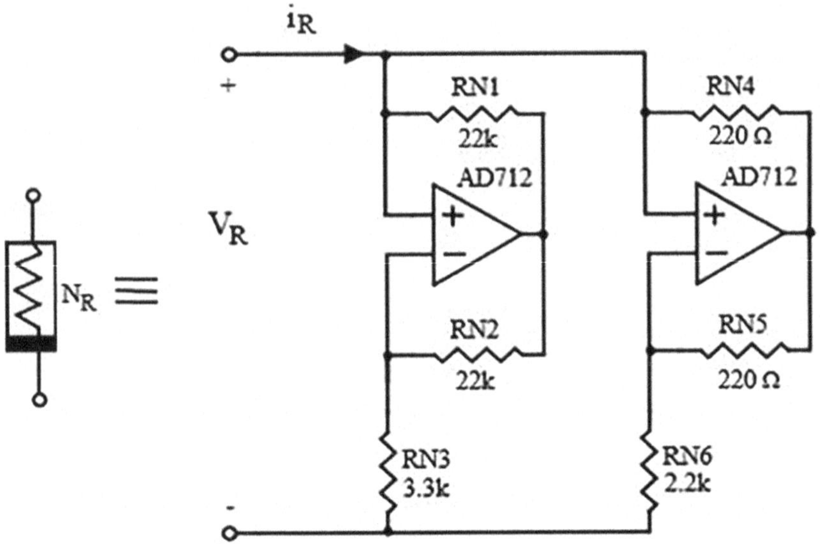

(a)

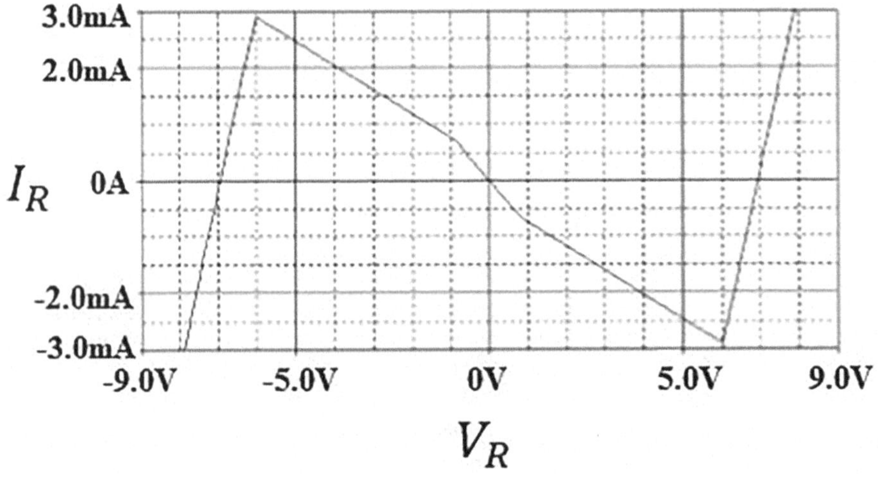

(b)

Fig. 5.17 (**a**) The op-amp-based nonlinear resistor structure implemented by Kennedy. (**b**) Simulated v–i characteristic of the nonlinear resistor of (**a**). (Kılıç et al. 2004)

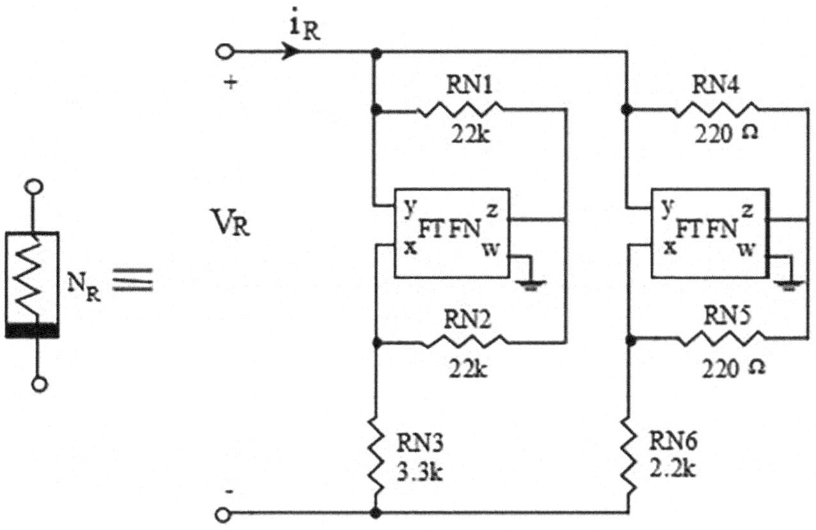

(a)

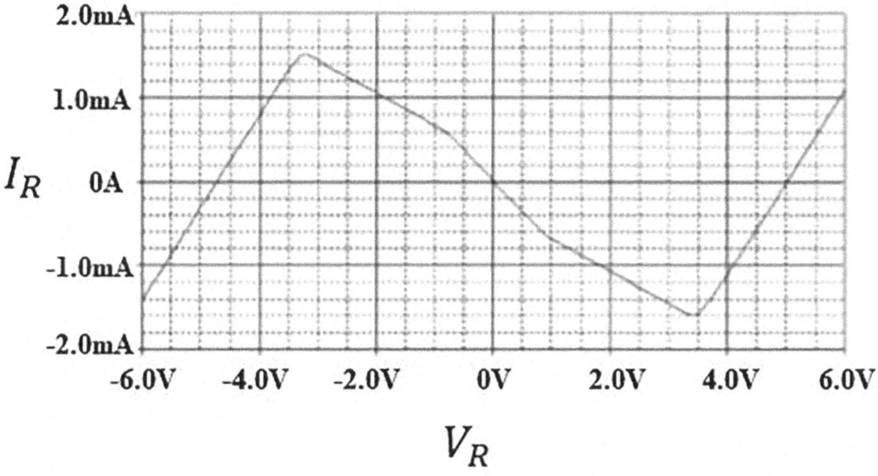

(b)

Fig. 5.18 (**a**) FTFN-based nonlinear resistor structure. (**b**) Simulated v–i characteristic of the nonlinear resistor of (**a**). (Kılıç et al. 2004)

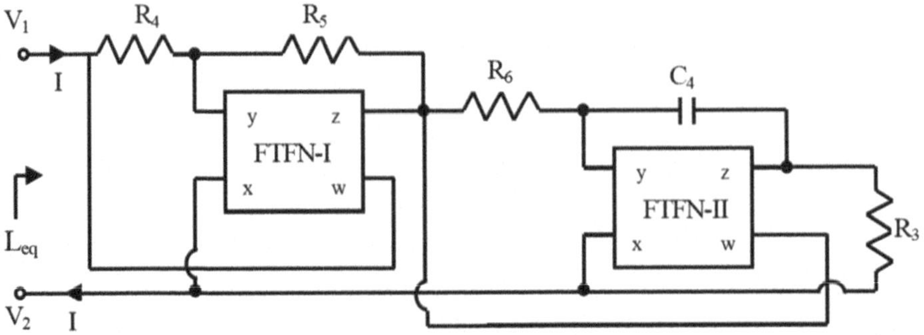

Fig. 5.19 FTFN-based inductance simulator. (Kılıç et al. 2004)

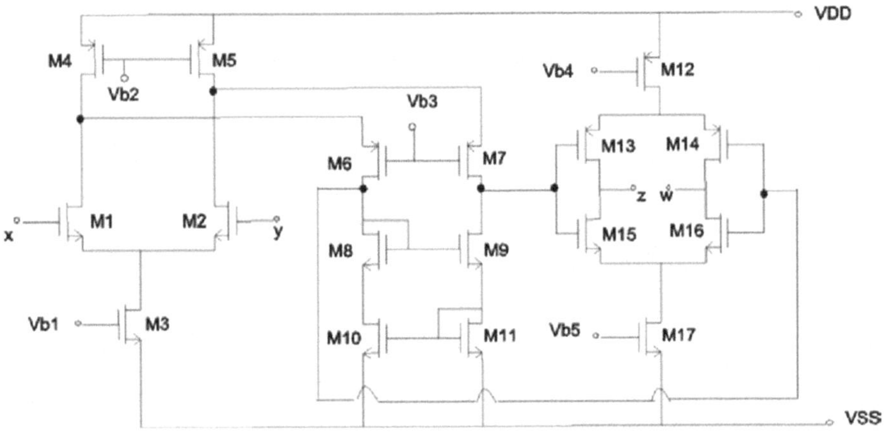

Fig. 5.20 A CMOS realization of the FTFN. (Kılıç et al. 2004)

$$L_{eq} = \frac{C_4 R_3 R_4 R_6}{R_5} \tag{5.20}$$

In Fig. 5.14a, the ideal grounded inductor is replaced by simulated inductance. In Fig. 5.19, the following element values are chosen: $R_3 = R_4 = R_5 = R_6 = 1$ kΩ, $C_4 =$ 18 nF to simulate $L = 18$ mH. Here the floating inductance is used as a grounded inductor by connecting one port of the floating inductance to ground for simplicity.

The PSPICE simulations were performed using a CMOS realization of the FTFN, shown in Fig. 5.20, with the same parameters as in the literature (Kılıç et al. 2002). While the chaotic waveforms and the chaotic spectrum of the proposed inductorless FTFN-based version of Chua's circuits are shown in Fig. 5.21, the double-scroll attractors are illustrated in Fig. 5.22.

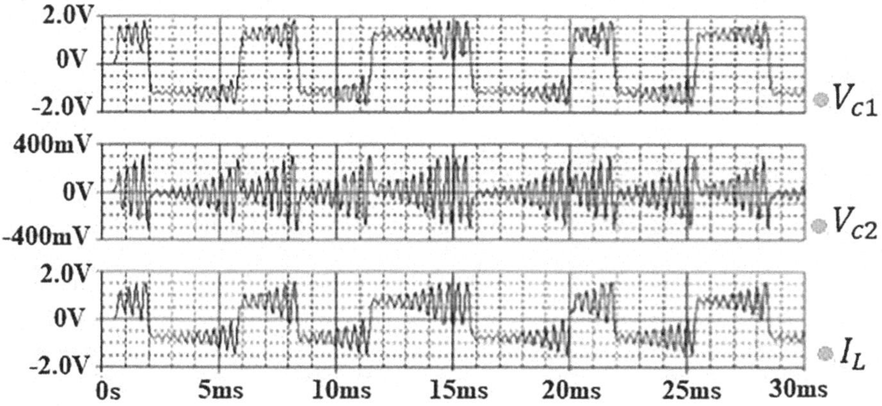

(a)

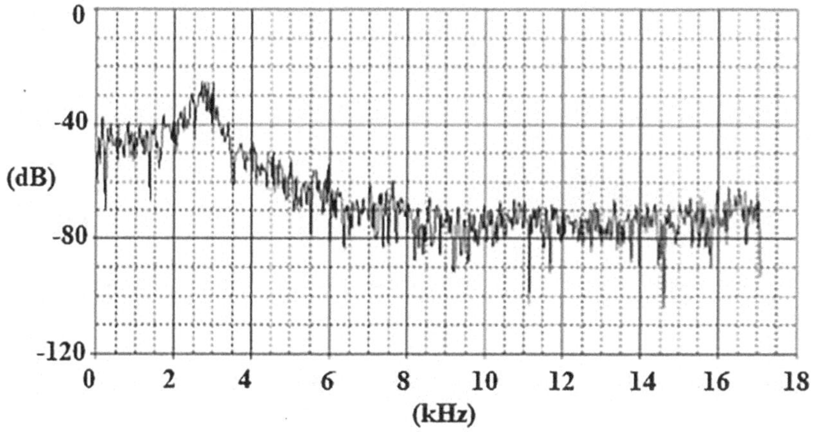

(b)

Fig. 5.21 (**a**) Simulations of chaotic circuit dynamics $V_{C1}(t)$, $V_{C2}(t)$, $iL(t)$. (**b**) The chaotic spectrum for simulated time waveform $V_{C2}(t)$ from FTFN-based Chua's circuit. (Kılıç et al. 2002, 2004)

Concluding Remarks

New inductorless realization of the autonomous Chua's circuit using the FTFN-based inductance simulator and Chua's diode has been introduced. The simulation results indicate that the proposed FTFN-based inductorless version of Chua's circuit exhibits its original chaotic behaviors. The CMOS implementation of Chua's circuit using FTFN-based circuit topologies for inductance simulator and Chua's diode provides new possibilities to the designer for the integrated circuit realization of chaotic communication systems.

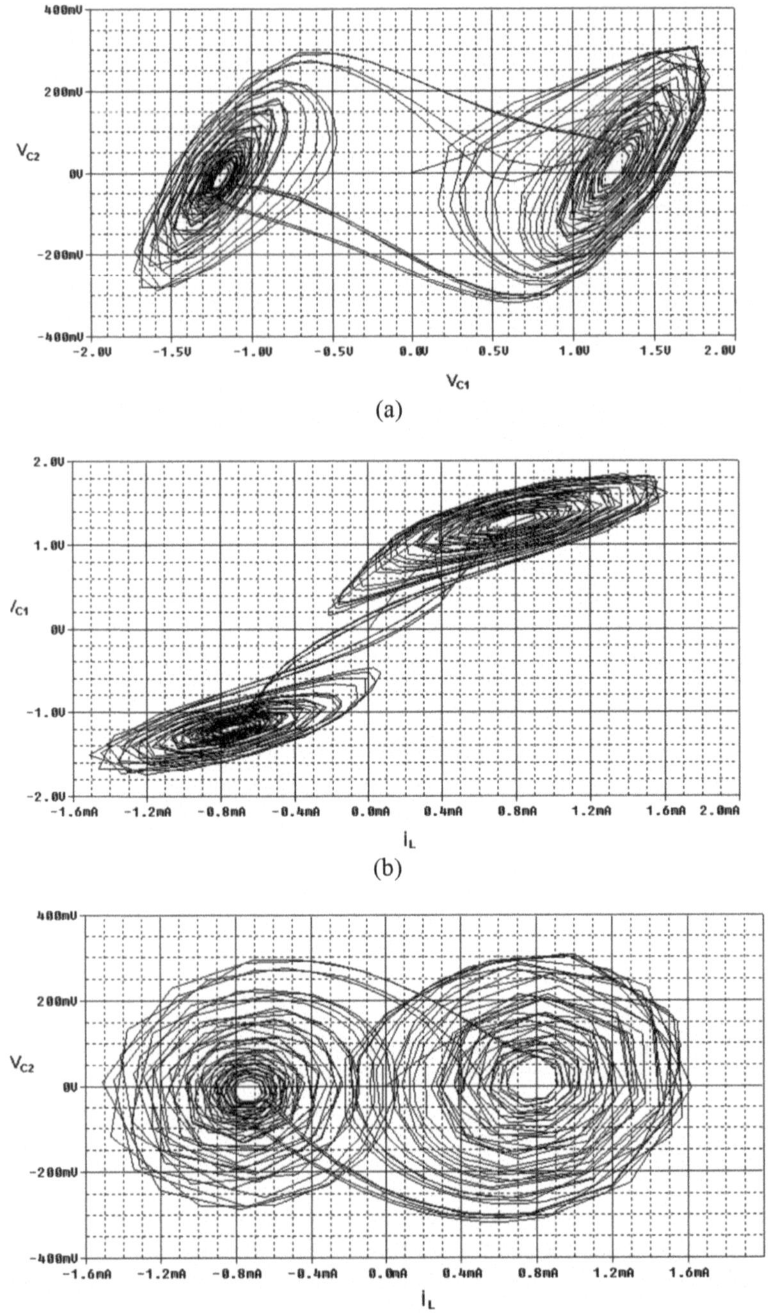

Fig. 5.22 The double-scroll attractors observed in FTFN-based Chua's circuit. (**a**) Projection in the ($V_{C2} - V_{C1}$) plane, (**b**) projection in the ($V_{C1} - iL$) plane, and (**c**) projection in the ($V_{C2} - iL$) plane. (Kılıç et al. 2002, 2004)

References

Abuelma'Atti, M. T. and Al-Zaher, H. A. Universal two-input two-output current-mode active biquad using FTFNs. International Journal of Electronics, 86(2), 181–188, 1999.

Arbel, A. F. and Goldminz, L. Output stage for current-mode feedback amplifiers, theory and applications. Analog Integrated Circuits and Signal Processing, 2(3), 243–255, 1992.

Cabeza, R., Carlosena, A. and Serrano, L. Unified approach to the implementations of universal active devices. Electronics Letters, 30(8), 618–620, 1994.

Çam, U., Kuntman, H. A new CMOS realization of four terminal floating nullor (FTFN). Int. Journal of Electronics, 87, 7, 809– 817, 2000.

Çam, U., Çiçekoğlu, O. and Kuntman, H. A new four terminal floating nullor based single-input three-output current-mode multifunction filter. Microelectronics Journal, 30(2), 115–118, 1999.

Cam, U., Çiçekoğlu, O., Gülsoy, M. and Kuntman, H. (2000). New voltage and current mode first-order all-pass filters using single FTFN. Frequenz, 54(7–8), 177–179.

Çam, U., Çiçekoğlu, O. and Kuntman, H. Universal series and parallel immittance simulators using four terminal floating nullors. Analog Integrated Circuits and Signal Processing, 25(1), 59–66, 2000.

Gray, R., Hurst, P.J., Lewis, S.H., Meyer, R.G. Analysis and design of analog integrated circuits, John Wiley & Sons, Inc. (2001)

Higashimura, M. Current-mode allpass filter using FTFN with grounded capacitor. Electronics Letters, 27(13), 1182–1183, 1991a.

Higashimura, M. Realisation of current-mode transfer function using four-terminal floating nullor. Electronics Letters, 27(2), 170–171, 1991b.

Kilic, R., Çam, U., Kuntman, H. and Uzunhisarcikli, E. Realization of inductorless Chua's circuit using FTFN-based nonlinear resistor and inductance simulator, 2004.

Kılıç, R., Alçı, M., Çam, U., Kuntman, H. Improved realization of Mixed-mode chaotic circuit. Int. Journal of Bifurcation and Chaos, 12, 6, 1429–1435, 2002.

Kılıç, R., Alçı, M., Tokmakçı, M. Mixed-mode chaotic circuit. Electronic Lett., 36, 2, 103–104, 2000.

Laopoulos, T. H., Siskos, S., Bafleur, M., Givelin, P. H. and Tournier, E. Design and applications of an easily integrable CMOS operational floating amplifier for the megahertz range. Analog Integrated Circuits and Signal Processing, 7(2), 103–111, 1995.

Liu, S. I. Single-resistance-controlled sinusoidal oscillator using two FTFNs. Electronics letters, 33(14), 1185–1186, 1997.

Liu, S. I. Cascadable current-mode filters using single FTFN. Electronics Letters, 31(23), 1965–1966, 1995.

Liu, S. I. and Liao, Y. H. Current-mode quadrature sinusoidal oscillator using single FTFN. International Journal of Electronics, 81(2), 171–175, 1996.

Liu, S. T. and Lee, J. L. Insensitive current/voltage-mode filters using FTFNs. Electronics Letters, 32(12), 1079–1080, 1996.

Senani, R., Gupta, S. S. Implementation of Chua's chaotic circuit using current feedback op-amps. Electronic Lett., 34, 9, 829–830, 1998.

Senani, R. A novel application of four-terminal floating nullors. Proceedings of the IEEE, 75(11), 1544–1546, 1987.

Toumazou, F. D. Lidgey, and DG Haigh, Analog IC Design: The Current Mode Approach. London, UK: Peregrinus, 1990.

Torres, L. A. B., Aguirre, L. A. Inductorless Chua's circuit. Electronic Lett., 36, 23, 1915–1916, 2000.

Chapter 6
Current Operational Amplifiers

Basic Concept

In recent years, the current-mode approach has been extensively investigated. Compared to the voltage-mode circuits, the current-mode circuits have some significant advantages such as inherent wide bandwidth, wide dynamic range, and simple circuitry with lower voltage supplies. Current-mode operational amplifier (COA) is one of the useful current-mode integrated building blocks. The main advantage of using COA is its ability to replace the voltage operational amplifier (VOA) when applying the adjoint network theorem in voltage-mode to current-mode transformation. The circuit symbol of fully differential COA is reported in Fig. 6.1. COA ideally exhibits zero input resistance and infinite output resistance and current gain.

It is not difficult to increase output resistance and current gain to reasonable values. On the other hand, to get very low input resistance, some complicated negative feedback configurations have been suggested and it generally worsens the frequency response of the amplifier. Moreover, these structures cause bigger problems in fully differential COAs because of its double input. Positive feedback is another solution for getting better input resistance. This paper presents a new approach dependent on positive feedback to decrease input resistance. Traditional current output stage (Arbel–Goldminz output stage) is the most commonly used transconductance stage (Arbel and Goldminz, 1992), but it suffers from its output resistance. A novel current output stage with improved output resistance is proposed in this study. In the literature, most of the COA designs are single ended. However, fully differential analog signal processing provides several advantages over single-ended analog signal processing. Some CMOS implementations of fully differential COA have been reported (Cheng and Wang 1997; Jun and Kim 1998; Kaulberg 1993). Unfortunately, these COAs do not offer low input resistance and most of them do not have proper frequency response. The proposed fully differential COA has simple circuitry and offers high performance. Some useful current amplifier-

H. Kuntman, D. Özenli, *Trends in Circuit Design for Analog Signal Processing*, Analog Circuits and Signal Processing,
https://doi.org/10.1007/978-3-030-96836-6_6

Fig. 6.1 Circuit symbol of fully differential COA. (Altun and Kuntman 2006, 2007, 2008)

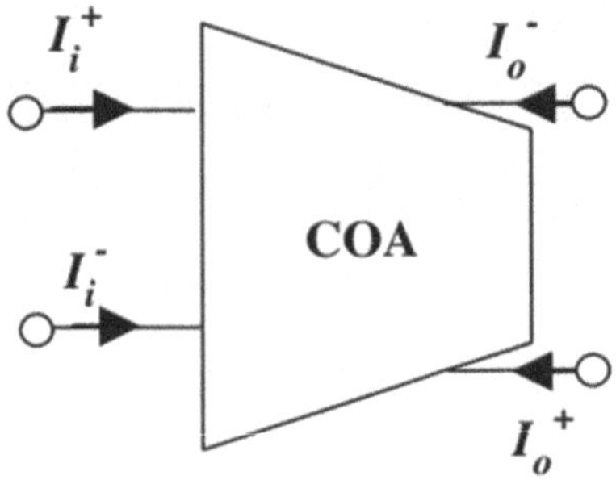

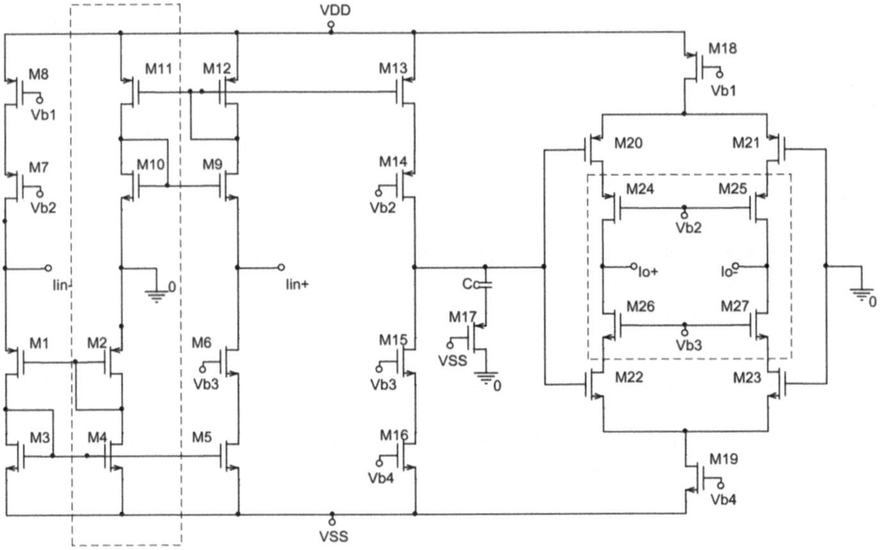

Fig. 6.2 Schematic of the described COA. (Altun and Kuntman 2008)

based filter applications have been presented in the literature (Souliotis et al. 2001; Kilinc and Cam 2004). To demonstrate the performance of the proposed COA, a new fourth-order band-pass (BP) filter that enables maximal flatness around the center frequency is proposed.

A typical COA is shown in Fig. 6.2; the described amplifier has very simple structure. W/L of transistors and DC values of the circuit are reported in Tables 6.1 and 6.2, respectively. The amplifier is configured from a differential input transimpedance stage followed by a differential output transconductance stage. Shown in dashed lines at the input stage, M_2, M_4 and M_{10}, M_{11} compose positive feedback loops to reduce positive and negative input resistances, respectively. Because only four more transistors are used for input resistance improvements, it is noteworthy the frequency response of the amplifier is not affected by these

Table 6.1 Transistor dimensions

Transistors	W(μm)/L(μm)
M1, M2	20/0.7
M3. M4, M5, M6	20/1
M7	10/0.7
M8	27/0.7
M9, M10	15/0.7
M11, M12, M13	20/1.4
M14	40/1.4
M15	20/0.7
M16	17.8/1.4
M17	12.5/1
M18	118/1
MI9	47/1
M20, M21	75/1
M22, M23	40/0.7
M24, M25	100/1
M26, M27	60/0.7

Table 6.2 DC values of the COA

Parameter	Value
V_{DD}–V_{SS}	±1.5 V
V_{b1}, V_{b2}	0.6 V, –0.3 V
V_{b3}, V_{b4}	0.3 V, –0.8 V
I_{D8}, I_{D16}	30 μA
I_{D18}, I_{D19}	100 μA

additional transistors. Equations (6.1) and (6.2) show input resistances without positive feedback loops and generally their values are not low enough.

$$r_{in-} \cong \frac{1}{g_{m1}}, \tag{6.1}$$

$$r_{in+} \cong \frac{1}{g_{m9}}. \tag{6.2}$$

Input resistance equations of the proposed COA are shown in Eqs. (6.3) and (6.4). The second terms of Eqs. (6.3) and (6.4) mainly affect input resistance value. If we select its value close to zero, it also goes near zero. Moreover, its value must be bigger than zero to overcome the stability problem. If we choose $g_{m3} = g_{m4}$, $g_{m1} = g_{m2}$ and $g_{m11} = g_{m12}$, $g_{m9} = g_{m10}$, then we will both overcome the stability problem and obtain very low input resistances. Consequently we come to such a decision that $(W/L)_{M3}$=$(W/L)_{M4}$, $(W/L)_{M1}$=$(W/L)_{M2}$, $(W/L)_{M11} = (W/L)_{M12}$, $(W/L)_{M9} = (W/L)_{M10}$. Furthermore, because positive feedback is increasing the stability problem, MOS transistor matching parameters are also considered in this study. To analyze the MOS transistor mismatches, firstly we need to calculate variances of relative current gain factor (K) mismatch for transistors M_1–M_4 and M_9–M_{12}. As it will be reported in the "Simulation Results" part, input resistances stay positive while considering the matching parameters in the worst case.

$$r_{in-} \cong \frac{1}{g_{m1}g_{m3}}\left[(g_{ds1} + g_{m3} + g_{ds3}) - \frac{g_{m1}g_{m4}}{g_{ds4} + g_{m2} + g_{ds2}}\right], \tag{6.3}$$

$$r_{in+} \cong \frac{1}{g_{m9}g_{m12}}\left[(g_{ds9} + g_{m12} + g_{ds12}) - \frac{g_{m1}g_{m4}}{g_{ds11} + g_{m10} + g_{ds10}}\right]. \tag{6.4}$$

Output resistance of traditional current output stage (Arbel–Goldminz output stage) is shown in Eq. (6.5), (Arbel and Goldminz, 1992).

$$r_{out+} = r_{out-} \cong \left[\left(\frac{g_{m20}g_{ds21}}{g_{m21} + g_{m20}}\right) + \left(\frac{g_{m22}g_{ds23}}{g_{m23} + g_{m22}}\right)\right]^{-1} \tag{6.5}$$

M_{24}, M_{25}, M_{26}, and M_{27} in dashed line are added to the conventional current output stage for getting very big output resistance values. The output resistance equation of the proposed COA is represented in Eq. (6.6). Compared to the conventional one, the proposed output stage approximately offers $g_m r_o$ times higher output resistance.

$$r_{out+} = r_{out-} \cong \left[\left(\frac{g_{m20}g_{ds21}g_{ds25}}{g_{m25}(g_{m21} + g_{m20})}\right) + \left(\frac{g_{m22}g_{ds23}g_{ds27}}{g_{m27}(g_{m23} + g_{m22})}\right)\right]^{-1} \tag{6.6}$$

Transistor M17 operates like a resistance (R_c) and improves the frequency response of the COA. C_c is inserted at high impedance node for compensation. DC current gain and gain-bandwidth product are given by Eqs. (6.7) and (6.8), respectively:

$$A_i(0) \cong \frac{g_{m20}g_{m22}}{2}\left[\frac{g_{ds14}g_{ds13}}{g_{m14}} + \frac{g_{ds15}g_{ds16}}{g_{m15}}\right]^{-1} \tag{6.7}$$

$$f_{GBW} \cong \frac{1}{2\pi}\frac{g_{m20} + g_{m22}}{2Cc} \tag{6.8}$$

COA-Based Filter Realizations

The second-order low-pass (LP) and high-pass (HP) filter configurations are shown in Fig. 6.3. Single COA is used for each realization and matching conditions are $R_1 = R_3 = R_{\mathrm{LP}}$, $C_1 = C_2 = C_{\mathrm{LP}}$ for LP filter and $R_4 = R_5 = R_{\mathrm{HP}}$, $C_4 = C_6 = C_{\mathrm{HP}}$ for HP filter. The node analysis of the circuit of Fig. 6.3a yields current transfer function as follows:

$$\frac{i_{\text{out-LP}}}{i_{\text{in}}} = \frac{\frac{1}{2R_{\mathrm{LP}}R_2C_{\mathrm{LP}}^2}}{s^2 + s\frac{1}{R_2C_{\mathrm{LP}}} + \frac{1}{2R_{\mathrm{LP}}R_2C_{\mathrm{LP}}^2}} \tag{6.9}$$

The natural angular frequency ω_0 and the quality factor Q of the LP filter can be expressed as

$$w_0 = \sqrt{\frac{1}{2R_{\mathrm{LP}}R_2C_{\mathrm{LP}}^2}}, \quad Q = \sqrt{\frac{R_2}{2R_{\mathrm{LP}}}} \tag{6.10}$$

It should be noted that ω_0 and Q are orthogonally adjustable. It means that ω_0 can be adjusted without disturbing the Q by changing either C_{LP} or R_{LP} and R_2 simultaneously while R_2/R_{LP} is constant. Sensitivity analysis of the LP filter shows that

$$S_{R_{\mathrm{LP}}}^{w_0} = S_{R_2}^{w_0} = -1/2, S_{C_{\mathrm{LP}}}^{w_0} = -1, S_{R_2}^{Q} = 1/2, S_{R_{\mathrm{LP}}}^{Q} = -1/2 \tag{6.11}$$

which are equal less than unity in magnitude. The node analysis of the circuit of Fig. 6.3b yields current transfer function as follows:

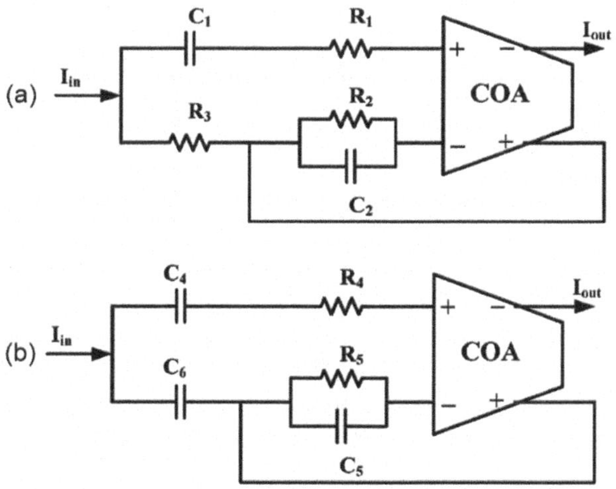

Fig. 6.3 (**a**) COA-based low-pass filter topology. (**b**) COA-based high-pass filter topology. (Altun and Kuntman 2008)

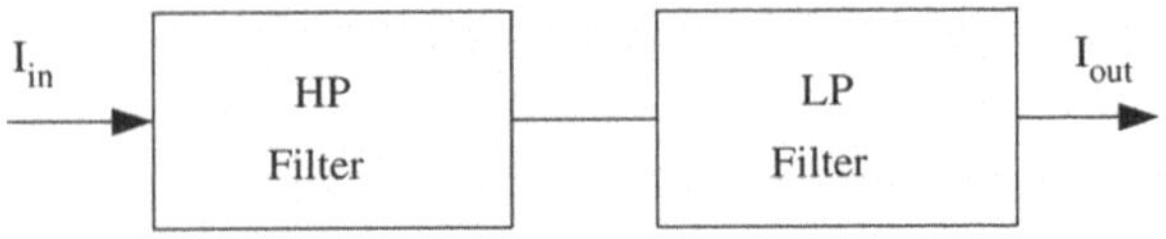

Fig. 6.4 Band-pass filter structure formed by high-pass and low-pass filters. (Altun and Kuntman 2008)

$$\frac{i_{\text{out}-\text{HP}}}{i_{\text{in}}} = \frac{s^2}{s^2 + s\frac{1}{R_{\text{HP}}C_{\text{HP}}} + \frac{2}{R_{\text{HP}}^2 C_{\text{HP}} C_5}} \tag{6.12}$$

The natural angular frequency ω_0 and the quality factor Q of the HP filter can be expressed as

$$w_0 = \sqrt{\frac{2}{R_{\text{HP}}^2 C_{\text{HP}} C_5}}, \quad Q = \sqrt{\frac{C_{\text{HP}}}{2C_5}} \tag{6.13}$$

It should be again noted that ω_0 and Q are orthogonally adjustable. It means that ω_0 can be adjusted without disturbing the Q by changing either R_{HP} or C_{HP} and C_5 simultaneously while C_{HP}/C_5 is constant. Sensitivity analysis of the HP filter shows that

$$S_{C_{\text{HP}}}^{w_0} = S_{C_5}^{w_0} = -1/2, S_{R_{\text{HP}}}^{w_0} = -1, S_{C_{\text{HP}}}^{Q} = 1/2, S_{C_5}^{Q} = -1/2 \tag{6.14}$$

which are equal-less than unity in magnitude. As shown in Fig. 6.4, it is possible to realize BP filter by using HP and LP filters connecting one's output to another's input. With using this BP filter structure, we can arrange the passband width.

Simulation Results

SPICE is used for simulation with process parameters of a 0.35 μm CMOS technology. BSIM3v3 parameter sets are used for modeling transistors, of which threshold voltages are nearly 0.5 V for NMOS and −0.7 V for PMOS. The transistor widths range from 5 to 120 μm.

Table 6.3 summarizes the performance of the COA. The COA provides 96 dB DC gain, 92 MHz UGBW, and 60° phase margin guaranteeing single pole behavior throughout the UGBW. Input and output resistance values are also fairly good. For transistors M_1–M_4 and M_9–M_{12}, K/K values are calculated between 0.16% and 0.27%. As a result, input resistances become +52 Ω (n) and +47 Ω (p) for the worst case (still positive). Because of the class A operation, the speed of the COA is limited by quiescent current. An open-loop frequency characteristic is seen from Fig. 6.5. The transient response to a step input current of ±5 μA is shown in Fig. 6.6.

Table 6.3 Performance parameters of the COA

Parameter	Value
Power dissipation	0.66 mW
Open-loop gain	96 dB
GBW	92 MHz
Phase margin ($C_c = 1.2$ p $R_c = 2.4$ kΩ)	60°
Output voltage range	±0.6 V
Slew rate	4 μA/ns
Input resistance (n)	124Ω
Input resistance (p)	109Ω
Output resistance	30MΩ
Input voltage offset (n)	≈ 1.6 mV
Input voltage offset (p)	≈ −3.5 m

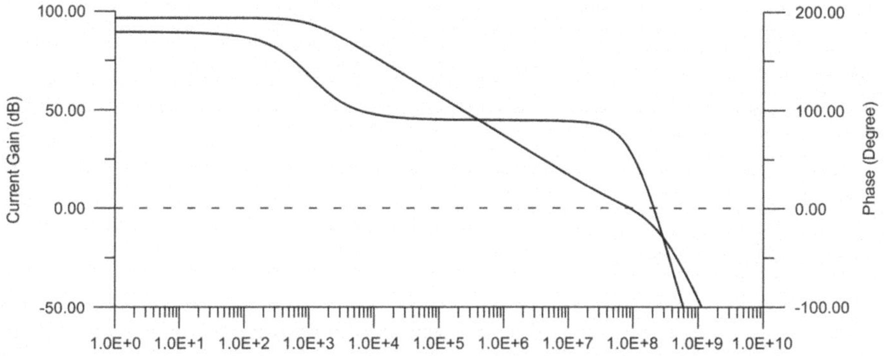

Fig. 6.5 Open-loop frequency response of the COA. (Altun and Kuntman 2008)

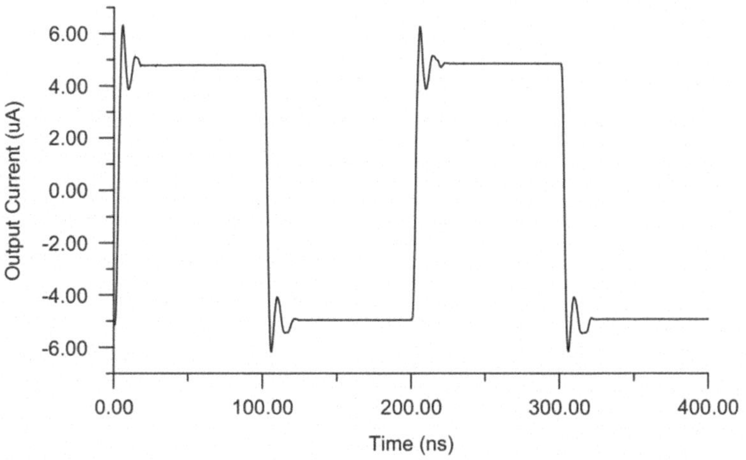

Fig. 6.6 Response of the COA in unity-gain feedback to a ±5 μA input step (f = 5 MHz). (Altun and Kuntman 2008)

Note that if we did not use impedance improvements, input resistance and output resistance values would be nearly 2.5 and 340 k, respectively. On the other hand, output voltage range would be better, nearly ±0.9 V.

Simulation Results of the COA-Based Filter Shown in Fig. 6.4

BP filter configuration is simulated by choosing the passive element values of LP filter $C_{LP} = 7.5$ pF, $R_{LP} = R_2 = 5$ k which results in a natural frequency of $f_0 \approx 3.0$ MHz and a quality factor of $Q = 0.707$ and for BP filter $C_{HP} = C_5 = 15$ pF, $R_{HP} = 50$ k which results in a natural frequency of $f_0 \approx 300.1$ kHz and a quality factor of $Q = 0.707$. So, the center frequency of the BP filter is 1 MHz and this configuration enables maximally flat passband. From Fig. 6.7 it can be understood that the theoretical and simulation results are in good agreement. The dependence of the output harmonic distortion of BP filter on input current amplitude is illustrated in Fig. 6.8. From Fig. 6.8, we can see that the harmonic distortion rapidly increases if the input signal is increased beyond 180 μA for the chosen fully differential COA implementation.

Concluding Remarks

In this section, very accurate, fully differential COA is described and simulation results are presented. A novel approach is used in input resistance improvement and also very high output resistance is achieved by modifying traditional current output

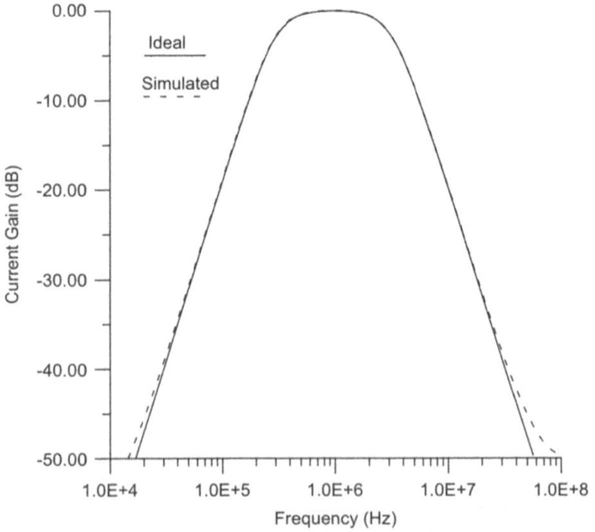

Fig. 6.7 Simulated and ideal band-pass filter responses. (Altun and Kuntman 2008)

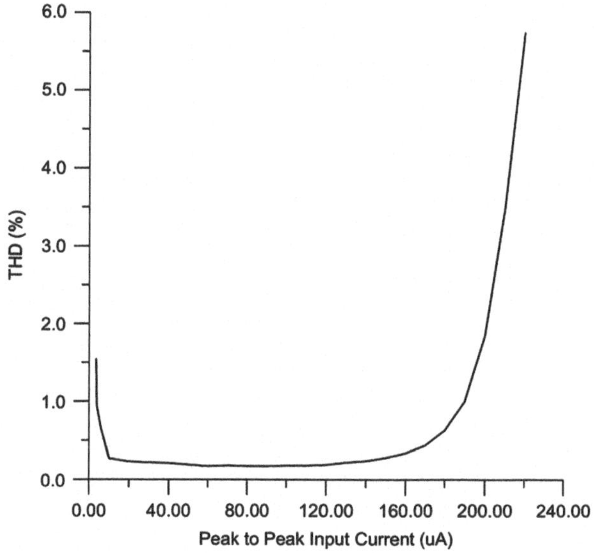

Fig. 6.8 Total harmonic distortion (THD) values of the filter versus input peak-to-peak current at 1 MHz frequency. (Altun and Kuntman 2008)

stage. Due to the simple circuitry of the proposed COA, higher than 90 MHz GBW is obtained. It also offers very high DC gain and ±0.6 V output voltage swing. As an application example, a new COA-based fourth-order BP filter is proposed. Simulation results of the filter are the evidences of the accuracy of the proposed fully differential COA.

A High-Drive Fully Differential Current-Mode Operational Amplifier Providing High Output Impedance with a Filter Application

In this section, fully differential CMOS current-mode operational amplifier (COA) which provides high drive capability and output impedance is presented. The introduced COA is operated under ±1.5 V voltage supplies and designed with 0.35 μm CMOS process. Using class AB input and output stages, the amplifier can drive a 1 kΩ load resistance with a maximum bipolar output current of about 350 μA while the quiescent current of all the branches are only 20 μA. Furthermore, it exhibits 100 dB DC gain, 85 MHz gain-bandwidth product, and 6.1 GΩ output resistance. Finally, a second-order notch filter is realized to demonstrate the performance of the described COA.

Figure 6.9 shows the CMOS realization of the two-stage COA with the transistor dimensions and DC values reported in Tables 6.4 and 6.5, respectively. Both input

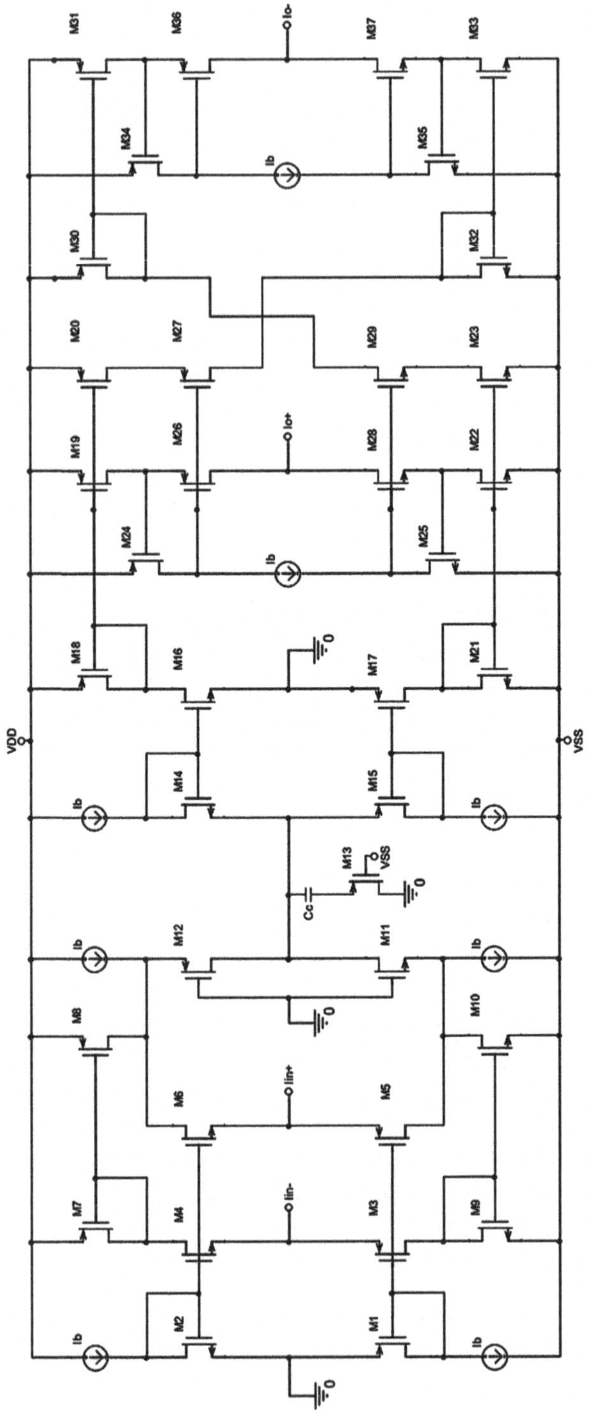

Fig. 6.9 Schematic of the COA. (Altun and Kuntman 2007)

Table 6.4 Transistor dimensions

Transistors	W(μm)/L(μm)
M2, M4, M6, M14, M15, M16, M17, M28, M28, M37	20/0.7
M1, M3, M5, M18, M19, M20, M24, M26, M27, M30, M31, M34, M36	30/0.7
M21, M22, M23, M25, M32, M33, M35	10/0.7
M7, M8	30/1
M9, M10	10/1
M11	20/1.4
M12	40/1.4

Table 6.5 Performance parameters of the COA

Parameter	Value
Power dissipation	0.72 mW
Open-loop gain	100 dB
GBW	85 MHz
Phase margin ($C_c = 1.5$ p $R_c = 2.2$ kΩ)	62°
Output voltage range	±1 V
Output current range	±250 μA
Slew rate	100 u A/ns
Input resistance (n, p)	1.6 kΩ
Output resistance	6.1 GΩ
Input voltage offset (n, p)	≈ 0.1 mV

and output stages operate in class AB which enables high input–output current swing and drive capability. Input resistance equations of the proposed COA are:

$$r_{\text{in}+} \cong \frac{1}{g_{m5} + g_{m6}} \tag{6.15}$$

$$r_{\text{in}-} \cong \frac{1}{g_{m3} + g_{m4}} \tag{6.16}$$

Compensation capacitance (C_c) is inserted at the high impedance node series with M_{13} which operates like a resistance. M_{13} is modeled with the resistance R_c.

At the output stage, by using the regulated cascode structure, output resistance values are improved and obtain nearly gmro times higher values compared to those of simple cascode ones. The output resistance equation of the proposed COA is:

$$r_{\text{out}+} = r_{\text{out}-} \cong \left[\frac{g_{ds26}g_{ds19}g_{ds24}}{g_{m26}g_{m24}} + \frac{g_{ds28}g_{ds22}g_{ds25}}{g_{m28}g_{m25}} \right]^{-1} \tag{6.17}$$

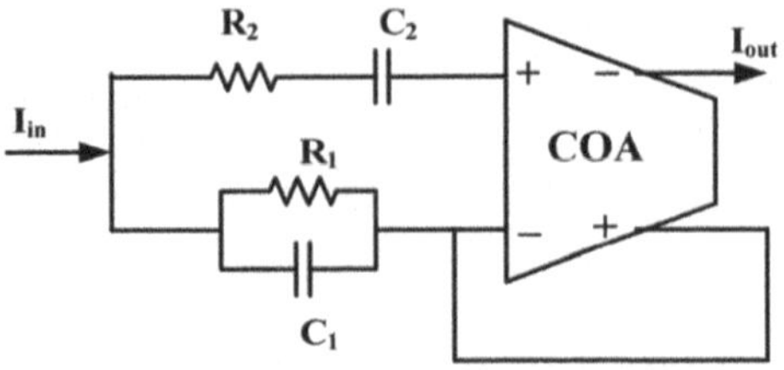

Fig. 6.10 COA-based notch filter topology. (Altun and Kuntman 2007)

The DC current gain and gain-bandwidth product of the proposed COA are given by

$$A_i(0) \cong \left[g_{m16} + g_{m17} \left[\frac{g_{ds12} g_{ds8}}{g_{m12}} + \frac{g_{ds11} g_{ds10}}{g_{m11}} \right] \right]^{-1} \tag{6.18}$$

$$f_{\mathrm{GBW}} \cong \frac{1}{2\pi} \frac{g_{m16} + g_{m17}}{Cc} \tag{6.19}$$

COA-Based Notch Filter Realization

As an application example, COA-based notch filter (band-stop) is offered. Shown in Fig. 6.10, a single COA is used for realization and matching conditions are selected $C_2 = 2C_1$, $R_1 = 2R_2$.

The node analysis of the circuit of Fig. 6.10 yields current transfer function as follows:

$$A(s) = \frac{i_{\mathrm{out}}}{i_{\mathrm{in}}} = \frac{s^2 + \frac{1}{4R_2^2 C_1^2}}{s^2 + s\frac{2}{R_2 C_1} + \frac{1}{4R_2^2 C_1^2}} \tag{6.20}$$

Simulation Results of the Described COA

Simulations were performed with SPICE using the BSIM3v3 model parameters of a 0.35 μm n-well CMOS process. Transistor threshold voltages are 0.5 V for NMOS and −0.7 V for PMOS. The frequency response of the loop gain of the amplifier is reported in Fig. 6.11. The DC loop gain is 100 dB, the gain-bandwidth product exceeds 80 MHz, and the phase margin is greater than 60°.

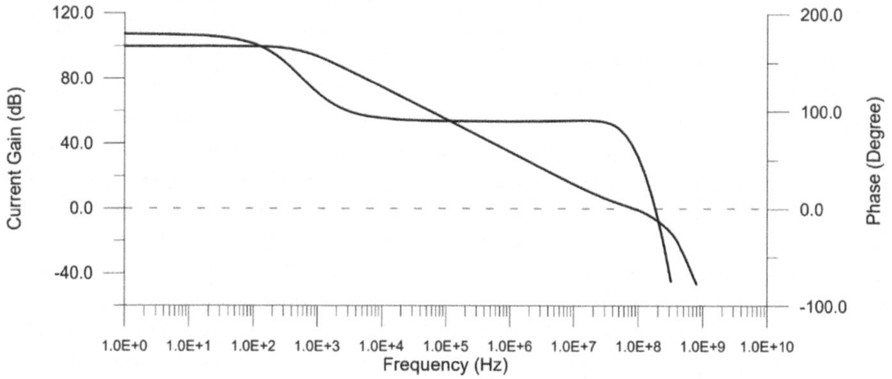

Fig. 6.11 Open-loop frequency response of the COA. (Altun and Kuntman 2007)

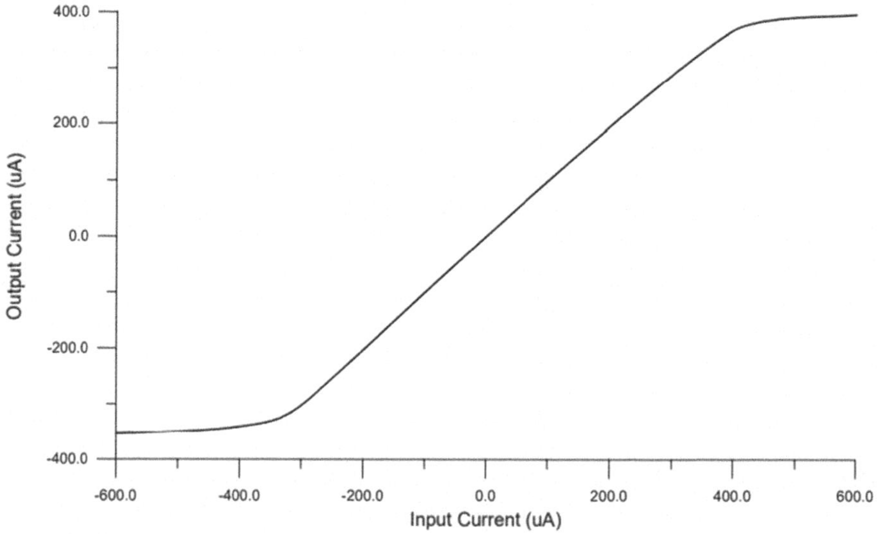

Fig. 6.12 Input--output transcharacteristic of the amplifier in unity-gain configuration. (Altun and Kuntman 2007)

The input–output transcharacteristic in unity-gain configuration, with a load resistance of 1 kΩ, is depicted in Fig. 6.12, where a maximum achievable output current of about 350 μA is clearly shown. A good power conversion is therefore obtained since the quiescent current in each branch is only of 20 μA ($I_b = 20$ μA). The transient response to a step input current of ±200 μA is shown in Fig. 6.13. As we neglect the outer capacitances, compensation capacitance (C_c) dominates the

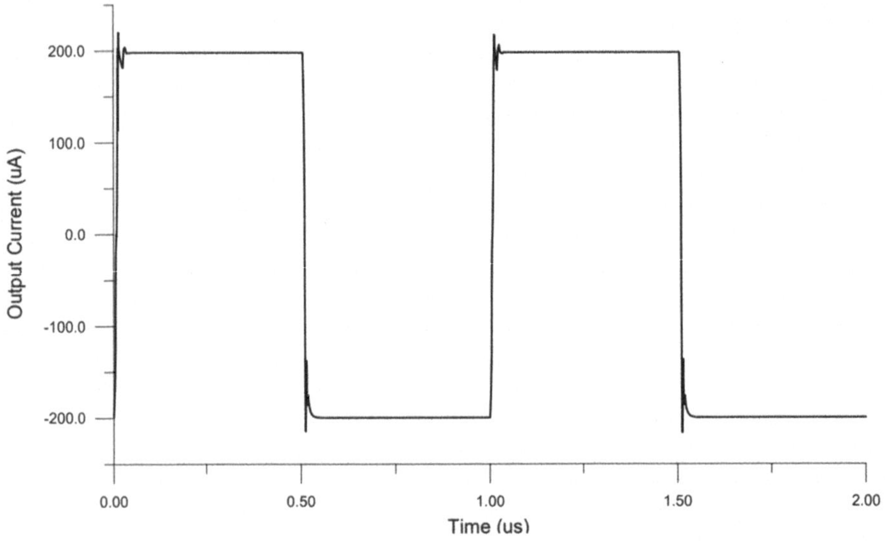

Fig. 6.13 Response of the COA in unity-gain feedback to a ±200 μA input step ($f = 1$ MHz). (Altun and Kuntman 2007)

slew-rate performance of the COA. As C_c is selected as 1.5 pF, it also means that adding output capacitance not bigger than 1.5 pF does not considerably worsen the speed of the proposed COA.

The main performance of the COA is summarized in Table 6.5. Note that with using impedance improvement, output resistance values become very high—6.1 GΩ.

Simulation Results of the COA-Based Filter

The following values are selected to realize a COA-based band-stop filter. The quality factor (Q) is 1/4, center frequency is ≈ 400 kHz, and element values in the circuit are chosen as $R_2 = R_1/2 = 10$ kΩ, $C_1 = C_2/2 = 20$ pF. As shown in Fig. 6.14, simulated and ideal filter responses are almost the same, up to ≈ 10 MHz. Figure 6.15 explains the transient characteristic of the filter. With up to 700 μA peak-to-peak input signal value, the THD is small enough to allow band-stop filter to work properly.

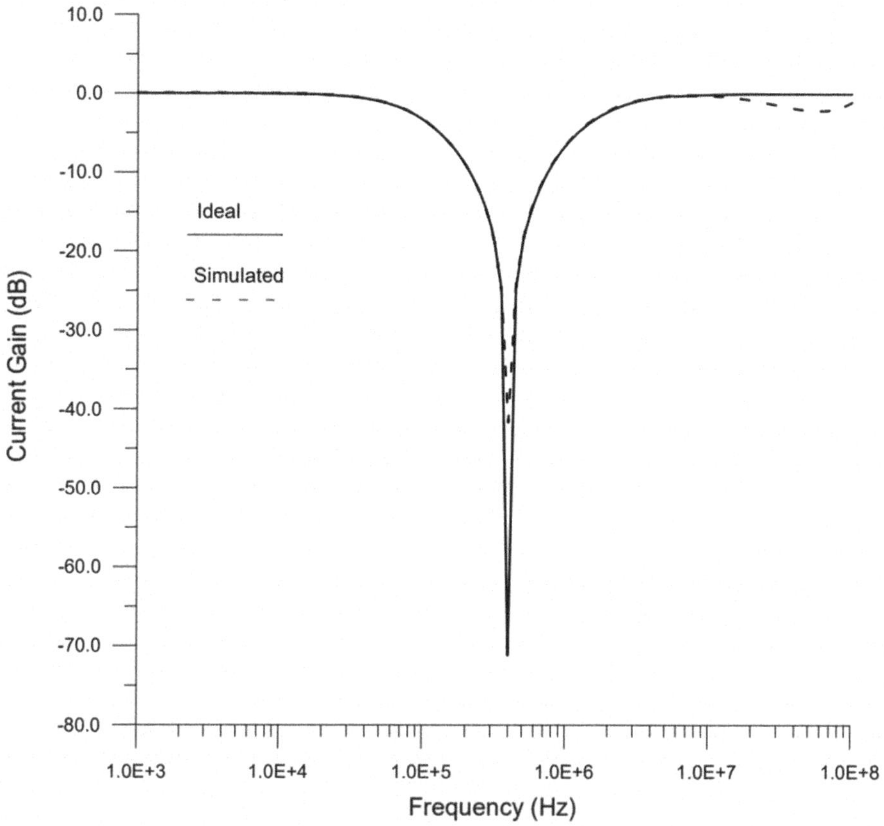

Fig. 6.14 Simulated and ideal band-pass filter responses. (Altun and Kuntman 2007)

Concluding Remarks

In this section, a high-performance fully differential COA is described. High current swing range is achieved both for inputs and outputs of the proposed COA which means high drive capability. Furthermore, very high output resistance is obtained by modifying the simple cascode output stage. The described COA also offers 85 MHz GBW and 100 dB DC gain. As an application example, a new COA-based second-order band-stop filter is proposed. Simulation results of the filter agree with the theoretical analysis demonstrating accuracy of the proposed fully differential COA.

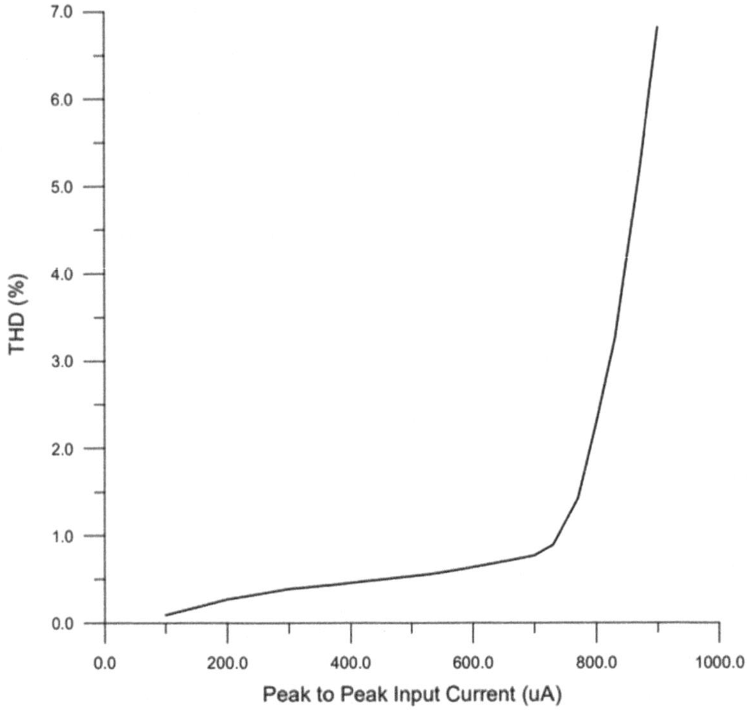

Fig. 6.15 Total harmonic distortion (THD) values of the filter versus input peak-to-peak current at 10 kHz frequency with RL = 1 kΩ. (Altun and Kuntman 2007)

References

Altun, M., & Kuntman, H. (2006, September). A wideband CMOS current-mode operational amplifier and its use for band-pass filter realization. In 2006 International Conference on Applied Electronics (pp. 3–6). IEEE.

Altun, M., & Kuntman, H. (2008). Design of a fully differential current mode operational amplifier with improved input–output impedances and its filter applications. AEU-International Journal of Electronics and Communications, 62(3), 239–244.

Altun, M., & Kuntman, H. (2007). A high drive fully differential current mode operational amplifier providing high output impedance and filter application. In Proceedings of the Fifth International Conference on Electrical and Electronics Engineering, Electronics (pp. 44–47).

Arbel, A. F., & Goldminz, L. (1992). Output stage for current-mode feedback amplifiers, theory and applications. Analog Integrated Circuits and Signal Processing, 2(3), 243–255.

Cheng, K. H., & Wang, H. C. (1997, June). Design of current mode operational amplifier with differential-input and differential-output. In 1997 IEEE International Symposium on Circuits and Systems (ISCAS) (Vol. 1, pp. 153–156). IEEE.

Jun, S., & Kim, D. M. (1998). Fully differential current operational amplifier. Electronics Letters, 34(1), 62–63.

Kaulberg, T. (1993). A CMOS current-mode operational amplifier. IEEE journal of solid-state circuits, 28(7), 849–852.

Kilinç, S., & Çam, U. (2004). Current-mode first-order allpass filter employing single current operational amplifier. Analog Integrated Circuits and Signal Processing, 41(1), 47–53.

Palmisano, G., & Pennisi, S. (2000). Dynamic biasing for true low-voltage CMOS class AB current-mode circuits. IEEE transactions on circuits and systems II: Analog and digital signal processing, 47(12), 1569–1575.

Palmisano, G., Palumbo, G., & Pennisi, S. (1999). CMOS current amplifiers (Vol. 499). Boston, MA: Kluwer Academic Publishers.

Souliotis, G., Chrisanthopoulos A, & Haritantis L., (2001), Current differential amplifiers: new circuits and applications, International Journal of Circuit Theory Application, 29(6):553–574.

Chapter 7
Current Feedback Operational Amplifiers, CFOA, and Applications

Basic Concept

The conventional operational amplifiers were successfully used over the years for the design of analog signal processing circuits. The maximum operation frequency of operational amplifier-based circuits` is determined primarily by the limited gain-bandwidth product and by the slew rate of the operational amplifier. Since their introduction in 1985, the popularity of current feedback amplifiers has increased considerably as they were found to be able to overcome the limitations arising from conventional operational amplifiers. A current feedback amplifier is equivalent to a plus-type second-generation current conveyor with a voltage buffer. The term current feedback is used because the error signal entering at the feedback node of the op-amp is in the form of a current and this gives to the amplifier a constant closed-loop bandwidth capability (Wong 1989). Ideally, the bandwidth of the current feedback op-amp is independent of the closed-loop gain. Therefore, the closed-loop gain-bandwidth product increases linearly with the closed-loop gain. This is a major advantage over the voltage-feedback op-amp architecture, which exhibits a constant GBW. Since the current feedback operational amplifier (CFOA) has a larger bandwidth and a higher slew rate than the conventional operational amplifier, analog signal processing circuits built around the CFOA are expected to operate at higher frequencies than the op-amp-based circuits. The current feedback amplifiers are gaining popularity as alternative building blocks for analog signal processing because they offer the following advantages over the conventional op-amps:

(i) Wide bandwidth which is relatively independent of the closed-loop gain
(ii) Very high slew rate
(iii) Simplicity of realization of various functions with the least possible number of external passive components

H. Kuntman, D. Özenli, *Trends in Circuit Design for Analog Signal Processing*,
Analog Circuits and Signal Processing,
https://doi.org/10.1007/978-3-030-96836-6_7

Consequently, there is a growing interest employing CFOAs for the realization of active filters, immittance simulators, single frequency, as well as single-element-controlled variable frequency sinusoidal oscillators and single/multiphase oscillators using CFOA pole. Recently, several current conveyor-based and CFOA-based oscillators are proposed in the literature. The CFOA-based topologies offer the following advantages compared to the current conveyor-based circuits (Abdullah Al-Walaie and Alturaigi, 1997; Abuelma'atti, 1987, 1993; Abuelma'atti and Al-Shahrani, 1996; Abuelma'atti and Humood, 1987; Abuelma'atti and Khan, 1997; Abuelma'atti et al. 1995, 1996; Abuelma'atti and Al-Shahrani, 1998; Aksoy et al., 2001; Awad and Soliman, 1999; Bhaskar and Senani, 1993; Celma et al., 1992; Celma et al. 1994a; Celma et al., 1994b; Chang, 1994; Çam et al. 2003; Fabre et al., 1995; Ferri et al., 2003; Gupta and Senani 1998; Harold, 1988; Incekaraoglu and Çam, 2005; Jana and Nandi, 1984; Kaçar, 2010), (Kaçar and Kuntman, 2009), (Keskin and Hancioglu, 2005; Kuntman et al., 2000; Liu, 1995; Liu and Tsay, 1996; Liu et al., 1994a, 1994b; Manetakis and Toumazou, 1996; Martinez et al., 1996; Martinez et al., 1997; Minaei et al., 2003; Senani and Gupta, 1997; Senani and Singh, 1992; Single, 1996; Soliman, 1998; Vanisri and Toumazou, 1992; Wang and Lee, 2000; Wu et al., 1995; Yuce, 2006, 2007, 2008, 2009; Yuce and Minaei, 2008; Yuce et al., 2006; Yuce et al., 2006a, 2006b; Zeki and Toker, 2005). As mentioned above the CFOA has an additional low-impedance terminal, which buffers the z terminal of the current conveyor with a unity gain; thus, CFOA-based oscillators exhibit low-impedance voltage-mode output. In previous publications new topologies were presented employing two CFOAs. Some of them employ two capacitors and three resistors where both capacitors are grounded. Another topology employing two CFOAs, two capacitors, and four resistors with both capacitors grounded and allowing the control of the oscillation frequency by a resistance ratio is given in the related literature. On the other hand, three circuits for high frequency applications are presented which make use of the CFOA pole, two circuits employ two CFOAs but no capacitors, and one circuit employs one CFOA and only one capacitor. Numerous single CFOA-based topologies are also presented in the literature. Most presented circuits are derived from a given configuration. The most general configuration with four nodes plus ground is proposed by Senani and Singh; the analysis and the presented results however are restricted to canonical case (employing two capacitors) and variable frequency single-resistance-controlled oscillators (SRCO). The Abuelmaatti configuration is interesting which introduces an extra node in addition to the four nodes of the CFOA and allows additional circuits. CFOA-based multiphase oscillators are also presented. Finally, Senani and Singh introduced new oscillators employing two or three CFOAs using the state variable synthesis approach.

In the following part, using a systematic circuit generation method (Cicekoglu and Kuntman 1998), new single CFOA-based oscillator circuits are derived. In the frame of this study, we propose two topologies with six nodes (five nodes plus ground), starting from six nodes plus ground configuration which is more general than the previously reported general configurations. From these topologies five SRCOs are derived. Four of them are found to be new; they cannot be derived

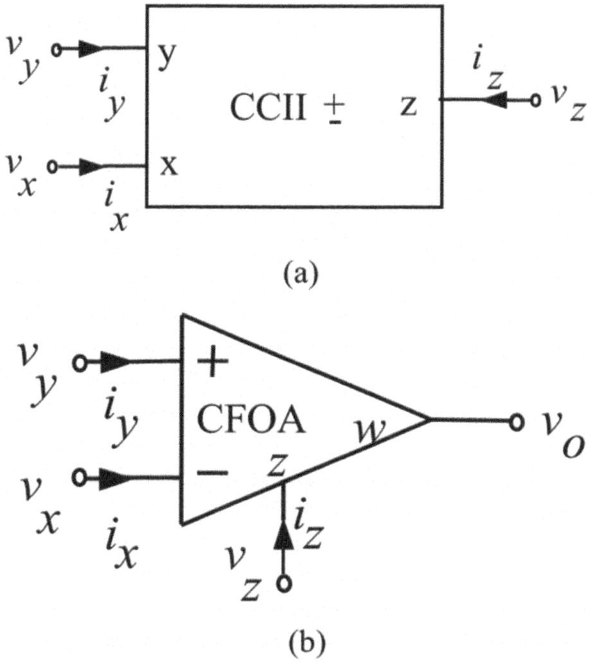

Fig. 7.1 (**a**) The current conveyor and (**b**) the CFOA symbols

from Senani and Singh (1996) and Abuelma'atti and Al-Shahrani (1997), and one of them is identical to the previously reported one (Abuelma'atti and Al-Shahrani 1997). Furthermore, the frequency stability of these oscillators is discussed. Moreover, an overview of the single-CFOA-based oscillators is also given.

The Current Conveyor and the Current Feedback Op-amp

The symbolic representation of second-generation current conveyor (CCII) and CFOA is shown in Fig. 7.1a, b, respectively. For the CFOA the terminals denoted with plus and minus signs are respectively the noninverting (infinite impedance) and inverting (zero impedance) input, w is the output, and z is the compensation terminal. The output terminal is low impedance (voltage source) and the compensation terminal is high impedance (current source). The main difference between the CFOA and the CCII is the unity gain voltage buffering of the z terminal of the CC to obtain the w terminal of the CFOA. For the CFOA the relation between the input/output voltages and currents can be given by the following matrix equation:

$$\begin{bmatrix} i_y \\ v_x \\ i_z \\ v_o \end{bmatrix} = \begin{bmatrix} 0 & 0 & 0 \\ 1 & 0 & 0 \\ 0 & 1 & 0 \\ 0 & 0 & 1 \end{bmatrix} \begin{bmatrix} v_y \\ i_x \\ v_z \end{bmatrix} \tag{7.1}$$

Non-ideal characteristics: The matrix representation given with Eq. (7.1) is valid only for the ideal case. In fact the voltage and the current copying abilities of the CC and the voltage buffer are with errors ev, ei, and eo. Taking these errors into account the terminal relations of the CFOA for the non-ideal case are given as follows:

$$i_y(t) = 0, v_x(t) = \beta v_y(t), i_z(t) = \alpha i_x(t) \ v_o = \gamma v_z \tag{7.2}$$

$$\beta = 1 - \varepsilon_v(|\varepsilon_v| \ll 1), \alpha = 1 - \varepsilon_i(|\varepsilon_i| \ll 1) \tag{7.3}$$

are voltage and current gains of the CFOA where ε_v and ε_i are voltage and current tracking errors, respectively.

The Realization Circuits, Application Examples

The n-nodes plus ground general configuration are shown in Fig. 7.2. In this work six nodes plus ground network are chosen as the starting point. All nodes are connected by admittances except certain node pairs for which a connection is redundant. These node pairs are nodes where x–y terminals are connected directly and nodes where z–w terminals are connected directly and the admittance from the low-impedance w terminal to ground. Thus, the total number of admittances is 18. From this parent configuration two five nodes plus ground general configurations are derived and they are shown in Figs. 7.3 and 7.4. Considering that an admittance may be the parallel connection of a resistance and capacitance, the total number of possibilities is enormous. Through an exhaustive search many circuits are derived and a subset with certain properties is given here for space reasons. The starting configuration is more general than the Senani and Singh and Abuelmaatti and

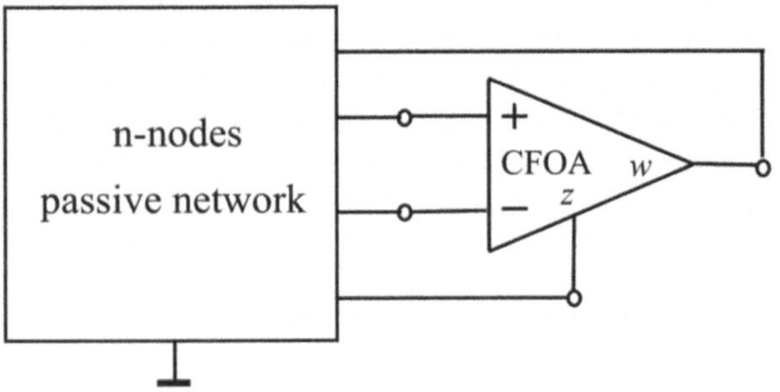

Fig. 7.2 The general single CFOA n-nodes plus ground network. (Toker et al. 2002)

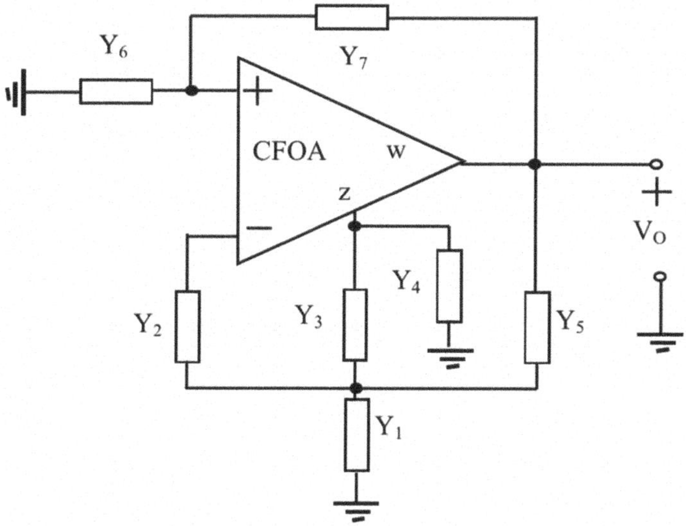

Fig. 7.3 The derived first oscillator topology. (Toker et al. 2002)

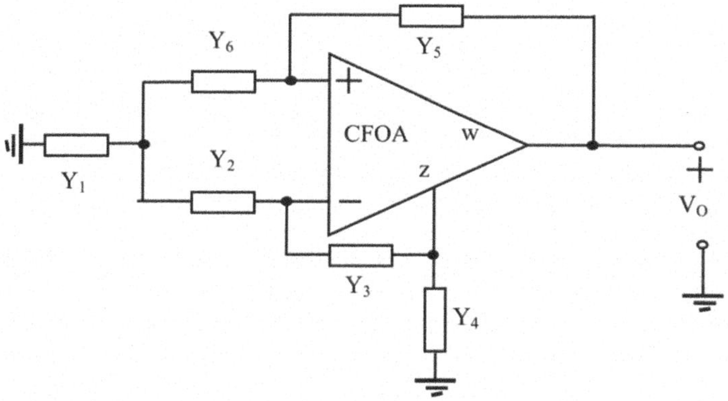

Fig. 7.4 The derived second oscillator topology. (Toker et al. 2002)

Al-Sharani circuits. Therefore, new circuits are found in addition to the existing circuits in the literature. Senani and Singh derived their circuit from the most general five nodes (four nodes plus ground) configuration by deleting the redundant admittances. Their circuit is shown in Fig. 7.5. Our two presented general configurations are six nodes (five nodes plus ground) circuits. For the circuit in Fig. 7.5 the x, y, and z terminals of the CFOA are directly connected by an admittance to ground. Since this admittance can either be a capacitor or a resistor or a parallel combination, a realization with a series combination of a capacitor and a resistor is not observed, which would mean the introduction of an extra node. From the presented topologies in Figs. 7.3 and 7.4, new SRCOs are obtained selecting the passive elements

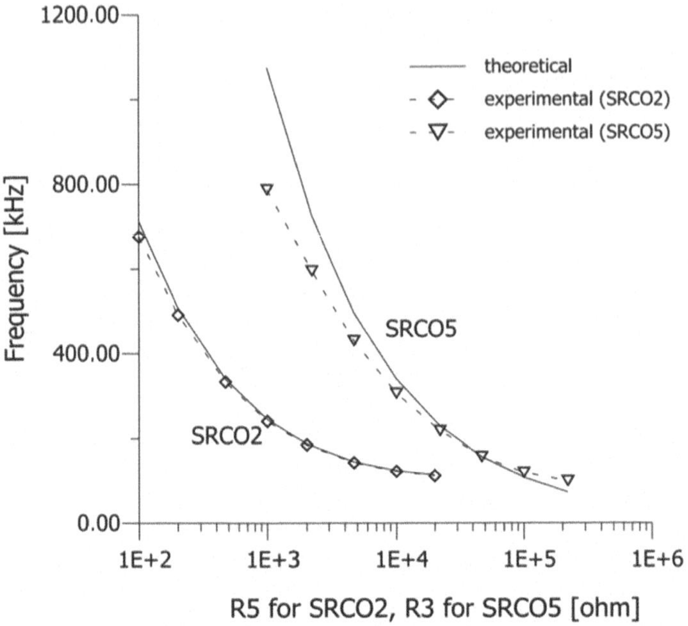

Fig. 7.5 The oscillation frequency versus R_5 (for SRCO2) and R_3 (for SRCO5) with passive element values $R_2 = 4.7$ kOhm, $R_4 = 10$ kOhm, $C_1 = C_3 = 220$ pF (for SRCO2) and $R_1 = R_5 = R_6 = 1$ kOhm, $C_2 = 390$ pF and $C_4 = 120$ pF (for SRCO5). (Toker et al. 2002)

appropriately, where the oscillation frequency and oscillation condition can independently be adjustable which results in variable frequency oscillators.

(i) The first oscillator topology (Fig. 7.3): Using the relationship given in Eq. (7.1), which characterizes the CFOA and routine analysis for the circuit topology illustrated in Fig. 7.3, the characteristic polynomial is calculated. If we set the characteristic polynomial equal to zero, the following equation is obtained:

$$Y_1Y_3(Y_6 + Y_7) + Y_2Y_6(2Y_3 + Y_5) + Y_4(Y_1 + Y_2 + Y_3 + Y_5)(Y_6 + Y_7) = Y_1Y_2Y_7 \tag{7.4}$$

(a) Case 1: (SRCO1)

Choosing Y4 = 0, Y5 = 0

$$Y_1 = G_1, Y_2 = sC_2, Y_3 = Y_3, Y_6 = sC_6 \textit{ and } Y_7 = G_7.$$

$$C_2G_7 = C_6G_3,\ \omega_0^2 = \frac{G_1G_7}{2C_2C_6} \tag{7.5}$$

(b) Case 2: (SRCO2)

Choosing Y6 = 0 and selecting the admittances as follows:

$$Y_1 = sC_1, Y_2 = G_2, Y_3 = sC_3, Y_4 = G_4 \textit{ and } Y_5 = G_5$$

$$C_1G_2 = C_1G_4 + C_3G_4, \quad \omega_0^2 = \frac{G_4(G_2 + G_5)}{C_1C_3} \tag{7.6}$$

(ii) The second oscillator topology (Fig. 7.4)

$$Y_1Y_2Y_5 = Y_1Y_6(2Y_3 + Y_4) + (Y_1 + Y_2 + Y_6)Y_4Y_5 \tag{7.7}$$

(c) (SRCO3). Selecting the admittances as

$$Y_1 = G_1, Y_2 = G_2, Y_3 = G_3, Y_4 = sC_4, Y_5 = sC_5,$$
$$Y_6 = G_6$$
$$G_2C_5 = C_4G_6, \quad \omega_0^2 = \frac{2G_1G_3G_6}{C_4C_5(G_1 + G_2 + G_6)} \tag{7.8}$$

(d) (SRCO4):

$$Y_1 = sC_1, Y_2 = G_2, Y_3 = G_3, Y_4 = G_4, Y_5 = G_5 \textit{ and } Y_6 = sC_6$$

$$G_2C_1 = (C_1 + C_6)G_4, \quad \omega_0^2 = \frac{G_2G_4G_5}{C_1C_6(2G_3 + G_4)} \tag{7.9}$$

(e) (SRCO5). Selecting the admittances as (Table 7.1)

$$Y_1 = G_1, Y_2 = sC_2, Y_3 = G_3, Y_4 = sC_4, Y_5 = G_5 \text{ and } Y_6 = G_6$$

$$G_1G_5C_2 = C_4(G_1G_6 + G_1G_5 + G_5G_6), \quad \omega_0^2 = \frac{2G_1G_3G_6}{C_2C_4G_5} \tag{7.10}$$

Frequency Stability

Frequency stability analysis is an important topic in the design stage of sinusoidal oscillators. The classical frequency stability factor SF is defined as

Table 7.1 Oscillation frequencies and conditions for the presented single-resistance-controlled oscillator's non-ideal case

Circuit	Frequency of oscillation $\left(\omega_0^2\right)$	Condition of oscillation
SRCO1	$\omega_0^2 = \frac{G_1 G_2}{C_2 C_6 (1+\alpha)}$	$\alpha\beta\gamma C_2 G_1 G_7 = C_6 G_1 G_3 + (1+\alpha)(1+\beta\gamma) C_2 G_3 G_7$
SRCO2	$\omega_0^2 = \frac{G_4(G_2+G_5)+(1-\beta)\alpha\gamma G_2 G_5}{C_1 C_3}$	$\alpha\beta\gamma C_2 G_2 = C_2 G_4 + C_3 G_4 + (1+\alpha)(1-\beta\gamma) C_3 G_2 + (1-\gamma) C_3 G_5$
SRCO3	$\omega_0^2 = (1+\alpha)\frac{G_3 G_6 [G_1+(1-\beta)] G_2}{C_4 C_5 (G_1+G_2+G_6)}$	$\alpha\beta\gamma C_5 G_1 G_2 = C_4 G_1 G_6 (1-\beta) G_2 G_6 (C_4 + \alpha\gamma C_5) + (1+\alpha)(1-\beta\gamma) C_5 G_3 (G_1 + G_2 + G_6)$
SRCO4	$\omega_0^2 = \frac{G_2 G_5 [G_4+(1+\alpha)(1-\beta\gamma) G_3]}{C_1 C_6 [(1+\alpha) G_3 + G_4]}$	$\alpha\beta\gamma C_1 G_2 G_5 = G_4 G_5 (C_1 + C_6) + (1+\alpha)(1-\beta\gamma) G_3 G_5 (C_1 + C_6) +$ $(1-\beta) C_6 G_2 [(1+\alpha) G_3 + G_4 + \alpha\gamma G_5]$
SRCO5	$\omega_0^2 = \frac{(1+\alpha) G_3 [G_1 G_6 + (1-\beta\gamma) G_5 (G_1+G_6)]}{C_2 C_4 (G_5 + (1-\beta) G_6)}$	$\alpha\beta\gamma C_2 G_1 G_5 = C_4 (G_1 G_6 + G_1 G_5 + G_5 G_6) + \alpha\gamma (1-\beta) C_2 G_5 G_6 +$ $(1+\alpha) C_2 G_3 [(1-\beta\gamma) G_5 + G_4 + (1-\beta) G_6]$

$$S_F = \left. \frac{\mathrm{d}\phi(u)}{\mathrm{d}u} \right|_{u=1} \tag{7.10}$$

where ω/ω_o and $\Phi(u)$ represent the phase function of the open-loop transfer function of the circuit given in Figs. 7.3 and 7.4. Good frequency stability is obtained for high values of SF. The results are illustrated in Table 7.2. Note that as a result of the above analysis, the effect of the frequency-controlling resistor on the stability factor can be stated as:

(a) SRCO1: Good stability can be obtained by choosing large values of n, namely, for high R1 values.
(b) SRCO2: For small R5 values good stability factor is obtained.
(c) SRCO3: For small R3 values good stability factor is obtained.
(d) SRCO4: For large R5 values good stability factor is obtained.
(e) SRCO5: For small R3 values good stability factor is obtained.

Therefore, SRCO1 and SRCO4 are not suitable for high frequency operation because at high frequencies their frequency stability factors are reduced. On the other hand, SRCO2, SRCO3, and SRCO5 are suitable for high frequency operation.

Experimental Results and Discussion

All the proposed oscillators are experimentally tested using the commercially available current feedback amplifier AD844 from analog devices. As two examples the experimental results of the second and fifth single-resistance-controlled oscillators (SRCO2 and SRCO5) are illustrated in Fig. 7.5. The passive component values of SRCO2 and SRCO5 are chosen as $R_2 = 4.7$ kOhm, $C_1 = 220$ pF, $C_3 = 220$ pF, $R_4 = 10$ kOhm (for SRCO2) and $R_1 = 1$ kOhm, $C_2 = 390$ pF, $C_4 = 120$ pF, $R_5 = 1$ kOhm, $R_6 = 1$ kOhm (for SRCO5). The frequency of oscillation is controlled with R_5 for SRCO2 and with R_3 for SRCO5: The supply voltages are chosen as $V_{DD} = 12$ V and $V_{SS} = 12$ V. The experimental results are in good agreement with the theoretical results. The deviation at high frequencies in oscillation frequency (for SRCO5) from theoretical values is caused by the parasitics of the CFOA and the stray capacitances. The output waveform of the SRCO2 is illustrated in Fig. 7.6 for an oscillation frequency of 300 kHz.

In this section CFOA oscillators are investigated starting from a generalized parent configuration. From this configuration all the previously reported oscillators are obtained. By investigation of six-node general configuration, two circuit topologies are found to be interesting. Oscillator circuits are obtained from these configurations. Four of these oscillator circuits presented are new and one of them is reported previously. Two of them can be operated also with CFOA pole. Furthermore, the frequency stability of the presented circuits is also investigated and some of these circuits are found to be suitable for high frequency operation. Moreover, the

Table 7.2 Frequency stability of the proposed oscillators

Oscillator	Frequency stability	Capacitor selection	Definition of n	Frequency stability
SRCO1	$S_F = 2\sqrt{\frac{2C_2G_7}{G_1C_6}}$	$C_2 = C_6 = C$	$\frac{G_7}{G_1} = \frac{R_1}{R_7} = n$	$S_F = \sqrt{8n}$
SRCO2	$S_F = 2\sqrt{\frac{C_1C_3(G_2+G_5)}{G_4(C_1+C_3)^2}}$	$C_1 = C_3 = C$	$\frac{G_5}{G_4} = \frac{R_4}{R_5} = n$	$S_F = \sqrt{2+n}$
SRCO3	$S_F = 2\sqrt{\frac{2C_5G_3(G_1+G_2+G_6)}{C_4G_1G_6}}$	$C_4 = C_5 = C$	$\frac{R_6+2R_1}{R_3} = n$	$S_F = \sqrt{8n}$
SRCO4	$S_F = 2\sqrt{\frac{C_1C_6G_2(2G_3+G_4)}{(C_1+C_6)^2G_4G_5}}$	$C_1 = C_6 = C$	$R_5\left(\frac{2}{R_3} + \frac{1}{R_4}\right) = n$	$S_F = \sqrt{2}$
SRCO5	$S_F = 2\sqrt{2\frac{C_2G_1G_3G_5G_6}{C_4(G_1G_6+G_1G_5+G_5G_6)^2}}$	$C_2 = 3C_4 = C$	$\frac{R}{R_3} = n for$ $G_1 = G_5 = G_6 = G$	$S_F = \sqrt{8n/3}$

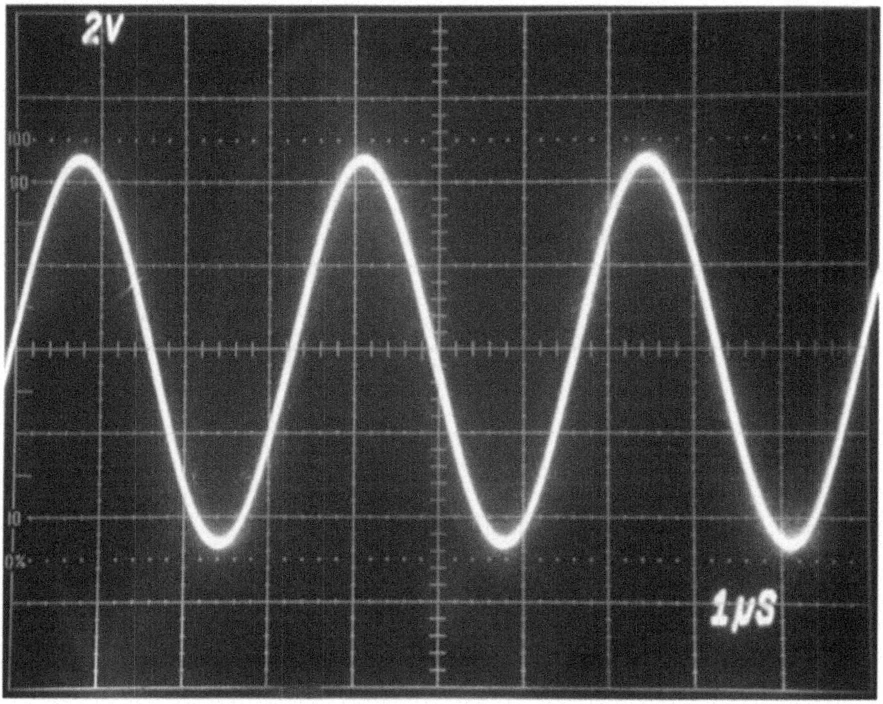

Fig. 7.6 The waveform of SRCO2 at 300 kHz. (Toker et al. 2002)

comparison of the previously reported and in this work proposed SRCOs is given. To verify the theoretical analysis experimental tests for the presented circuits are performed.

CFOA-Based Lossless and Lossy Inductance Simulators

Inductance simulator is a useful component in the circuit synthesis theory especially for analog signal processing applications such as filter, chaotic oscillator design, analog phase shifters, and cancellation of parasitic element. In this section, new four inductance simulator topologies employing a single current feedback operational amplifier are presented (Kacar and Kuntman 2011). The presented topologies require few passive components. The first topology is intended for negative inductance simulation, the second topology is for lossy series inductance, the third one is for negative lossy parallel inductance (−R) (−L), and the fourth topology is for negative parallel (−R) (−L) (−C) simulation. The performance of the proposed CFOA-based inductance simulators is demonstrated on both a second-order low-pass filter and inductance cancellation circuit. PSPICE simulations are given to verify the theoretical analysis.

The first topology is intended for negative inductance simulation, the second topology is for lossy series inductance, the third one is for negative lossy parallel inductance (−R) (−L), and the fourth topology is for negative parallel (−R) (−L) (−C) simulation. Finally, a second-order low-pass filter and inductance cancellation circuit are constructed using the proposed series inductance and negative inductance simulator. Simulation results are included to verify the theory (Tables 7.3 and 7.4).

Deviations from the ideal characteristics will affect the performance of the circuits realized with CFOA. Taking the current and voltage tracking errors into account, the current–voltage terminal characteristic of the non-ideal CFOA becomes

$$\begin{bmatrix} I_y \\ I_z \\ V_x \\ V_w \end{bmatrix} = \begin{bmatrix} 0 & 0 & 0 \\ \alpha & 0 & 0 \\ 0 & \beta & 0 \\ 0 & 0 & \gamma \end{bmatrix} \begin{bmatrix} I_x \\ V_y \\ V_z \end{bmatrix} \tag{7.11}$$

where α, β, and γ are the non-ideal current and voltage gains.

Applications and Simulation Results

All of the circuits illustrated in Fig. 7.7 are tested with SPICE simulations. The test circuits were constructed with AD844. The circuit in Fig. 7.7(b) was supplied with symmetrical voltages of 10 V. The element values are chosen as $C_1 = 10$ nF, $R_1 = R_2 = 1$ kΩ to obtain an inductance of $L_{eq} = 10$ mH and a series resistance of $R_{eq} = 0.5$ kΩ. The frequency response of the topology in Fig. 7.7(b) is chosen as an example to demonstrate the performance of the derived inductance simulators. Figure 7.8 shows that the magnitudes of the impedances of an ideal inductor and a lossy inductor that is a series R_L circuit can be made very close for a set of selected values over many decades.

Negative inductance is especially important if the parasitic effects have to be cancelled. The magnitude of the negative inductance increases with frequency in the same way as for positive inductances. However, a negative inductance provides a negative 90° phase shifting. Typical waveforms of the voltage and current through the proposed negative inductance simulator are shown in Fig. 7.9.

To illustrate an application of the negative inductance simulator, it is employed in an inductance cancellation circuit shown in Fig. 7.10 (Jaikla and Siripruchyanun 2006), where $V_{in} = 1$ V, $R = 100$ kΩ and $L = 5$ mH. Applying a sinusoidal input signal of 1.59 MHz with an amplitude of $V_{IN} = 1$ V, the transient responses of output current IR and input voltage are obtained and illustrated in Fig. 7.11. It is shown that the inductance in the circuit has been cancelled by the negative inductance simulator.

In order to demonstrate the performances of the proposed circuits, the second circuit with traditional CFOA is chosen as an example. Furthermore, a low-pass filter

Table 7.3 Actively realizable inductance forms

Circuit	Z_{eq}	L_{eq}	G_{eq}	C_{eq}	Type
Fig. 2 (a)	$Z_{eq} = -\frac{sC}{2G_1G_2}$	$L_{eq} = -\frac{C}{2G_1G_2}$	–	–	Pure-L
Fig. 2 (b)	$Z_{eq} = -\frac{sC+G_1+G_2}{G_1G_2}$	$L_{eq} = \frac{C}{G_1G_2}$	$G_{eq} = G_1 + G_2$	–	L with series R
Fig. 2(c)	$Z_{eq} = -\frac{sC}{sCG_2+G_1G_2}$	$L_{eq} = -\frac{C}{G_1G_2}$	$G_{eq} = -G_2$	–	–L with parallel – R
Fig. 2(d)	$Z_{eq} = -\frac{sC_1}{G_1G_2+s^2C_1C_2+sC_1G_2+sC_2G_1}$	$L_{eq} = -\frac{C_1}{G_1G_2}$	$G_{eq} = -\left(\frac{C_1G_2+C_2G_1}{C_1}\right)$	$-C_2$	–L and –C parallel with -R

Table 7.4 CFOA-based grounded inductance circuits for the non-ideal case

Circuit	Matching	Z_{eq}
Fig. 2 (a)	No	$Z_{eq} = -\frac{sC\alpha\beta\gamma}{G_1(1+\alpha)(G_2+sC(1-\beta\gamma))}$
Fig. 2(b)	No	$Z_{eq} = \frac{sC\alpha\gamma+G_1+G_2}{G_1(G_2+sC\alpha(-\beta+\gamma))}$
Fig. 2(c)	No	$Z_{eq} = \frac{sC}{sC(-1+\gamma)G_1+sC\alpha\beta G_2+\alpha\beta\gamma G_1G_2}$
Fig. 2(d)	No	$Z_{eq} = \frac{sC_1}{G_1G_2\alpha\beta\gamma+s^2C_1C_2\alpha\beta+sC_1(G_2\alpha\beta+G_1(-1+\gamma))+sC_2G_1\alpha\beta\gamma}$

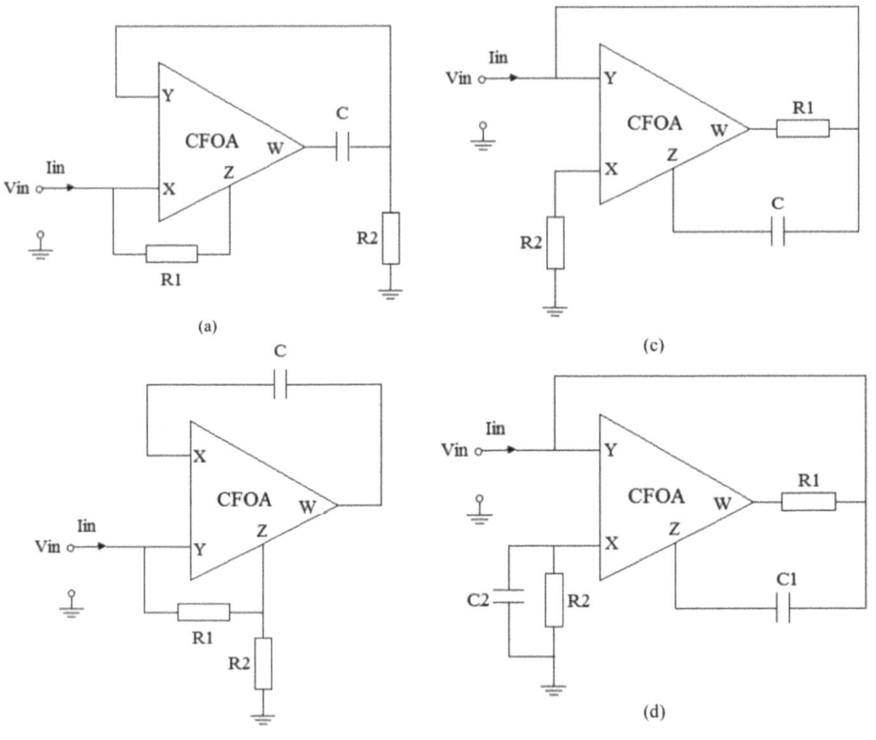

Fig. 7.7 Inductance simulators realized using single CFOA. (Kacar and Kuntman 2011)

circuit was realized as an application example by connecting a series capacitor to the circuit illustrated in Fig. 7.7(b). The proposed inductance simulator was used for constructing a second-order voltage-mode low-pass filter shown in Fig. 7.12. The circuit is designed to simulate a series R–L combination consisting of an inductor with $L_{eq} = 0.1$ mH and a resistance $R_{eq} = 0.5$ kΩ. To obtain these values, the following values are chosen: $R_1 = R_2 = 1$ kΩ and C = 0.1 nF. Choosing the capacitor with $C_L = 100$ pF and driving the resulting RLC circuit with voltage of 1 V, the frequency responses of the ideal and simulated low-pass filter circuits are

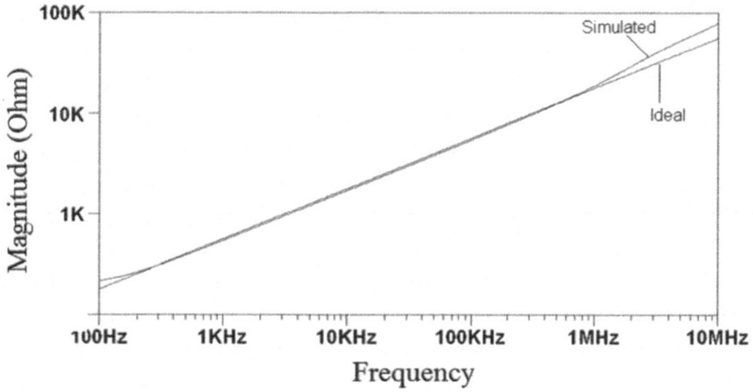

Fig. 7.8 Ideal and simulated frequency responses of the series (RL) inductor simulator. (Kacar and Kuntman 2011)

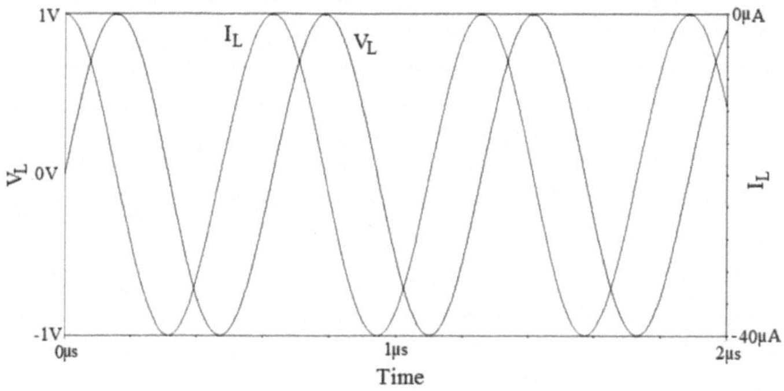

Fig. 7.9 Waveforms of voltage and current of the proposed negative inductance. (Kacar and Kuntman 2011)

Fig. 7.10 Inductance cancellation circuit using negative inductance simulator. (Kacar and Kuntman 2011)

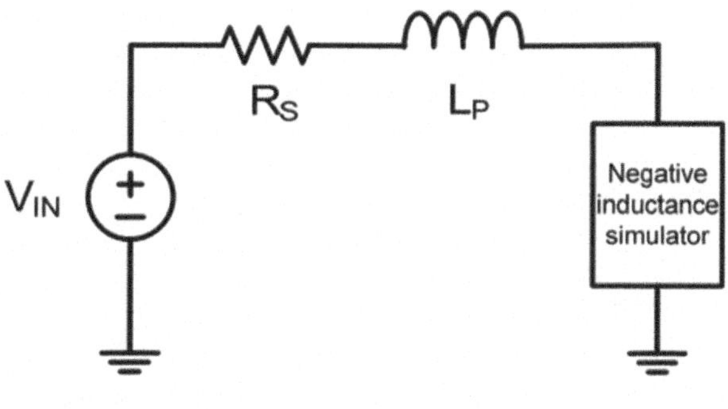

investigated. The circuit was supplied with symmetrical voltages of ±10 V. PSPICE simulations were performed using AD844.

The frequency response of the actual circuit obtained from SPICE simulations is given with the response of the ideal circuit in Fig. 7.13. The simulation results show

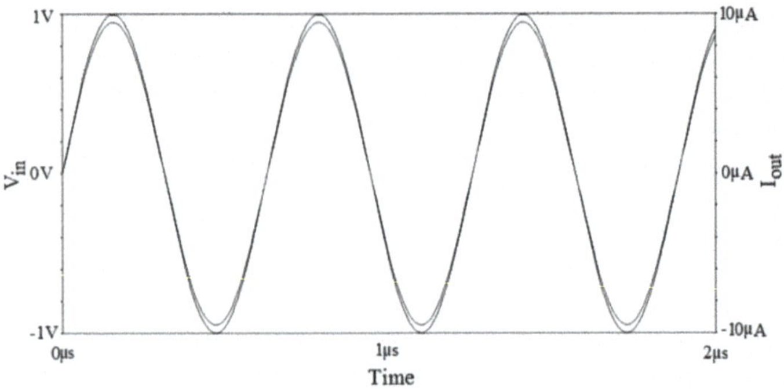

Fig. 7.11 Transient responses of the inductance cancellation circuit. (Kacar and Kuntman 2011)

Fig. 7.12 Voltage-mode low-pass filter application of the series +L with +R lossy inductor simulators. (Kacar and Kuntman 2011)

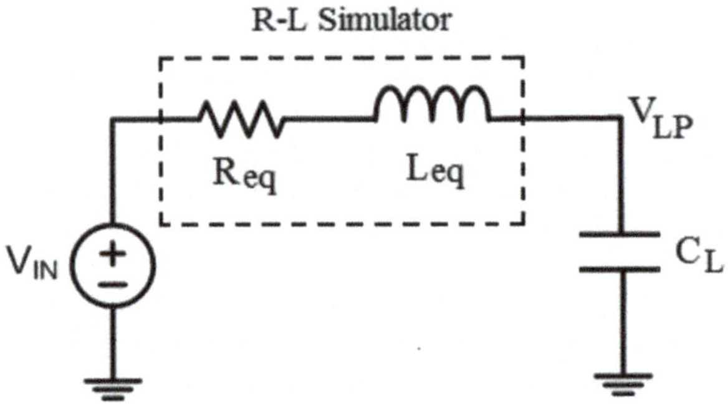

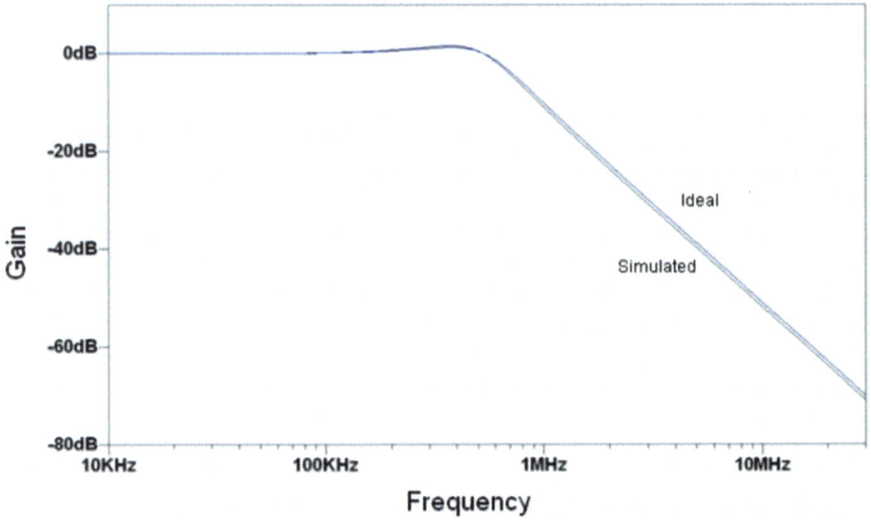

Fig. 7.13 Ideal and simulated frequency response of low-pass filter. (Kacar and Kuntman 2011)

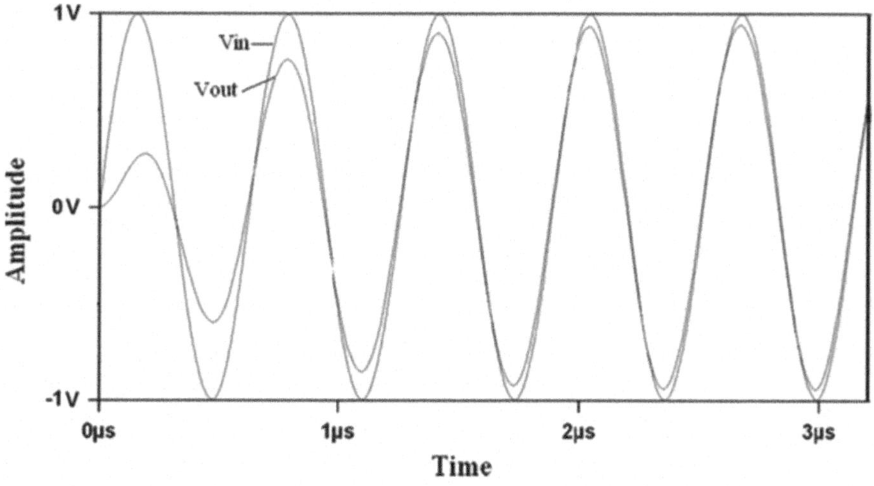

Fig. 7.14 Time domain response of circuit in Fig. 7.12 for 2 V peak-to-peak 1 MHz sine wave input for low-pass filter configuration. (Kacar and Kuntman 2011)

that filter characteristics are in good agreement with the predicted theoretical values. There is little difference between ideal and non-ideal responses in the high frequency region of the frequency response of this filter. Time domain analysis result is given in Fig. 7.14 for peak-to-peak 2 V, 1 MHz sine wave input for low-pass filter configuration for the circuit in Fig. 7.12.

Concluding Remarks

In this section four different inductance simulating topologies employing a current feedback operational amplifier are proposed. Each presented topology employs only one CFOA. Examples of applications are given in order to illustrate the practical use of the topologies. Simulation results are included to verify the theory. It is expected that the proposed series and parallel inductance simulators will be useful in analog signal processing applications such as filter and chaotic oscillator design.

References

Abdullah Al-Walaie, Soliman. and Alturaigi, M. A. (1997). Current mode simulation of lossless floating inductance. International Journal of Electronics, 83(6), 825–830.
Abuelma'atti, M. T. (1987). Two-minimum-component single current conveyor RC oscillators.

Abuelma'atti, M. T. (1993). *New current-controlled ccii-based RC oscillators with grounded capacitors and resistors. International journal of circuit theory and applications, 21(5), 487–491.*

Abuelma'atti, M. T. and Al-Shahrani, S. A. M. (1996). A novel low-component-count single-element-controlled sinusoidal oscillator using the CFOA pole. International Journal of Electronics, 80(6), 747–752.

Abuelma'atti, M. T. and Al-Shahrani, S. A. M. (1997). New CFOA-based sinusoidal oscillators. International Journal of Electronics, 82(1), 27–32.

Abuelma'atti, M. T. and Humood, N. A. (1987). Research note: Two new minimum-component Wien-bridge oscillators using current-conveyors. International journal of electronics, 63(5), 669–672.

Abuelma'atti, M. T. and Khan, M. H. (1997). New sinusoidal oscillators employing the CCII internal pole. International journal of electronics, 83(6), 817–824.

Abuelma'atti, M. T., Al-Ghumaiz, A. A. and Khan, M. H. (1995). Novel CCII—based single-element controlled oscillators employing grounded resistors and capacitors. International Journal of Electronics, 78(6), 1107–1112.

Abuelma'atti, M. T., Farooqi, A. A. and Alshahrani, S. M. (1996). Novel RC oscillators using the current-feedback operational amplifier. IEEE Transactions on Circuits and Systems I: Fundamental Theory and Applications, 43(2), 155–157.

Abuelma'atti, M.T. and Al-Shahrani, S. A. M. (1998). Novel CFOA-based sinusoidal oscillators. International journal of electronics, 85(4), 437–441.

Aksoy, M., Özcan, S., Kuntman, H. and Çiçekoğlu, O. (2001). Transresistance Multifunction Filter Using Current Feedback Amplifier. In Proceedings of the International Conference on Electrical and Electronics Engineering, ELECO (pp. 36–39).

Awad, I. A. and Soliman, A. M. (1999). Inverting second generation current conveyors: the missing building blocks, CMOS realizations and applications. International Journal of Electronics, 86(4), 413–432.

Bhaskar, D. R. and Senani, R. (1993). New current-conveyor-based single-resistance-controlled/voltage-controlled oscillator employing grounded capacitors. Electronics letters, 29(7), 612–614.

Celma, S., Martinez, P. A. and Carlosena, A. (1992). Minimal realisation for single resistor controlled sinusoidal oscillator using single CCII. Electronics letters, 28(5), 443–444.

Celma, S., Martinez, P. A. and Carlosena, A. (1994a). Approach to the synthesis of canonic RC-active oscillators using CCII. IEE Proceedings-Circuits, Devices and Systems, 141(6), 493–497.

Celma, S., Martinez, P. A. and Carlosena, A. (1994b). Current feedback amplifier based sinusoidal oscillators. IEEE Transactions on Circuits and Systems I: Fundamental Theory and Applications, 41(12), 906–908.

Chang, C. M. (1994). Novel current-conveyor-based single-resistance-controlled/voltage-controlled oscillator employing grounded resistors and capacitors. Electronics letters, 30(3), 181–183.

Cicekoglu, O. and Kuntman, H. H. (1998). Circuit design techniques: AD-HOC methods or systematic generation methods. In Proceedings of the Tenth International Conference on Microelectronics (Cat. No. 98EX186) (pp. 187–190). IEEE.

Çam, U., Kaçar, F., Cicekoglu, O., Kuntman, H. and Kuntman, A. (2003). Novel grounded parallel immittance simulator topologies employing single OTRA. AEU-International Journal of Electronics and Communications, 57(4), 287–290.

Fabre, A., Saaid, O. and Barthelemy, H. (1995). On the frequency limitations of the circuits based on second generation current conveyors. Analog Integrated Circuits and Signal Processing, 7(2), 113–129.

Ferri, G., Guerrini, N. C. and Diqual, M. (2003). CCII-based floating inductance simulator with compensated series resistance. Electronics Letters, 39(22), 1560–1562.

Gupta, S. S. and Senani, R. (1998). State variable synthesis of single resistance controlled grounded capacitor oscillators using only two CFOAs. IEE Proceedings-Circuits, Devices and Systems, 145(2), 135–138.

Harold, P. (1988). Current-feedback op amps ease high-speed circuit design. EDN, July, 7.

Incekaraoglu, M. and Çam, U. (2005). Realization of series and parallel RL and CD impedances using single differential voltage current conveyor. Analog Integrated Circuits and Signal Processing, 43(1), 101–104.

Jaikla, W. and Siripruchyanan, M. (2006). Floating positive and negative inductance simulators based on OTAs. In 2006 International Symposium on Communications and Information Technologies (pp. 344–347). IEEE.

Jana, P. B. and Nandi, R. (1984). Single current conveyor tunable sinewave RC oscillator. Electronics letters, 20(1), 44–45.

Kacar, F. and Kuntman, H. (2011). CFOA-based lossless and lossy inductance simulators. Radioengineering, 20(3), 627–631.

Kaçar, F. (2010). New lossless inductance simulators realization using a minimum active and passive components. Microelectronics Journal, 41(2–3), 109–113.

Kaçar, F. and Kuntman, H. (2009). On the realization of the FDNR simulators using only a single current feedback operational amplifier. In 2009 International Conference on Electrical and Electronics Engineering-ELECO 2009 (pp. II–223). IEEE.

Keskin, A. Ü. and Hancioglu, E. (2005). CDBA-Based Synthetic Floating Inductance Circuits with Electronic Tuning Properties. ETRI journal, 27(2), 239–242.

Kuntman, H., Gülsoy, M. and Çiçekoğlu, O. (2000). Actively simulated grounded lossy inductors using third generation current conveyors. Microelectronics Journal, 31(4), 245–250.

Liu, S. I. (1995). Single-resistance-controlled/voltage-controlled oscillator using current conveyors and grounded capacitors. Electronics letters, 31(5), 337–338.

Liu, S. I. and Tsay, J. H. (1996). Single-resistance-controlled sinusoidal oscillator using current-feedback amplifiers. International Journal of Electronics, 80(5), 661–664.

Liu, S. I., Chang, C. C. and Wu, D. S. (1994a). Active-R sinusoidal oscillators using the CFA pole. International journal of electronics, 77(6), 1035–1042.

Liu, S. I., Shih, C. S. and Wu, D. S. (1994b). Sinusoidal oscillators with single element control using a current-feedback amplifier. International Journal of Electronics, 77(6), 1007–1013.

Manetakis, K. and Toumazou, C. (1996). Current-feedback op amp suitable for CMOS VLSI technology. Electronics Letters, 32(12), 1090–1092.

Martinez, P. A., Celma, S. and Sabadell, J. (1996). Designing sinusoidal oscillators with current-feedback amplifiers. International Journal of Electronics, 80(5), 637–646.

Martinez, P. A., Sabadell, J. and Aldea, C. (1997). Grounded resistor controlled sinusoidal oscillator using CFOAs. Electronics Letters, 33(5), 346–348.

Minaei, S., Cicekoglu, O., Kuntman, H. and Türköz, S. (2003). Electronically tunable, active only floating inductance simulation. International Journal of Electronics, 89(12), 905–912.

Senani, R. and Gupta, S. S. (1997). Synthesis of single-resistance-controlled oscillators using CFOAs: simple state-variable approach. IEE Proceedings-Circuits, Devices and Systems, 144(2), 104–106.

Senani, R. and Singh, V. K. (1992). Single-element-controlled sinusoidal oscillator employing single current conveyor IC. Electronics letters, 28(4), 414–415.

Senani, R. and Singh, V. K. (1996). Novel single-resistance-controlled-oscillator configuration using current feedback amplifiers. IEEE Transactions On Circuits And Systems Part 1 Fundamental Theory And Applications, 43, 698–699.

Single, O. U. A. (1996). Synthesis of canonic single-resistance-controlled-oscillators using a single current-feedback-amplifier. IEE Proc.-Circuits Devices Syst, 143(1), 71.

Soliman, A. M. (1998). Current mode CCII oscillators using grounded capacitors and resistors. International Journal of Circuit Theory and Applications, 26(5), 431–438.

Toker, A., Çiçekoğlu, O. and Kuntman, H. (2002). On the oscillator implementations using a single current feedback op-amp. Computers & Electrical Engineering, 28(5), 375–389.

Vanisri, T. and Toumazou, C. (1992). Wideband and high gain current-feedback opamp. Electronics Letters, 28(18), 1705–1707.

Wang, H. Y. and Lee, C. T. (2000). Systematic synthesis of RL and CD immittances using single CCIII. International Journal of Electronics, 87(3), 293–301.

Wong, J. (1989). Current-feedback op-amps extend high-frequency performance. Electronic Design News, 211–215.

Wu, D. S., Liu, S. I., Hwang, Y. S. and Wu, Y. P. (1995). Multiphase sinusoidal oscillator using the CFOA pole. IEE Proceedings-Circuits, Devices and Systems, 142(1), 37–40.

Yuce, E. (2006). On the realization of the floating simulators using only grounded passive components. Analog Integrated Circuits and Signal Processing, 49(2), 161–166.

Yuce, E. (2007). On the implementation of the floating simulators employing a single active device. AEU-International Journal of Electronics and Communications, 61(7), 453–458.

Yuce, E. (2008). Grounded inductor simulators with improved low-frequency performances. IEEE Transactions on Instrumentation and Measurement, 57(5), 1079–1084.

Yuce, E. (2009). Novel lossless and lossy grounded inductor simulators consisting of a canonical number of components. Analog Integrated Circuits and Signal Processing, 59(1), 77–82.

Yuce, E. and Minaei, S. (2008). A modified CFOA and its applications to simulated inductors, capacitance multipliers, and analog filters. IEEE Transactions on Circuits and Systems I: Regular Papers, 55(1), 266–275.

Yuce, E., Cicekoglu, O. and Minaei, S. (2006). Novel floating inductance and FDNR simulators employing CCII+ s. Journal of Circuits, Systems, and Computers, 15(01), 75–81.

Yuce, E., Minaei, S. and Cicekoglu, O. (2006a). Limitations of the simulated inductors based on a single current conveyor. IEEE Transactions on Circuits and Systems I: Regular Papers, 53(12), 2860–2867.

Yuce, E., Minaei, S. and Cicekoglu, O. (2006b). Resistorless floating immittance function simulators employing current controlled conveyors and a grounded capacitor. Electrical Engineering, 88(6), 519–525.

Zeki, A. and Toker, A. (2005). DXCCII-based tunable gyrator. AEU-International Journal of Electronics and Communications, 59(1), 59–62.

Chapter 8
Active-Only Circuits

Basic Concept: Realizations of Current-Mode and Voltage-Mode Multifunction Filters Without External Passive Elements

The interest on analog continuous-time active filters without using external passive elements is growing. In these types of filters active devices such as operational amplifiers (op-amps) and operational transconductance amplifiers (OTAs) are employed. In these filters the pole of the operational amplifier that gives the integrating nature in the frequency range of interest and the operational transconductance amplifier that can implement the necessary resistance dimension is utilized. One advantage of these active-only filters over the gm-C filters is that a low output impedance signal output is available at the outputs of the op-amp if the signal is taken from the corresponding terminal. Also elimination of the passive elements may result in a reduction of chip area for integrated circuit implementations. Since filter parameters can be controlled by the OTA current, such a filter circuit would enable easy programmability. Erdogan et al. investigated the problem in their published papers (Erdogan et al. 2004, 2006). Furthermore, Minaei, Cicekoglu, Kuntman, Dündar, and Cerid described new realizations of current-mode and voltage-mode multifunction filters without external passive elements in Minaei et al. (2003b).

Analog continuous-time active filters utilizing an operational amplifier (op-amp) pole and the transconductance control property of the operational transconductance amplifier (OTA) have received considerable attention recently. These filters do not need to employ additional passive elements and are therefore sometimes called active-only filters. The major advantage of these circuits is the elimination of passive elements that may result in a reduction of chip area for integrated circuit implementations. Having multiple functions in a single circuit is especially useful since the same topology can be used for different filter functions. A growing number

H. Kuntman, D. Özenli, *Trends in Circuit Design for Analog Signal Processing*,
Analog Circuits and Signal Processing,
https://doi.org/10.1007/978-3-030-96836-6_8

of publications exist in the literature on op-amp and OTA-only filters (Tsukutani et al. 1996, 1999, 2000a, b, 2001; Abuelma'atti and Alzaher 1997; Singh and Senani 1998). In active-only filters, current-mode circuits are becoming more popular since they have many advantages over their voltage-mode counterparts. One of these advantages is that they have higher bandwidths. Another advantage is their higher dynamic ranges. For these types of filters, easy voltage or current control of filter parameters is important because, due to the integration process tolerances, filter characteristics deviate from the desired values. In fact, all integrated filters must be tuned after fabrication. The electronic tuning capability of these filters is achieved by adjusting them through the compensation capacitor (bandwidth) of the op-amps (Singh and Senani 1998) and/or the transconductance gm of the OTAs (Tsukutani et al. 1996, 1999, 2000a, b, 2001; Abuelma'atti and Alzaher 1997). This study by Erdogan, Topaloglu, Cicekoglu, Kuntman, and Morgul offers new topologies in current-mode active-only multifunction filter implementation (Erdogan et al. 2006). Three new current-mode active-only filters are proposed. By cascading the proposed filters, which implement LPN and HPN functions, higher-order band-pass and band-stop filter functions can be obtained. An AP function, which is implemented by the third filter circuit, is also introduced to give the chip designer more choices, depending on the application, where a phase correction may be needed. All of the circuits are tested with SPICE simulations using both ideal and MOSFET-based op-amp and OTA models. The layouts of the MOSFET-based op-amp and OTA cells are composed of SCMOS 2 μm technology. The filter chips are implemented with these cells. Using the SPICE parameters of the layout technology, post-layout simulations are performed, with the parasitics extracted from the chip layouts. All the results are included for verification. Simulation results show that filter characteristics are in quite good agreement with the theory.

The OTA and op-amp models: Ideally, the OTA is assumed as an ideal voltage-controlled current source. The gm (transconductance gain), which is used to relate output current to input voltage, is a function of the bias current, I_A. For the case of OTAs using MOS transistors in the saturation region, the gms are proportional to $(I_A)^{1/2}$; for MOS transistors operating in the weak inversion region, or for bipolar transistors, the g_{ms} are directly proportional to I_A. DO-OTAs (double-output OTAs) are used in the circuit schematics. They differ from OTAs in that they have two outputs with separately adjustable g_{ms}.

The multifunction type filter is especially versatile since the same topology can be used for different filter functions. A number of active-only voltage-mode or current-mode multifunction filters are reported in the literature, which realize second-order filters. These filters can be electronically tuned through the transconductance (g_m) of the OTAs and/or the gain-bandwidth product (B) of the op-amps.

These filters suffer from one or more of the following drawbacks:

1. The need of matching condition to realize different filter responses.
2. The need of additional active components for realizing a band-stop response simultaneously with low-pass, band-pass, and high-pass responses in current-mode operation and use of dual-output OTAs.

3. The use of three inputs and one output, single input and single output, or multi-input and multi-output, which implies that only one filter response can be used at a time.

In this part we propose two new multifunction filters. The first proposed filter operates in voltage mode and uses only two single-output OTAs and two op-amps. For a second-order filter the minimum number of integrators (op-amps) is two. To the best knowledge of the authors there is no active-only filter structure that realizes three filtering functions simultaneously but employing minimum number of active elements with controllable parameters at the same time yet. This filter realizes low-pass, band-pass, and high-pass responses without any matching condition. The circuit exhibits high input impedance which enables easy cascadability.

The second proposed filter is a current-mode multifunction filter which uses five single-output OTAs and two op-amps and can realize low-pass, band-pass, high-pass, and band-stop responses all at high output impedance terminals which enables easy cascadability. Also no matching condition is required for this filter. Because of current-mode operation, the filter enjoys advantages in terms of bandwidth, linearity, and dynamic range with respect to their voltage-mode counterparts. Moreover, both of the presented filters employ only single-output OTAs in contrast to the topologies requiring multiple-output OTAs (Tsukutani et al. 1996; Abuelma'atti and Alzaher, 1997; Tsukutani et al., 2000a) which can be considered as a further advantage. For the proposed filters the filter characteristics can be electronically tuned through biasing current of the op-amps or transconductance of the OTAs. Sensitivity analysis shows that the proposed multifunction filters have low sensitivities with respect to the circuit active parameters. Simulation results show that filter responses are in good agreement with the theory.

The OTA and Op-Amp Models

The symbol of the OTA is shown in Fig. 8.1. Ideally, the OTA is assumed an ideal voltage-controlled current source and can be described by the following equation:

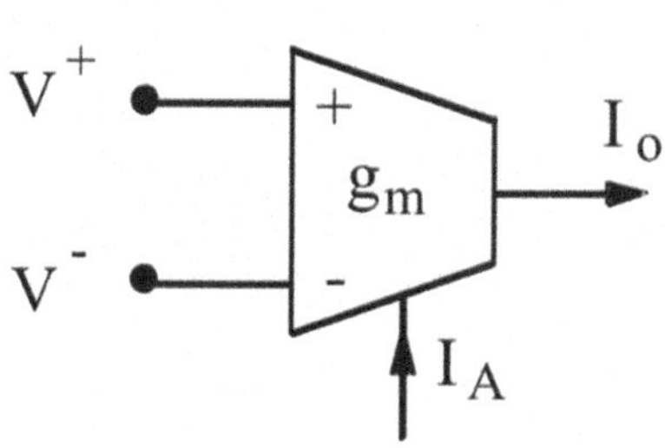

Fig. 8.1 Symbol of the OTA

$$I_o = g_m(V^+ - V^-) \tag{8.1}$$

where I_o is output current and V^+ and V^- denote noninverting and inverting input voltage of the OTA, respectively. Note that gm (transconductance gain) is a function of the bias current, *IA*. For the case of OTAs using MOS transistors in saturation, the gms are proportional to $\sqrt{I_A}$; for MOS transistors operating in weak inversion or bipolar transistors, the gms are directly proportional to *IA*. The op-amp on the other hand can be represented by the well-known first pole roll-off characteristic

$$A(s) = \frac{A_o\omega_{p1}}{s + \omega_{p1}} = \frac{B}{s + \omega_{p1}} \tag{8.2}$$

where A_o is open-loop DC gain, ω_{p1} is the first pole frequency, and B ($= A_o\omega_{p1}$) is the gain-bandwidth product of the operational amplifier. In the frequency range $\omega\omega_{p1}$, Eq. (8.2) reduces to

$$A(s) \cong \frac{B}{S} \tag{8.3}$$

This model of the op-amp is valid from a few kilohertz to a few hundred kilohertz. In this frequency range a monolithic OTA works as an ideal device.

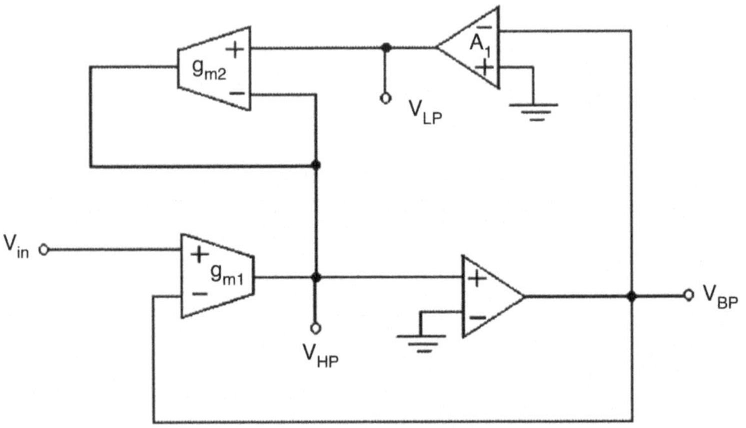

Fig. 8.2 The first proposed multifunction filter. (Minaei et al. 2003a, b)

The Proposed Active-Only Multifunction Filters

The first proposed multifunction type filter is shown in Fig. 8.2. The filter operates in voltage-mode and employs only two op-amps and two OTAs. Routine analysis of the circuit yields

$$\begin{aligned} T_{LP}(s) &= -\frac{g_{m1}}{g_{m2}} \cdot \frac{B_1 B_2}{\Delta} \\ T_{BP}(s) &= \frac{g_{m1}}{g_{m2}} \cdot \frac{B_2 s}{\Delta} \\ T_{HP}(s) &= \frac{g_{m1}}{g_{m2}} \cdot \frac{s^2}{\Delta} \\ \Delta = s^2 &+ \frac{g_{m1}}{g_{m2}} B_2 s + B_1 B_2 \end{aligned} \tag{8.4}$$

The angular resonant frequency ω_o, the parameter ω_o/Q, and the quality factor Q of the filter are given by

$$\omega_o = \sqrt{B_1 B_2} \quad \frac{\omega_o}{Q} = \frac{g_{m1}}{g_{m2}} B_2 \quad Q = \frac{g_{m2}}{g_{m1}} \sqrt{\frac{B_1}{B_2}} \tag{8.5}$$

The active sensitivities of the circuit are expressed as

$$\begin{aligned} &S_{B_1}^{\omega_o} = S_{B_2}^{\omega_o} = S_{B_1}^{Q} = -S_{B_2}^{Q} = \frac{1}{2}, \\ &S_{gm2}^{Q} = -S_{gm1}^{Q} = S_{gm1}^{\frac{\omega_o}{Q}} = -S_{gm2}^{\frac{\omega_o}{Q}} = S_{B2}^{\frac{\omega_o}{Q}} = 1, \\ &S_{gm1}^{\omega_o} = S_{gm2}^{\omega_o} = 0 \end{aligned} \tag{8.6}$$

Thus, all sensitivities are no more than unity. For this filter, B is used to adjust ω_o by means of the biasing current of the op-amp, whereas the OTA g_{ms} (transconductance values) are used to set the value of Q. No component matching constraints are imposed.

The second proposed multifunction type filter is shown in Fig. 8.3. The filter operates in current mode and uses five OTAs and two op-amps to realize low-pass, band-pass, high-pass, and band-stop responses simultaneously. The circuit is based on a transimpedance core which includes both of the op-amps and the first OTA denoted with transconductance gain g_{m2}. The other four OTAs operate in open-loop mode. Therefore, if not all functions are required, then the corresponding OTA can be removed. For example, if only a single function is required, then two OTAs and two op-amps are enough which is a minimum number for high output impedance active-only filters in current-mode operation. Routine analysis of the circuit yields

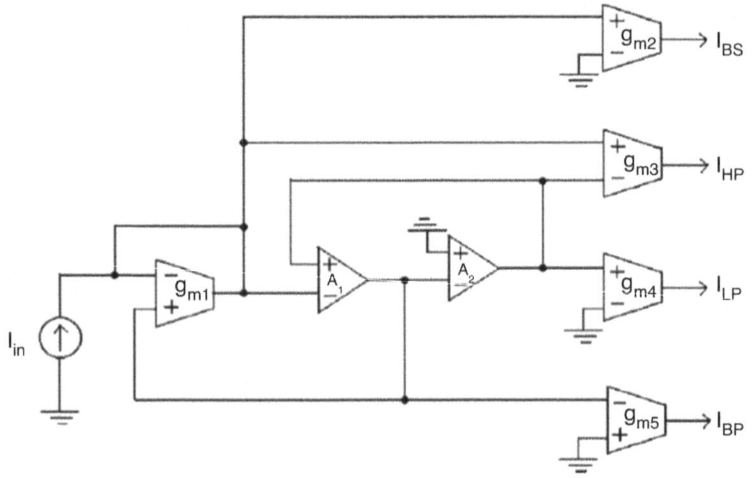

Fig. 8.3 The second proposed multifunction filter. (Minaei et al. 2003a, b)

$$
\begin{aligned}
T_{LP(s)} &= \frac{I_{LP}}{I_{in}} = \frac{g_{m4}}{g_{m1}} \cdot \frac{B_1 B_2}{\Delta} \\
T_{BP(s)} &= \frac{I_{BP}}{I_{in}} = \frac{g_{m5}}{g_{m1}} \cdot \frac{B_1 s}{\Delta} \\
T_{HP(s)} &= \frac{I_{HP}}{I_{in}} = \frac{g_{m3}}{g_{m1}} \cdot \frac{s^2}{\Delta} \\
T_{BS(s)} &= \frac{I_{BS}}{I_{in}} = \frac{g_{m2}}{g_{m1}} \cdot \frac{s^2 + B_1 B_2}{\Delta} \\
\Delta &= s^2 + B_1 s + B_1 B_2
\end{aligned}
\tag{8.7}
$$

B_1 and B_2 are the gain-bandwidth product of the first and second OP-AMP, respectively. Also gmi ($i = 1, 2\ldots5$) is the transconductance gain of the i'th OTA. The parameters ω_o, ω_o/Q, and Q can be found as follows:

$$\omega_o = \sqrt{B_1 B_2}, \quad \frac{\omega_o}{Q} = B_1, \quad Q = \sqrt{\frac{B_2}{B_1}} \tag{8.8}$$

It can be seen that the gains of the low-pass, band-pass, high-pass, and band-stop responses are equal to $G_{LP} = g_{m4} = g_{m1}$, $G_{BP} = g_{m5}\ g_{m1}$, $G_{HP} = g_{m3} = g_{m1}$, and $G_{BS} = g_{m2} = g_{m1}$, respectively. Therefore, the gain of the filter responses can be electronically tuned by adjusting g_{ms} of the OTAs, without disturbing the parameters ω_o, ω_o/Q, and Q. Sensitivity analysis of this filter shows that

$$S_{B_1}^{\omega_o} = S_{B_2}^{\omega_o} = S_{B_2}^{Q} = -S_{B_2}^{Q} = \frac{1}{2}, S_{B1}^{\frac{\omega_o}{Q}} = 1, S_{B2}^{\frac{\omega_o}{Q}} = 0$$
$$S_{gm1}^{\omega_o} = S_{gm1}^{\frac{\omega_o}{Q}} = S_{gm1}^{\omega_o} = 0 \quad (i = 1 - 5) \tag{8.9}$$

which all are no more than unity.

Effects of the Non-idealities of the Op-Amps

In this section we discuss the effects of non-idealities of the op-amps on the transfer functions of the proposed multifunction filters. Considering the parasitic pole of op-amp, the open-loop gain $A(s)$ is assumed to be

$$A(s) = \frac{B}{s\left(1 + \frac{s}{\omega_{p2}}\right)} = \frac{B}{s(1 + \tau s)} \tag{8.10}$$

where ω_{p2} denotes the second pole of op-amp and $\tau = 1/\omega_{p2}$. In the frequency range $\omega < \omega_{p2}$ of our interest, $A(s)$ is approximated by

$$A(s) \cong \frac{B}{S}(1 - \tau s) \tag{8.11}$$

Simulation Results and Discussion

To verify the theoretical analysis, the proposed circuits have been simulated using SPICE circuit simulation program. For both of the proposed circuits shown in Figs. 8.2 and 8.3, the OTAs are realized with the CMOS implementation shown in Fig. 8.4. The transconductance of this OTA is given by

$$g_m = N\sqrt{2K_n' \cdot I_B \cdot \left(\frac{W}{L}\right)_1} \tag{8.12}$$

in which $(W/L)_1$ is the aspect ratio of the input transistors; K_n is the transconductance parameter for an n MOS in saturation. The parameter N is the current multiplication factor of the OTA. The circuit was supplied with symmetrical voltages of ±2.5 V. The dimensions of the NMOS and PMOS transistors are given in Table 8.1. The MOS transistors are simulated using MIETEC 1.2 μ model.

The op-amps are realized with the same CMOS topology given in Fig. 8.4 where the OTAs are used as unbuffered op-amps. Note that this is provided by the CMOS technology where the input resistances of the OTAs can be assumed to be infinite

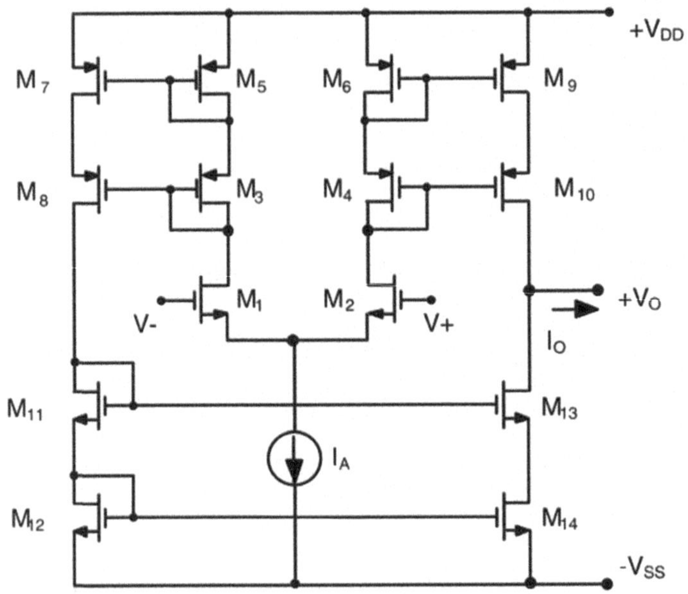

Fig. 8.4 Symmetrical CMOS cascode OTA. (Minaei et al. 2003a, b)

Table 8.1 Dimensions of MOS transistors in OTA structure

Transistor	*L* (μm)	*W* (μm)	Transistor	*L* (μm)	*W* (μm)
M1	3	60	M8	3	60
M2	3	60	M9	3	60
M3	3	12	M10	3	60
M4	3	12	M11	3	5
M5	3	12	M12	3	5
M6	3	12	M13	3	5
M7	3	60	M14	3	5

and therefore the outputs of the op-amps are not loaded by the next stage although the output resistance of the circuit is high. In this case, the output voltage of the unbuffered op-amp is related to the differential input voltage as

$$K_V = \frac{v_o}{v_+ - v_-} = gR_o \tag{8.13}$$

where g is the OTA transconductance and R_o is the OTA output resistance. The dominant pole, thus the gain-bandwidth product, is determined by a capacitor C_L connected to the output node and expressed as

$$f_d = \frac{1}{2\pi R_o C_L} \tag{8.14}$$

In this case, this load capacitor acts as a compensation capacitor. The output resistance of the op-amp is controlled by the biasing current. This provides the possibility of changing the gain-bandwidth product and the tuning of the filter by the current I_A. The dependence of the open-loop frequency response of the OP-AMP on the biasing current I_A is shown in Fig. 8.5. This property is important since integrated filters must be tuned. Current or voltage-controlled parameters make the filter suitable for on-chip tuning techniques. The first proposed filter is designed to realize a filter response with a natural frequency of $fo = 1.2$ MHz. To achieve this, the biasing currents of the CMOS OTAs are chosen as 200 μA and the biasing currents of the op-amps are taken as 100 μA. A compensation capacitor with a capacitance of $C_L = 50$ pF is connected between the outputs of the op-amps and the ground which results in a gain-bandwidth product (B) of 2π × 1.2 MHz. With these biasing current values the pole quality factor of the filter is obtained as $Q = 1$. Figure 8.6 shows the simulated frequency responses of the first proposed filter.

For the second proposed circuit shown in Fig. 8.3, the following setting has been selected to obtain the low-pass, band-pass, and high-pass and band-stop responses with natural frequency of $fo = 148.2$ kHz and a pole quality factor of $Qo = 0.707$: $B_1 = 2\pi \times 200$ kHz, $B_2 = 2\pi \times 109$ kHz, and $g_{mi} = 1$ mA/V ($i = 1$–5). Figure 8.7 shows the simulated frequency responses of the second proposed filter. As it can be seen from Figs. 8.6 and 8.7, the simulation results agree quite well with the theoretical analysis.

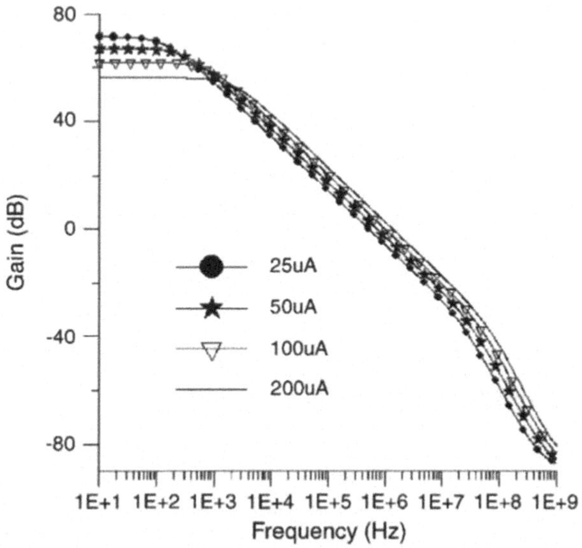

Fig. 8.5 Dependence of the open-loop frequency response of the op-amp on the biasing current. (Minaei et al. 2003a, b)

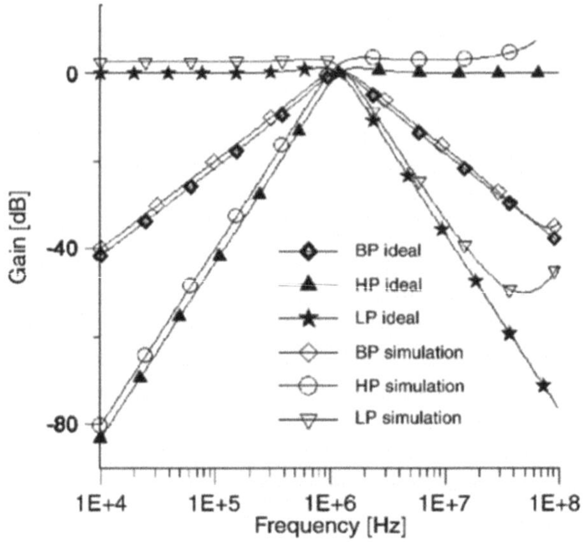

Fig. 8.6 Frequency responses of the first proposed multifunction filter. (Minaei et al. 2003a, b)

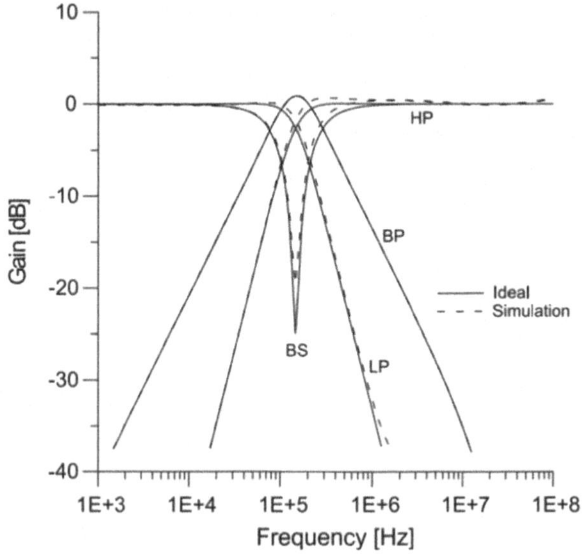

Fig. 8.7 Frequency responses of the second proposed multifunction filter. (Minaei et al. 2003a, b)

Concluding Remarks

New voltage-mode and current-mode multifunction filters using internally compensated op-amps and single-output OTAs have been proposed. The proposed voltage-mode circuit realizes low-pass, band-pass, and high-pass filter responses simultaneously, while the proposed current-mode circuit realizes low-pass, band-pass, high-pass, and band-stop filter responses simultaneously all at high output impedance, which enables easy cascading. The non-ideality effects of op-amps in the proposed circuits are investigated. The proposed filters enjoy the following advantages:

(i) Use of active elements without requiring any external passive components.
(ii) Provide electronically control of the parameters of filter through the bias current of the OTAs or op-amps, thus suitable for automatic on-chip tuning techniques.
(iii) Low active sensitivities.
(iv) Suitability for IC implementation.

Further Filter Applications Employing Active-Only Circuits

Further filter circuits employing active-only structures are shown in Figs. 8.8, 8.9, and 8.10 (Erdogan et al. 2004, 2006). The OTAs used in our circuits through Figs. 8.8, 8.9, and 8.10 are designed to give out current at the same phase angle as the differential input voltage.

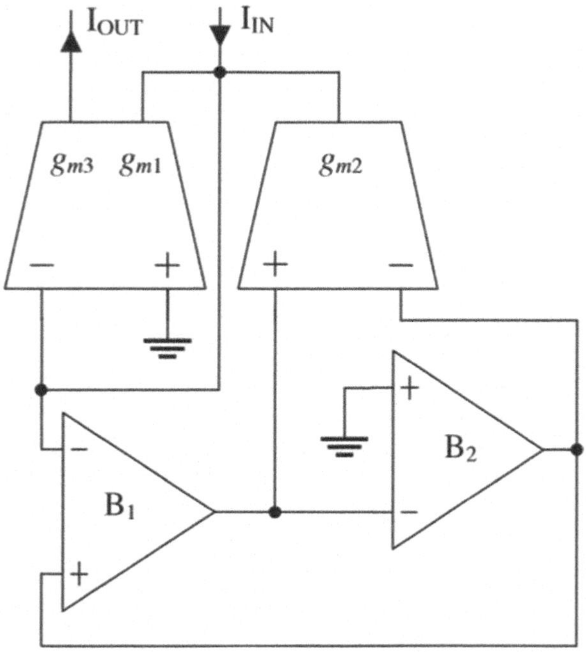

Fig. 8.8 Circuit 1, high-pass notch. (Erdogan et al. 2004)

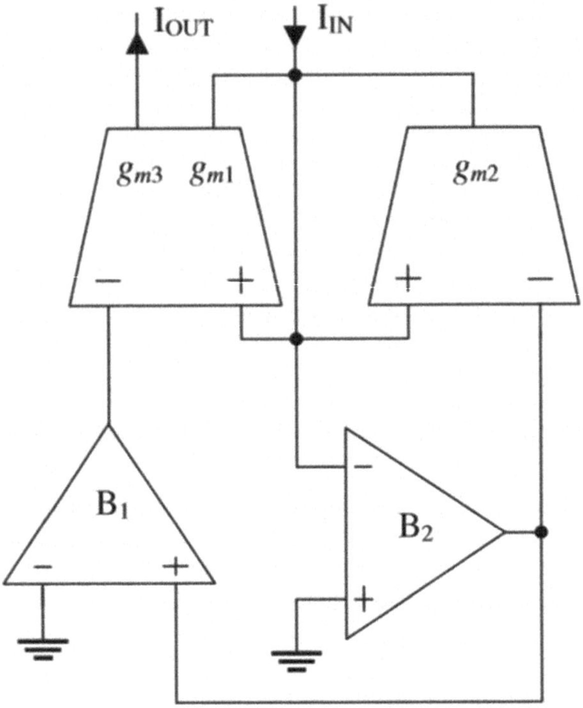

Fig. 8.9 Circuit 2, low-pass notch. (Erdogan et al. 2004)

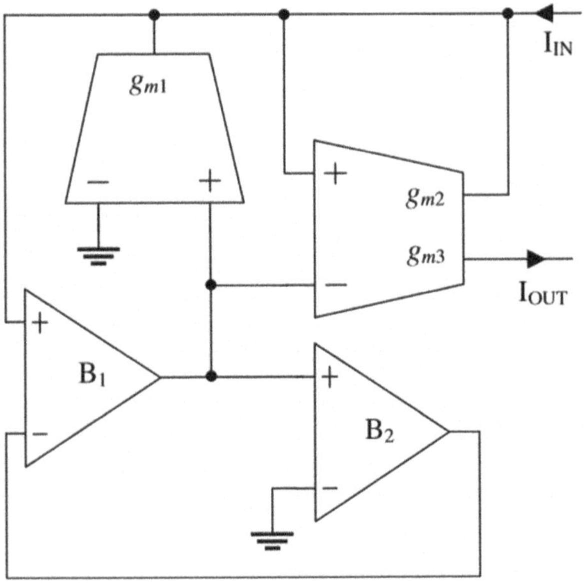

Fig. 8.10 Circuit 3, AP function. (Erdogan et al. 2004)

The op-amp, on the other hand, can be modeled by a single pole model, which can be written as B/s for the operating range of frequencies, that is to say, between the first and second poles in the frequency domain. This model of the op-amp is valid from a few kilohertz to a few megahertz. In this frequency range a bipolar monolithic OTA works as an ideal device.

The proposed filters. The proposed second-order filters are shown in Figs. 8.8, 8.9, and 8.10. Circuit 1 (shown in Fig. 8.8) implements the high-pass notch function, while circuit 2 (in Fig. 8.9) implements the low-pass notch function. Circuit 3 (in Fig. 8.10) realizes the all-pass function. Angular resonant frequency and quality factor, denoted by ω_P and Q_P, respectively, are independently adjustable by means of B_1, B_2, the gain-bandwidth products of both op-amps, assuming the open-loop gain $A(s)$ has the form of $A(s) = B/s$, and also by the g_m parameters belonging to the OTAs in the circuits. No component-matching constraints are imposed for the first two circuits unless one wants to have 0 dB gain for the responses at passbands. The filter transfer functions $T(s) = I\text{out}/I\text{in}$ are given by the following equations. Figure 8.8 (high-pass notch function, circuit 8.8):

$$\begin{aligned} T_{\text{HPN}}(s) = \frac{I_{\text{out}}}{I_{\text{in}}} &= \frac{A_{\text{HPN}}\left(s^2+\omega_0^2\right)}{s^2+\left(\omega_p/Q\right)s+\omega_p^2} \\ &= -\frac{g_{m3}(s^2+B_1B_2)}{g_{m1}s^2+g_{m2}B_1s+B_1B_2(g_{m1}+g_{m2})} \end{aligned} \tag{8.15}$$

$$\begin{aligned} \omega_p &= \sqrt{\frac{(g_{m1}+g_{m2})B_1B_2}{g_{m1}}} \quad \omega_0 = \sqrt{B_1B_2} \\ Q &= \frac{1}{g_{m2}}\sqrt{\frac{g_{m1}(g_{m1}+g_{m2})B_2}{B_1}} \quad A_{\text{HPN}} = -\frac{g_{m3}}{g_{m1}} \end{aligned} \tag{8.16}$$

The active sensitivities of the circuit are expressed as

$$\begin{aligned} S_{B_1}^{\omega_p} = S_{B_2}^{\omega_p} = S_{B_1}^{\omega_0} = S_{B_2}^{\omega_0} &= -S_{B_2}^{Q} = \frac{1}{2} \\ S_{gm1}^{\omega_p} = -S_{gm2}^{\omega_p} &= \frac{-g_{m2}}{2(g_{m1}+g_{m2})} \\ S_{gm1}^{Q} = S_{gm2}^{Q} &= \frac{2g_{m1}+g_{m2}}{2(g_{m1}+g_{m2})} \\ -S_{g_{m1}}^{A_{\text{HPN}}} = S_{g_{m3}}^{A_{\text{HPN}}} &= 1 \end{aligned} \tag{8.17}$$

Thus, all sensitivities are no more than unity.

Simulation Results, Discussion, and Design Example

To confirm the theoretical validity of the filters proposed in Fig. 8.8 (circuit 1), Fig. 8.9 (circuit 2), and Fig. 8.10 (circuit 3), a design example for each filter topology is performed. In the following section, the results obtained for circuit 1 are given as a chosen example. Simulations are performed according to the macromodels obtained by CMOS implementations (Laker and Sansen 1994) of op-amps and OTAs with the PSPICE simulation program. In addition, the layouts of OTAs and op-amps are drawn to construct the filter chips. Post-layout simulations are carried out with the parasitics extracted from the chip layout. The CMOS implementations of OTAs and op-amps are shown in Fig. 8.11.

The dimensions of the NMOS and PMOS transistors are given in Tables 8.2 and 8.3. Although we have used DO-OTAs in some of our circuits schematically, we have, for convenience, used single-output OTAs with parallel-connected inputs to simulate DO-OTAs. The filters are designed to realize a filter response with a resonant frequency f_0 of 417 kHz. To achieve this, the compensation capacitors of the op-amps are taken as 50 pF. A GBW of 417 kHz is obtained for both op-amps with these capacitances. If we were to build the filter for a higher frequency, we would need smaller capacitors and we could benefit from this property in IC

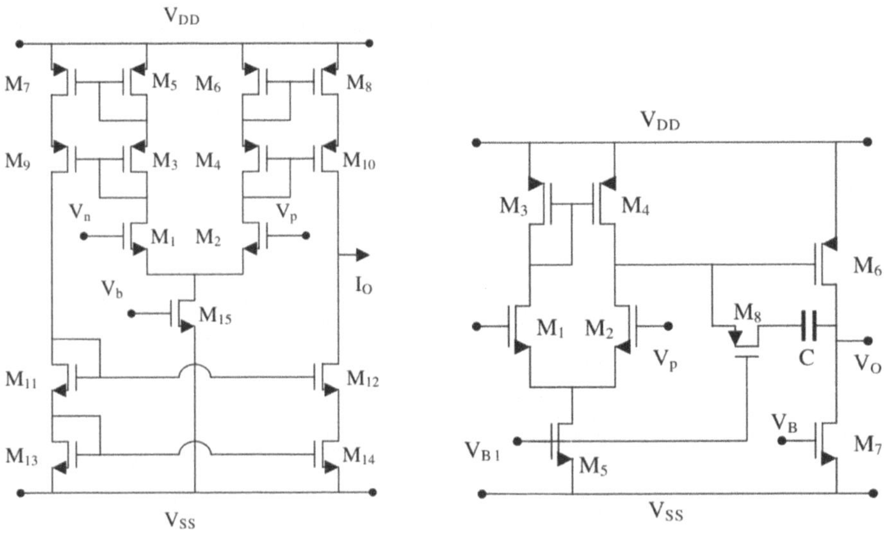

Fig. 8.11 CMOS OTA and CMOS op-amp OTA circuits. (Erdogan et al. 2004)

Table 8.2 Dimensions of transistors used in CMOS OTA

Transistor	L (μm)	W (μm)	Transistor	L (μm)	W (μm)
M1	3	60	M8	3	12
M2	3	60	M9	3	12
M4	3	12	M10	3	12
M5	3	12	M11	3	5
M6	3	12	M12	3	5
M7	3	12	M13	3	5
	3	12	M14	3	5
	3	12	M15	3	25

Table 8.3 Dimensions of transistors used in CMOS op-amp

Transistor	L (μm)	W (μm)	Transistor	L (μm)	W (μm)
M1	10	180	M5	32	12
M2	10	180	M6	10	392
M3	10	280	M7	10	232
M4	10	280	M8	10	39

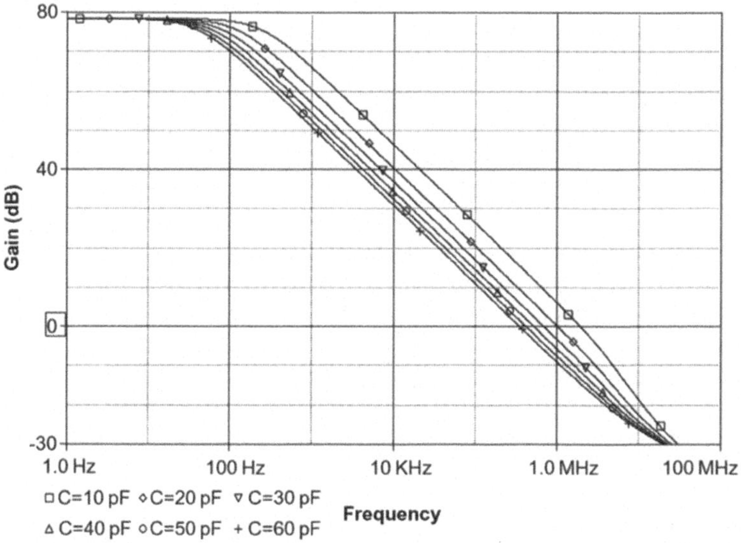

Fig. 8.12 Gain-bandwidth product dependence of CMOS op-amp on the compensation capacitor. (Erdogan et al. 2004)

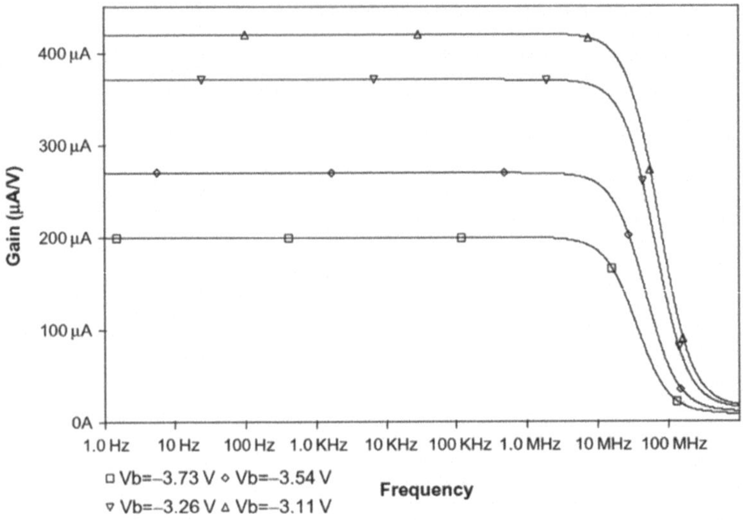

Fig. 8.13 Frequency response dependence of CMOS OTA on bias voltage. (Erdogan et al. 2004)

implementations. The dependence of the op-amp open-loop voltage gain on the biasing capacitor is obtained with the SPICE simulation program and is illustrated in Fig. 8.12. As can be seen from Fig. 8.12, the gain-bandwidth products are determined as 1.922 MHz, 1.026 MHz, 688 kHz, 519 kHz, 417 kHz, and 348 kHz for compensation capacitor values of 10 pF, 20 pF, 30 pF, 40 pF, 50 pF, and 60 pF, respectively, with a 78.3 dB gain at low frequencies. Figure 8.13 shows the transadmittance gain of OTAs for different bias voltages. It is observed that reducing gm reduces the bandwidth. Open-loop transconductance gains of 200 μA/V, 270 μA/V, 370 μA/V, and 420 μA/V are observed, respectively. With the chosen biasing voltages and compensation capacitance values, the pole quality factor of the filter is obtained as $Q_P = 1.41$. The zero frequency is found to be 416.87 kHz. Figure 8.14a, b shows respectively the simulated frequency and phase responses of the proposed filter. In addition, post-layout simulations were performed on the filter circuit with the parasitics extracted from the chip layout.

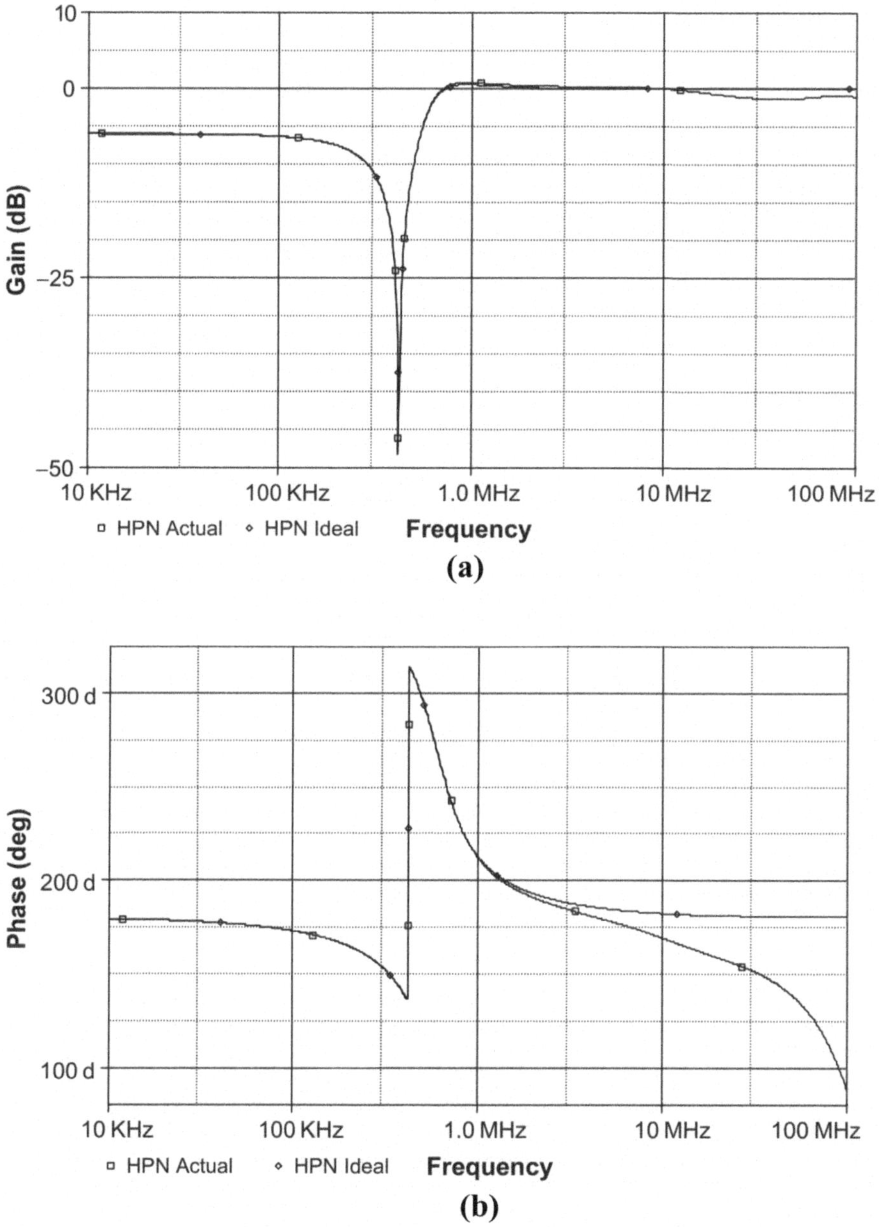

Fig. 8.14 (**a**) Frequency and (**b**) phase responses of the proposed HPN circuit. (Erdogan et al. 2004)

Concluding Remarks

This section reports actually three active-only current-mode filter structures. The filter structures are easily cascaded since they have high output impedances. This property can be easily used to achieve a circuit that can implement higher-order filters. Furthermore, current-mode operation is expected to overcome dynamic range limitation problems due to supply rails for cascaded sections. Simulation and post-layout simulation results (performed with parasitics extracted from the layouts) are included to verify the theory presented.

References

Abuelma'atti, M. T. and Ahzaher, H. A. (1998). Multi-function active-only current-mode filter with three inputs and one output. International Journal of Electronics, 85(4), 431–435.

Abuelma'atti, M. T. and Alzaher, H. A. (1997). Universal three input and one output current-mode filter without external passive elements. Electronics Letters, 33(4), 281–283.

Erdogan, E. S., Topaloglu, R. O., Kuntman, H. and Cicekoglu, O. (2004). New current–mode special function continuous-time active filters employing only OTAs and OPAMPs. International Journal of Electronics, 91(6), 345–359.

Erdogan, E. S., Topaloglu, R. O., Cicekoglu, O., Kuntman, H., and Morgul, A. (2006). Novel multiple function analog filter structures and a dual-mode multifunction filter International Journal of Electronics, 93 (9), 637–650.

Laker, K. R. and Sansen, W. M. (1994). Design of analog integrated circuits and systems (Vol. 1). New York: McGraw-Hill.

Minaei, S., Cicekoglu, O., Kuntman, H. and Türköz, S. (2003a). Electronically tunable, active only floating inductance simulation. International Journal of Electronics, 89(12), 905–912.

Minaei, S., Cicekoglu, O., Kuntman, H., Dündar, G. and Cerid, Ö. (2003b). New realizations of current-mode and voltage-mode multifunction filters without external passive elements. AEU-International Journal of Electronics and Communications, 57(1), 63–69.

Roberts, G. W. and Sedra, A. S. (1991). A general class of current amplifier-based biquadratic filter circuits. In 1991., IEEE International Symposium on Circuits and Systems (pp. 1821–1824). IEEE.

Singh, A. K., and Senani, R. (1998). Low-component count active-only immittances and their application in realizing simple multifunction biquads. Electronics Letters, 34, 718–719.

Topaloglu, R. O., Kuntman, H. and Cicekoglu, O. (2001), Novel Notch And Bandpass Filter Structures Using OTAs And OPAMPs. The 2nd International Conference on Electrical and Electronics Engineering (Electronics), pp.63–67, Bursa, 7–11 November 2001.

Tsukutani, T., Higashimura, M., Sumi, Y. and Fukui, Y. (2000a). Electronically tunable current-mode active-only biquadratic filter. International Journal of Electronics, 87(3), 307–314.

Tsukutani, T., Higashimura, M., Sumi, Y. and Fukui, Y. (2000b). Voltage-mode active-only biquad. International Journal of Electronics, 87(12), 1435–1442.

Tsukutani, T., Higashimura, M., Kinugasa, Y., Sumi, Y., and Fukui, Y. (1999). A general class of current-mode active-only filters. Proceedings of the International Technical Conference on Circuits/Systems, Computers and Communications, ITC-CSCC'99, Niigata, Japan, 297–300.

Tsukutani, T., Higashimura, M., Takahashi, N., Sumi, Y. and Fukui, Y. (2001). Novel voltage-mode biquad without external passive elements. International Journal of Electronics, 88(1), 13–22.

Tsukutani, T., Ishida, M., Tsuiki, S. and Fukui, Y. (1996). Current-mode biquad without external passive elements. Electronics Letters, 32(3), 197–198.

Chapter 9
MOS-Only and MOS-C Circuits

Basic Concept: Fundamentals of MOS-Only and MOS-C Circuits

The world of analog IC, the design of which does not include external passive components such as resistances, capacitances, or inductances, has attracted a great deal of interest over the last two decades. This kind of circuits, which consist of only MOSFETs, is called as MOSFET-only (MOS-only) structures. Rapid growth of demands on designs having low cost, low power, and small chip area requires new innovative solutions including MOS-only and MOS-C structures. In this respect, it is important to present new opportunities in MOS-only schemes to improve previously proposed structures in this area. In this chapter, novel MOS-only and MOS-C structures will be presented with promising performances regarding operating frequency range, power consuming, low cost, and small chip area issues.

A MOS-only active filter makes use of the intrinsic parameters, gate-to-source capacitance (C_{gs}), and transconductance (g_m) of a MOS transistor. Even though many filter designs, which incorporate active elements, are reported in the literature, they bring big chip area and restricted operating frequency owing to including high numbers of transistors. In this respect, MOS-only filters can be used to overcome these limits due to the fact that they enable to operate at higher frequencies and have smaller chip areas. However, the circuits suffer from linearity problems coming from MOSFET's intrinsic capacitances. To eliminate the nonlinearity issues, MOS-C filters include external capacitors to increase small valued gate-to-source capacitances. In this manner, the proposed MOS-C filters play a critical role to improve analog building blocks' performances (Fig. 9.1).

Before introducing the proposed circuits, let us emphasize that there are different solutions to the transistor-level sizing problem found in the literature, subdivided into three main categories (Daems et al. 2003):

H. Kuntman, D. Özenli, *Trends in Circuit Design for Analog Signal Processing*,
Analog Circuits and Signal Processing,
https://doi.org/10.1007/978-3-030-96836-6_9

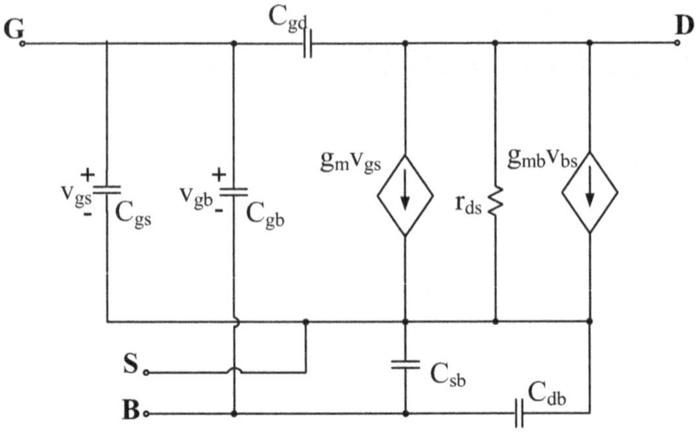

Fig. 9.1 Small signal model of MOSFETs in the proposed circuits (bulk source is shorted in general case and bulk capacitances are ignored). (Allen and Holberg 2002)

- *Knowledge-based sizing*: the knowledge of the designer is captured and modeled as a design plan or a set of rules, which are then executed during sizing.
- *Equation-based sizing*: the circuit's behavior is captured in a model that consists of a set of equations. These equations can be derived using automated or computer-assisted symbolic analysis. Sizing is performed using different optimization algorithms.
- *Simulation-based sizing*: the circuit's behavior is obtained by numerical transistor-level SPICE simulations and sizing is realized using optimization techniques.

In the last one, SPICE simulation-based algorithms can be organized to identify fundamental parameters of MOSFETs. Therefore, it will be possible to improve calculation accuracy and reduce design time by this way. Equation-based algorithms reduce computational burden and present fast optimization algorithms, whereas simulation-based ones bring more complexity with longer executional time. On the other hand, they include higher computational accuracy than the equation-based algorithms. In this chapter, to take advantage of the both techniques, most of the proposed circuits make use of simulation-based circuit sizing methods including polynomial regression-based algorithms.

Significant restriction is encountered when using conventional formulas in the manual design of analog circuits. Our proposed automation flow focuses on saturation region's small signal equations in order to eliminate that limits by means of polynomial models. To show manual design limits, Eqs. (9.1) and (9.2) are given as follows in conventional approach:

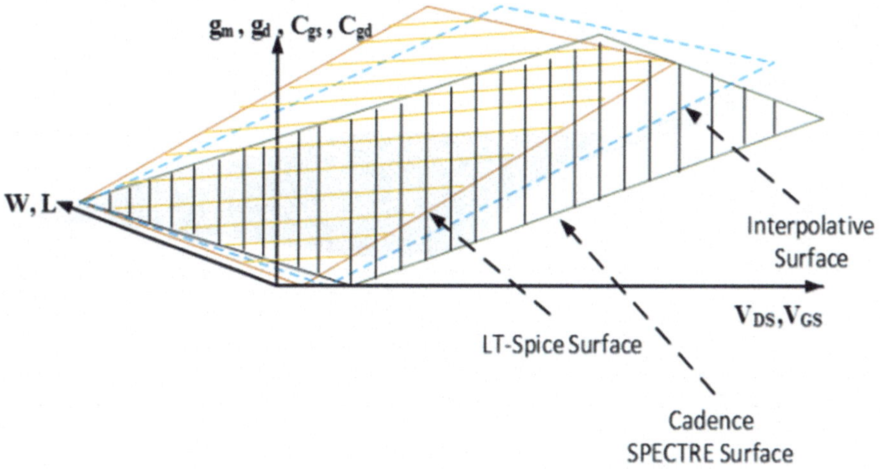

Fig. 9.2 Relaxation of simulation-based data coming from LT-SPICE and CADENCE with linear interpolation (Ozenli et al. 2018a, b).

$$C_{\mathrm{gs}} \approx \frac{2}{3} C_{\mathrm{ox}} WL \quad , \quad C_{\mathrm{gd}} \approx 0 \tag{9.1}$$

$$g_{\mathrm{m}} \approx \sqrt{2\mu C_{\mathrm{ox}} \left(\frac{W}{L}\right) I_{\mathrm{D}}} \quad , \quad g_{\mathrm{ds}} \approx \lambda I_{\mathrm{D}} \tag{9.2}$$

Design cycles considering conventional approach increase due to no direct relationship between fundamental design parameters ($V_{\mathrm{GS}}, V_{\mathrm{DS}}, \frac{W}{L}$ etc.) and small signal parameters. It is required to construct a robust link between design parameter terms and small signal formulas with high accuracy as much as possible. Equations (9.1) and (9.2) are regenerated based on LT-SPICE and CADENCE-SPECTRE data using approximately 350–400 samples of simulations as shown in Fig. 9.2. Regenerated formulas based on polynomial regression are reorganized with linear interpolation to alleviate mean error between simulator results and proposed automation flow. In the simulations, transistors' bulk and source terminals are shorted in order to avoid impact ionization. Furthermore, C_{ds} and junction capacitances consisting of bottom-plate and sidewall capacitances of the MOSFET are much more smaller than C_{gs} and C_{gd}. Especially, junction capacitances bring at the least 10 times smaller effect than C_{gs} to the center frequency or quality factor of the MOS-C or MOS-only filters in most cases. In this work, we include only g_{m}, g_{ds}, C_{gs}, C_{gd}, and I_{ds} in our automation flow algorithm and we avoid long computational cycles and high complexity problems. In this respect, polynomial regressive models can be given for C_{gs} and C_{gd} after linear interpolation as follows:

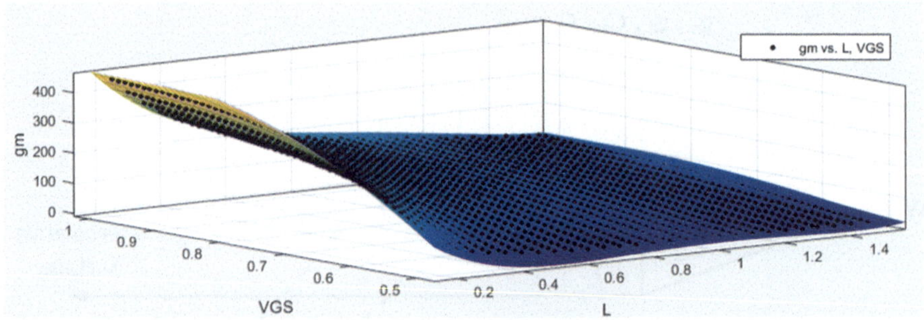

Fig. 9.3 g_m (μS) variation of NMOS device over channel length (μm) and V_{GS} (V) voltage (in this regression, $V_{DS} = 0.5$ V and dots are SPICE and CADENCE interpolated data, whereas the colored surface is a polynomial regressive model). (Ozenli et al. 2018a, b)

$$C_{gs} \approx 0.66 \cdot W \cdot \left(k_1 + k_2 L + k_3 V_{GS} + k_4 V_{GS} L + k_5 {V_{GS}}^2\right) \tag{9.3}$$

$$C_{gs} \approx 0.66 \cdot W \cdot \left(m_1 + m_2 L + m_3 V_{DS} + m_4 V_{DS} L + m_5 {V_{DS}}^2\right) \tag{9.4}$$

$$C_{gd} \approx 0.312 \cdot W \cdot \Big(n_1 + n_2 L + n_3 V_{DS} + n_4 L V_{DS} + n_5 {V_{DS}}^2 + n_6 L {V_{DS}}^2 + n_7 {V_{DS}}^3 + n_8 L {V_{DS}}^3 + n_9 {V_{DS}}^4\Big) \tag{9.5}$$

$$L \in [L_{\min}, L_{\min} + \Delta L \cdot 1, L_{\min} + \Delta L \cdot 2, L_{\min} + \Delta L \cdot 3, \ldots, L_{\max}],$$
$$W \in [W_{\min}, W_{\min} + \Delta W \cdot 1, W_{\min} + \Delta W \cdot 2, W_{\min} + \Delta W \cdot 3, \ldots, W_{\max}].$$

where L_{min} is minimum channel length as 0.18 μm, ΔL denotes step size as 0.01 μm, and L_{max} is 2 μm. Also, W_{min} is minimum channel width as 2 μm, ΔW is step size as 1 μm, and W_{max} is equal to 600 μm. In Eqs. (9.3) and (9.4), C_{gs} is defined by polynomial regressive models so that Eq. (9.3) specifies constant V_{GS} case, while Eq. (9.4) includes constant V_{DS} situation. Therefore, designers can prefer to reach to design constraints by keeping V_{GS} voltage of the devices constant. In the equations, "k," "n," and "m" are constants with different units coming from surface fitting. These constants are selected between ±0.1 and ±1.5. In addition to that, g_m and g_{ds} are reformulated as illustrated in Fig. 9.3 (Ozenli, 2018; Ozenli and Kuntman, 2017, 2018; Ozenli et al. 2018a, b).

The reformulations are given in terms of design parameters as follows:

$$g_m \approx W \cdot \left(\begin{array}{l} \left(\sum_{i=0}^{2} \sum_{j=0}^{2} p_{k,i,j} L^i {V_{GS}}^j\right) + \left(\sum_{i=0}^{1} p_{k,3i,3-3i} L^{3i} {V_{GS}}^{3-3i}\right) + \\ \left(\sum_{i=0}^{1} p_{k,4i,4-4i} L^{4i} {V_{GS}}^{4-4i}\right) + \left(\sum_{i=0}^{1} p_{k,2i+1,3-2i} L^{2i+1} {V_{GS}}^{3-2i}\right) \end{array} \right)$$
$$\boldsymbol{a} = \text{linspace}(0, V_{DD}, 9),\ \boldsymbol{a}(k) < V_{DS}(k) \le \boldsymbol{a}(k+1),\ k = 1, 2....8. \tag{9.6}$$

$$
\begin{aligned}
g_{ds} &\approx \frac{W}{L}\cdot\left(c_{i1}+c_{i2}V_{DS}+c_{i3}V_{GS}+c_{i4}{V_{DS}}^2+c_{i5}V_{GS}V_{DS}+c_{i6}{V_{DS}}^3\right.\\
&\left.+c_{i7}{V_{DS}}^2V_{GS}+c_{i8}{V_{DS}}^4+c_{i9}{V_{DS}}^3V_{GS}\right)\\
L &\in [L_{\min}, L_{\min}+\Delta L\cdot 1, L_{\min}+\Delta L\cdot 2, L_{\min}+\Delta L\cdot 3, \ldots, L_{\max}],\\
W &\in [W_{\min}, W_{\min}+\Delta W\cdot 1, W_{\min}+\Delta W\cdot 2, W_{\min}+\Delta W\cdot 3, \ldots, W_{\max}].\\
\boldsymbol{b} &= \mathrm{linspace}(0, V_{DD}, 9),\ \boldsymbol{b}(i) < V_{DS}(i) \le \boldsymbol{b}(i+1),\ i=1,2\ldots.8.
\end{aligned}
\tag{9.7}
$$

Equations (9.6) and (9.7) are calculated by using piecewise polynomial regressive model, whose coefficients are changed for each interval of V_{DS}. In this point of view, "c" and "p" are constants with different units having normalized values between ± 0.1 and ± 5. It should be noted that **linspace** is a built-in function of MATLAB and it can be described as follows:

$$
\begin{aligned}
\boldsymbol{a} &= \mathrm{linspace}(a_1, a_n, h)\\
&= \left\{\begin{array}{l}
\boldsymbol{a} = [a_1, a_2, a_3, \ldots, a_n], a \in \mathbb{R}^{1xn},\quad h, n \in Z^+\\
a_{i+1} - a_i = \dfrac{(a_n - a_1)}{h-1},\ i \in [1, 2, 3, \ldots, n]
\end{array}\right\}
\end{aligned}
\tag{9.8}
$$

In particular, bias circuit dimensioning in the analog design requires a great deal of computation and design cycle chains. To reduce their heavy computational cost and accuracy problems arising from standard square law model, I_{DS} must be reformulated. By using polynomial regressive models, I_{DS} may be calculated as shown in Eq. (9.9). Here, "d"s are regression constants and regenerated for each interval of V_{DS} with different unities and selected between $\pm 0.1 \times 10^{-5}$ and $\pm 5 \times 10^{-5}$.

$$
\begin{aligned}
I_{DS} &\approx \frac{W}{L}\cdot\left(d_{i1}+d_{i2}V_{GS}+d_{i3}{V_{GS}}^2+d_{i4}{V_{GS}}^3+d_{i5}{V_{GS}}^4\right)\\
\boldsymbol{e} &= \mathrm{linspace}(0, V_{DD}, 9), \boldsymbol{e}(i) < V_{DS}(i) \le \boldsymbol{e}(i+1),\ i=1,2\ldots.8.
\end{aligned}
\tag{9.9}
$$

The aforementioned polynomial regressive formulas allow the designer to use minimum channel length for the transistors, when chip area is very much limited. Also, it enables to increase the channel length in order to avoid channel length modulation or some small channel effects. In this regard, the proposed automation flow is illustrated in Fig. 9.4. The proposed automation flow algorithm runs as follows:

1. (a) Random netlist generation is performed by means of randomly changing of node numbers of the circuit components. In this step, it must be clarified which transistors are dominant ones and whose gate-to-source capacitances are much bigger than the others. It should be noted that the simplified AC model consisting of g_m, C_{gs}, and/or external C is used for the transfer functions. In this view, the obtained transfer functions between output ports and input are calculated and compared with standard second-order filter templates.

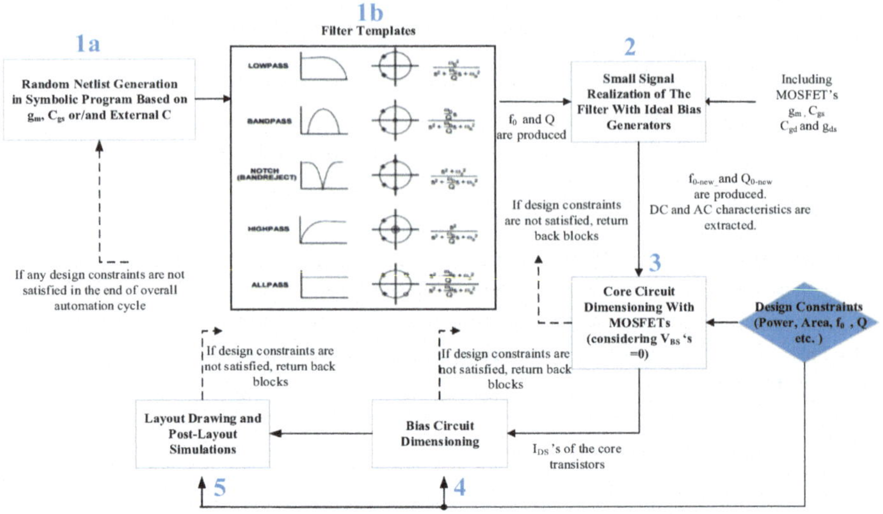

Fig. 9.4 Proposed MOSFET-only design automation flowchart (Part I). (Ozenli et al. 2018a, b)

(b) In comparison between generated transfer functions of 1a and templates, fundamental constraints are extracted in order to discard some components from numerator and/or denominator terms, which causes instability in SPICE simulation with ideal elements.

2. In the first step of the circuit dimensioning, small signal realization of the filter's core structure makes use of fundamental parameters of the devices (g_{m}, g_{ds}, C_{gs}, and C_{gd}). In this manner, detailed transfer functions are obtained with non-ideality effects. At the end of small signal realization of the filter with ideal bias generators, Routh–Hurwitz table is generated in order to check the stability of the core circuit. In addition to that, DC behavior of the circuit is examined by keeping all transistors in the saturation region. At the end of the second step, the relationship between node voltages, terminal voltages, and transistors is generated using the SPICE netlist. In this way, bias voltages can be produced.
3. After reformulation of the f_0 and Q (center frequency and quality factor) in the third block, core circuit is sized with real MOSFETs by taking into account designer-defined constraints such as power, area, operating speed, etc. In this step, bias voltages and transistor dimensions are extracted by using the aforementioned reformulations of the small signal parameters. Designer constraints also may be unified in one cost function to reach optimal solution or solution set as follows:

$$P_{\mathrm{cost}} = \alpha \cdot \mathrm{Power} + \beta \cdot \mathrm{Area} + f_0^{\gamma} + Q^{\delta} \quad \alpha, \beta \in \mathbb{R}^{+}, \gamma, \delta \in \mathbb{R}^{-} \tag{9.10}$$

At the end of this block, transistor types (NMOS or PMOS) of the core circuit which satisfy the user-defined design constraints are determined. After all solutions are obtained for each of the transistor types, the algorithm decides a unique solution (or solution list) which minimizes the cost function as given in Eq. (9.10).

4. Dimensioned core circuit in good agreement with design constraints is sent to the next block to be constructed as bias circuit by using core transistor currents. It should be noted that DC currents of transistors in the bias circuit can be easily calculated by collecting currents of the core transistors in the different branches. In the proposed model, bias transistors also should produce the required currents to satisfy the design constraints and DC conditions, which are produced in the second block. In this point of view, bias transistors' aspect ratios and terminal voltages can be calculated from Eq. (9.9) and DC conditions defined in the second block. (Checking of saturation condition for each MOSFET is made in this block and unsatisfied solutions are removed.)
5. At the end of Part I, post-layout simulations are performed with Monte-Carlo simulations containing process variations and transistors' mismatches. If design constraints defined in the third block are not satisfied in this step, bias circuit dimensioning is repeated with different transistor dimensions. If design constraints are still not met, automation flow returns back to the third and second blocks in order to re-generate new solutions for the same user-defined cost function and constraints. Overall, the loop is broken by either changing design constraints or repeating netlist generation in the first block. This step completes Part I of the proposed design automation flow.

As depicted in Fig. 9.5, the proposed automation model facilitates layout placement and drawing a skeleton. When bulk and source terminals are shorted in NMOS transistors, inevitably large chip area comes to the layout due to the isolated NMOS areas in the triple well technology. Even though the overall automation model is

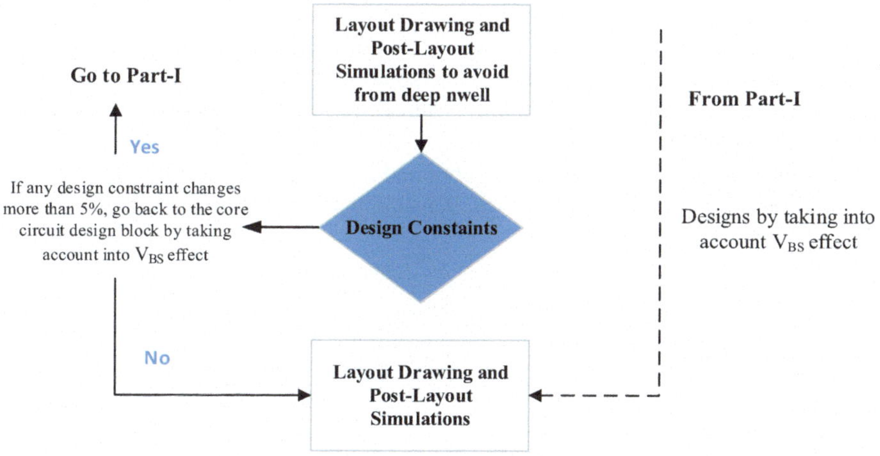

Fig. 9.5 Proposed MOSFET-only design automation flowchart (Part II). (Ozenli et al. 2018a, b)

processed with all bulks and sources shorted to eliminate the aforementioned impact ionization and body effect, to make the chip area as small as possible, NMOS's bulks and sources are disconnected where deep n-well is required. This mechanism is examined by controlling whether the design specs are satisfied or not by allowing 5% variation over target design constraints. If the disconnection of bulk and source terminals of NMOS causes to exceed 5% variation in designed circuit performance, change of V_{TH} is detected by operating point analysis in SPICE due to the body effect. In this regard, the aforementioned polynomial regressive models are changed by correction terms as follows in order to catch the same DC points:

$$C_{gs} \approx 0.66 \cdot W \cdot (k_1 + k_2L + k_3(V_{GS} + \Delta V_{TH}) + k_4(V_{GS} + \Delta V_{TH})L + k_5(V_{GS} + \Delta V_{TH})^2) \tag{9.11}$$

$$g_m \approx W \cdot \left(\begin{array}{l} \left(\sum_{i=0}^{2}\sum_{j=0}^{2} p_{k,i,j}L^i(V_{GS} + \Delta V_{TH})^j\right) + \left(\sum_{i=0}^{1} p_{k,3i,3-3i}L^{3i}(V_{GS} + \Delta V_{TH})^{3-3i}\right) + \\ \left(\sum_{i=0}^{1} p_{k,4i,4-4i}L^{4i}(V_{GS} + \Delta V_{TH})^{4-4i}\right) + \left(\sum_{i=0}^{1} p_{k,2i+1,3-2i}L^{2i+1}(V_{GS} + \Delta V_{TH})^{3-2i}\right) \end{array} \right) \tag{9.12}$$

According to the design flow of Part II, other aforementioned terms such as I_{DS}, g_{ds} and C_{gd} are corrected as well. At the end of this cycle, overall dimensions are revised with bias currents. After that, all of the circuit components are ready to be laid out.

Exemplary Circuit I

In Fig. 9.6, the core structure of the first proposed circuit is illustrated. This current-mode first example consists of three transistors and MOSFET capacitance (MOS-cap). If C_{mos}, $C_{gs2} >> C_{gs1}$, C_{gs3}, Eqs. (9.13), (9.14), and (9.15) can be obtained. In the equations, C_{mos} denotes capacitance of the MOS-cap. In this regard, small signal model of the core structure is shown in Fig. 9.7, whereas the complete circuit is given in Fig. 9.8.

The following transfer functions are obtained by considering a simplified form of the small signal model depicted in Figs. 9.7, 9.9, 9.10, and 9.11.

$$\frac{I_{BP}}{I_{IN}} = -\frac{sC_{gs2}g_{m1}}{s^2C_{mos}C_{gs2} + s(C_{gs2}g_{m1} + C_{gs2}g_{m3}) + g_{m2}g_{m3}} \tag{9.13}$$

$$\frac{I_{LP}}{I_{IN}} = \frac{g_{m2}g_{m3}}{s^2C_{mos}C_{gs2} + s(C_{gs2}g_{m1} + C_{gs2}g_{m3}) + g_{m2}g_{m3}} \tag{9.14}$$

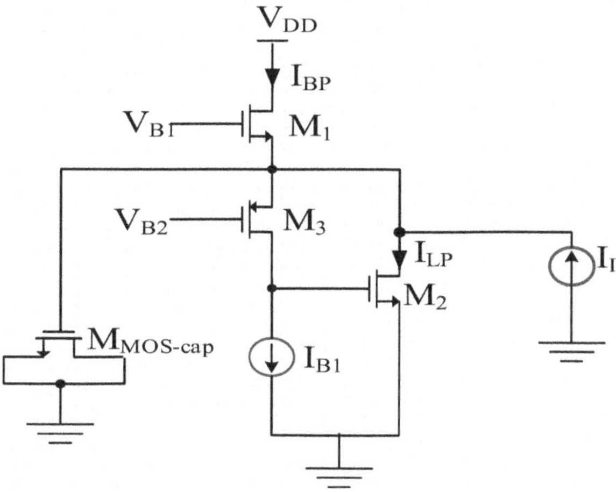

Fig. 9.6 The core structure of the proposed agile CM filter. (Ozenli et al. 2018a, b)

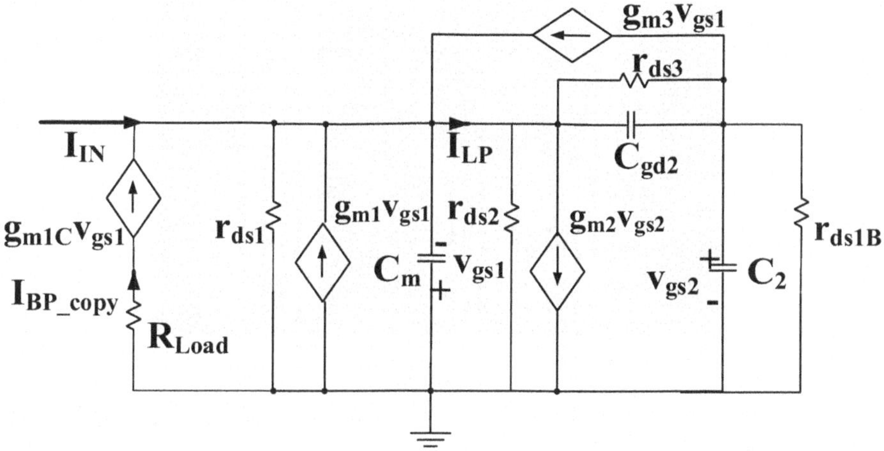

Fig. 9.7 AC small signal equivalent circuit of the proposed CM-BP filter. (Ozenli et al. 2018a, b)

$$\omega_0 = \sqrt{\frac{g_{m2}g_{m3}}{C_{mos}C_{gs2}}}, \quad Q = \sqrt{\frac{C_{mos}g_{m2}g_{m3}}{C_{gs2}(g_{m1}+g_{m3})^2}} \tag{9.15}$$

To indicate promising performance of the first design at the end of proposed automation model, total harmonic distortion (THD) analysis is given in Fig. 9.12a. With up to 100 μA peak-to-peak input current, the linearity of the BP output remains within acceptable limits. Also, noise simulation in Fig. 9.12b shows that the proposed first design has very low input referred noise because it has only 31pA/$\sqrt{Hz}$ of input referred noise at 213 MHz. In addition, Monte-Carlo analysis is depicted in Fig. 9.13.

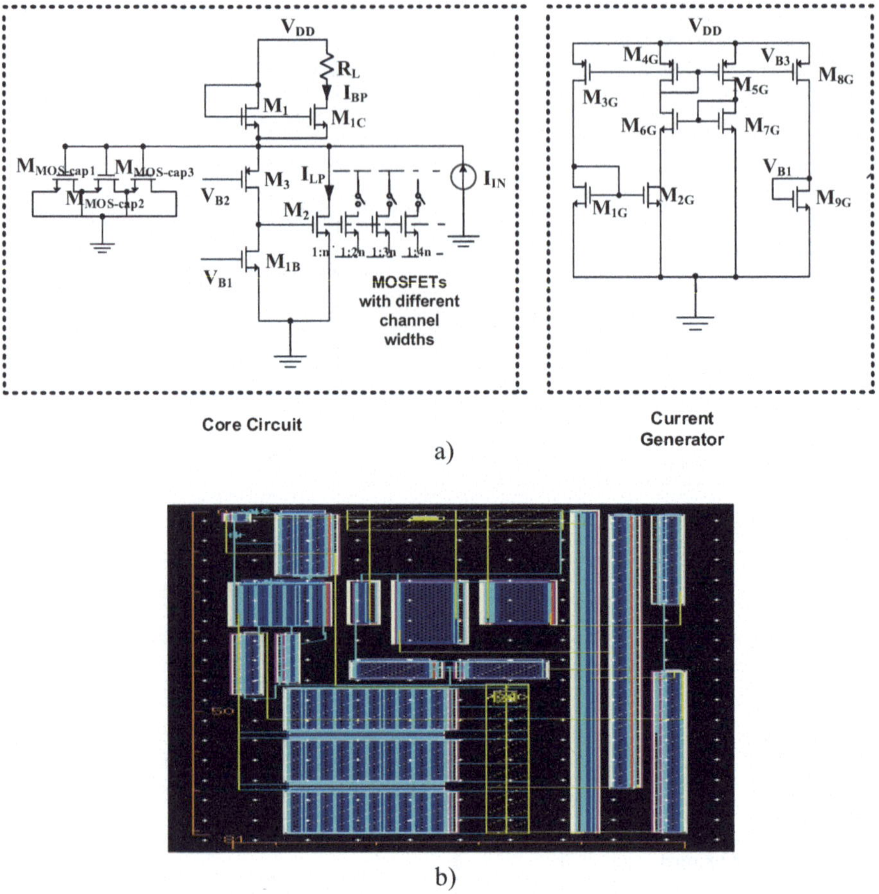

Fig. 9.8 Complete circuit of the first design (**a**) in schematic and (**b**) layout (3420 μm^2). (Ozenli et al. 2018a, b)

Temperature behavior over an extended range with overall mismatches and process variations' effects is depicted in Fig. 9.13, where the center frequency changes between 188 MHz and 225 MHz. This variation can be reduced by advanced current sources and temperature compensation techniques (Allen and Holberg 2002; Razavi 2017). In particular, the center frequency of the filter has a distribution with a mean of 207 MHz and the standard deviation value of 16 MHz over 1000 random samples of the Monte-Carlo analysis including overall mismatch and process variations' effects at 27 °C in part (b) of this figure.

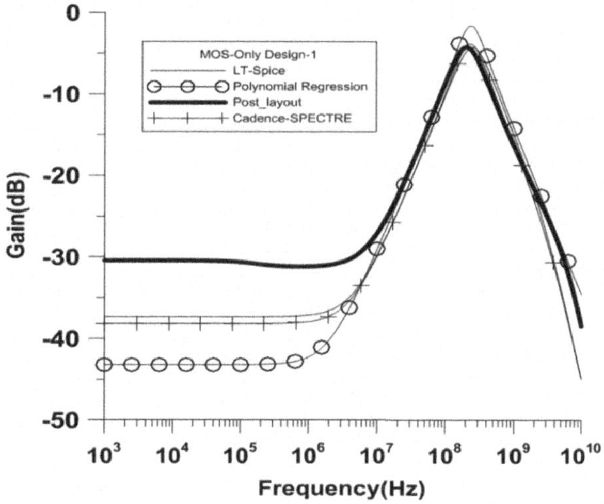

Fig. 9.9 Simulation results in comparison with the proposed automation model for the first circuit. (Ozenli et al. 2018a, b)

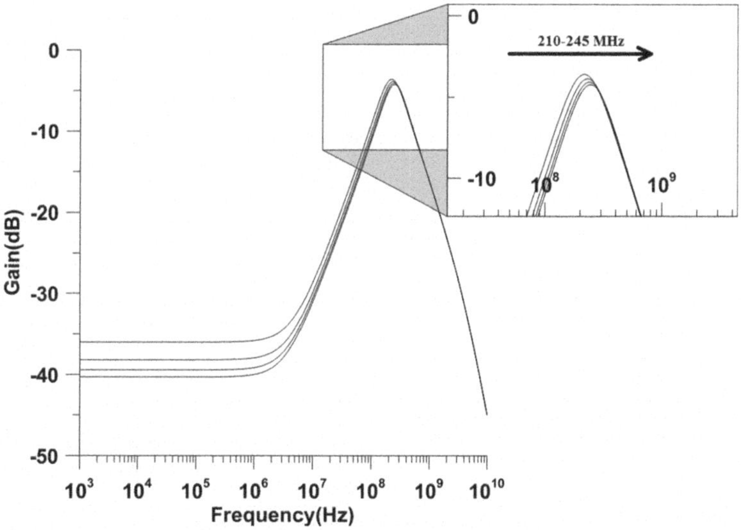

Fig. 9.10 Tunability of the center frequency of the first design. (Ozenli et al. 2018a, b)

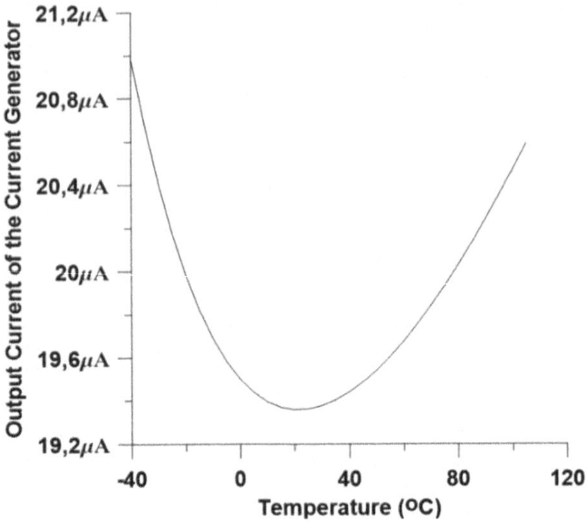

Fig. 9.11 Output current of the current generator in Figure 9.11 versus temperature. (Ozenli et al. 2018a, b)

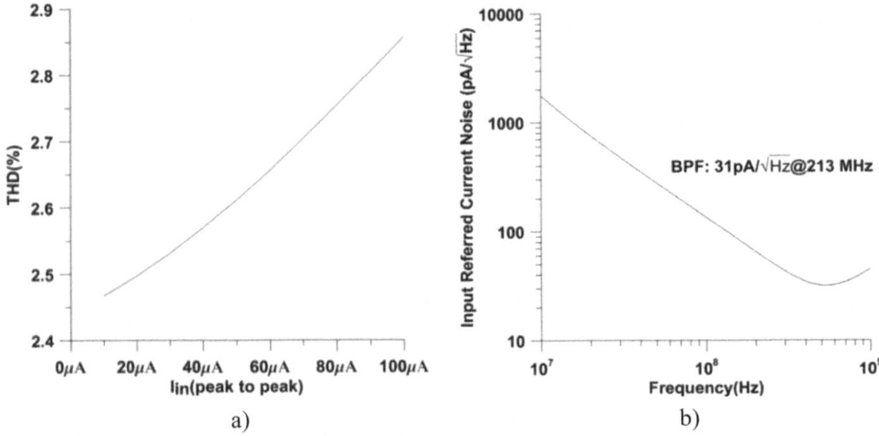

Fig. 9.12 (**a**) THD performance of the first design with regard to input current. (**b**) Input referred noise versus frequency between 100 MHz and 1 GHz. (Ozenli et al. 2018a, b)

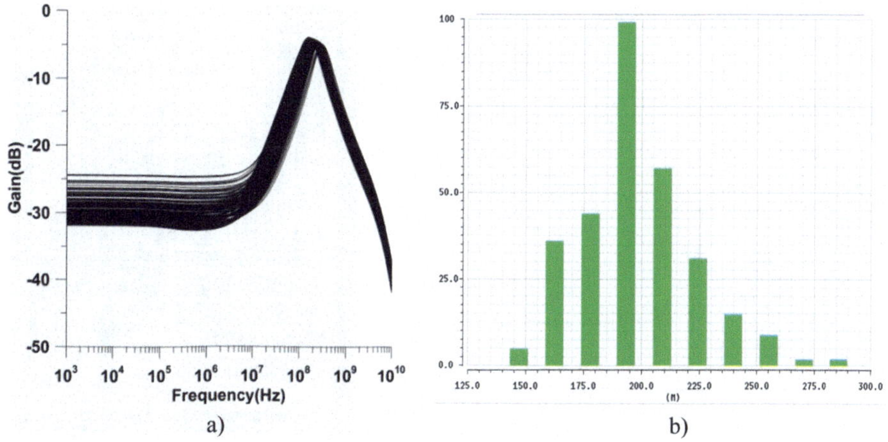

Fig. 9.13 Monte-Carlo analysis of the proposed structure. (**a**) BP frequency response variation from −40 °C to 120 °C with process variations and mismatches. (**b**) Monte-Carlo analysis at the corner of 27 °C. (Ozenli et al. 2018a, b)

Fig. 9.14 The core structure of the proposed second agile CM filter with BP/LP outputs. (Ozenli et al. 2018a, b)

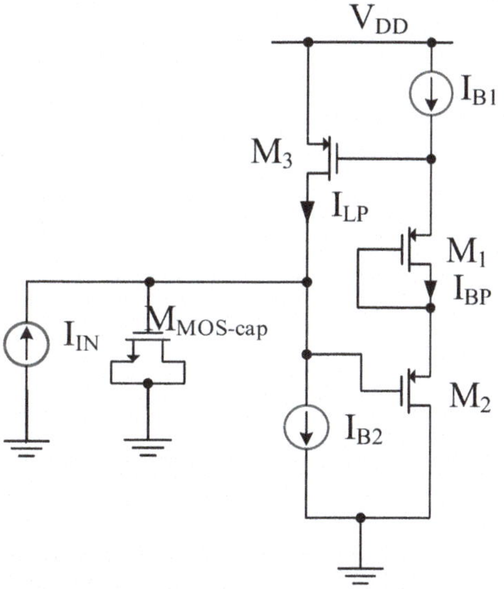

Exemplary Circuit II

The second example circuit is given in Fig. 9.14 with LP and BP outputs. This structure has agility property due to the feedback from LP terminal to input in order to control the operating frequency with the concept illustrated in the presented work

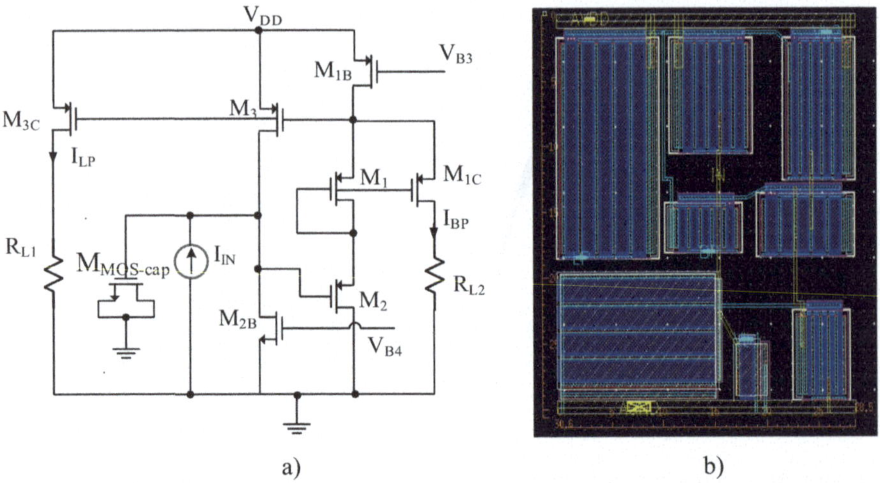

Fig. 9.15 Complete circuit of the second exemplary design (**a**) in schematic and (**b**) layout (852 μm²). (Ozenli et al. 2018a, b)

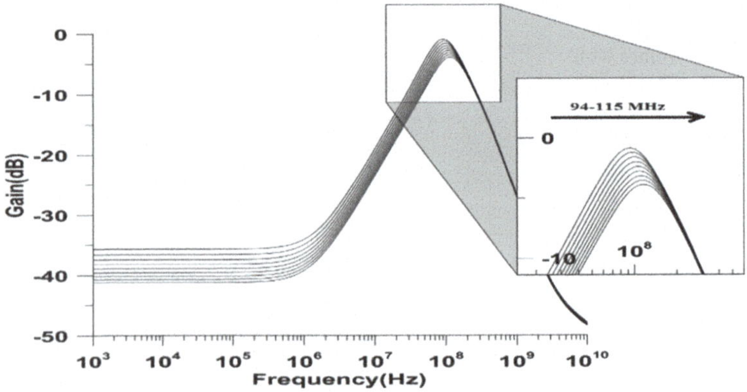

Fig. 9.16 Tunability of the center frequency of the second design. (Ozenli et al. 2018a, b)

(Lakys and Fabre 2013). The novel structure gives an opportunity to be used in FM band applications or aforementioned IF band-filtering implementations as seen in the first example design.

If C_{mos}, $C_{gs3} >> C_{gs1}$, C_{gs2}, Eqs. (9.16), (9.17), and (9.18) can be obtained. In these equations, C_{mos} represents capacitance of the MOS-cap (Figs. 9.15 and 9.16).

$$\frac{I_{BP}}{I_{IN}} = -\frac{sC_{gs3}g_{m1}g_{m2}}{s^2 C_{mos}C_{gs3}(g_{m2}+g_{m1}) + sC_{mos}g_{m1}g_{m2} + g_{m1}g_{m2}g_{m3}} \tag{9.16}$$

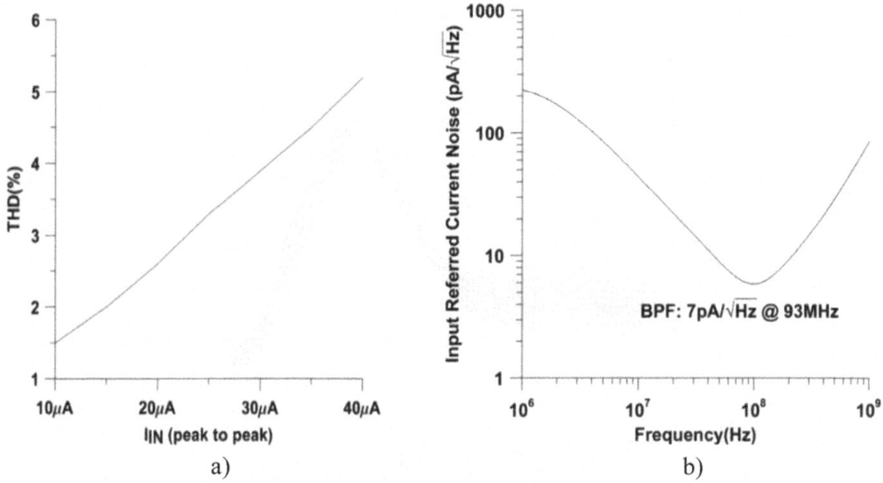

Fig. 9.17 (**a**) THD performance of the second design with regard to input current. (**b**) Input referred noise versus frequency between 1 MHz and 1 GHz. (Ozenli et al. 2018a, b)

$$\frac{I_{\mathrm{LP}}}{I_{\mathrm{IN}}} = -\frac{g_{\mathrm{m1}}g_{\mathrm{m2}}g_{\mathrm{m3}}}{s^2 C_{\mathrm{mos}}C_{\mathrm{gs3}}(g_{\mathrm{m1}}+g_{\mathrm{m2}}) + sC_{\mathrm{mos}}g_{\mathrm{m1}}g_{\mathrm{m2}} + g_{\mathrm{m1}}g_{\mathrm{m2}}g_{\mathrm{m3}}} \tag{9.17}$$

$$\omega_0 = \sqrt{\frac{g_{\mathrm{m1}}g_{\mathrm{m2}}g_{\mathrm{m3}}}{C_{\mathrm{mos}}C_{\mathrm{gs3}}g_{\mathrm{m1}} + C_{\mathrm{mos}}C_{\mathrm{gs3}}g_{\mathrm{m2}}}}, \quad Q = \sqrt{\frac{C_{\mathrm{gs3}}g_{\mathrm{m3}}(g_{\mathrm{m1}}+g_{\mathrm{m2}})}{C_{\mathrm{mos}}g_{\mathrm{m1}}g_{\mathrm{m2}}}} \tag{9.18}$$

To prove promising performance of the second design at the end of the section, THD analysis is given in Fig. 9.17a. With up to 40 μA peak-to-peak input current, the THD of the BP output remains within acceptable limits as 5%. It should be considered that Design 2 consumes moderately low power as 270 μW with the current generator. This is due to the fact that this circuit guarantees good linearity up to 25 μA peak-to-peak current with the THD of 3%. Moreover, noise simulation in Fig. 9.17b shows that the proposed second design has only $7\text{pA}/\sqrt{Hz}$ of input referred noise at 93 MHz.

Finally, Monte-Carlo analysis is illustrated in Fig. 9.18. In Fig. 9.18a, process variations and mismatches' effects are examined where temperature varies between −40 °C and 120 °C. It can be easily observed that operating frequency changes from 80 to 112 MHz. This variation can be easily reduced by advanced compensation techniques. In addition, the center frequency of the filter has a distribution with a mean of 94 MHz and the standard deviation value of 6.8 MHz over 100 random samples of the Monte-Carlo analysis including overall mismatch and process variations' effects in part (b) of this figure.

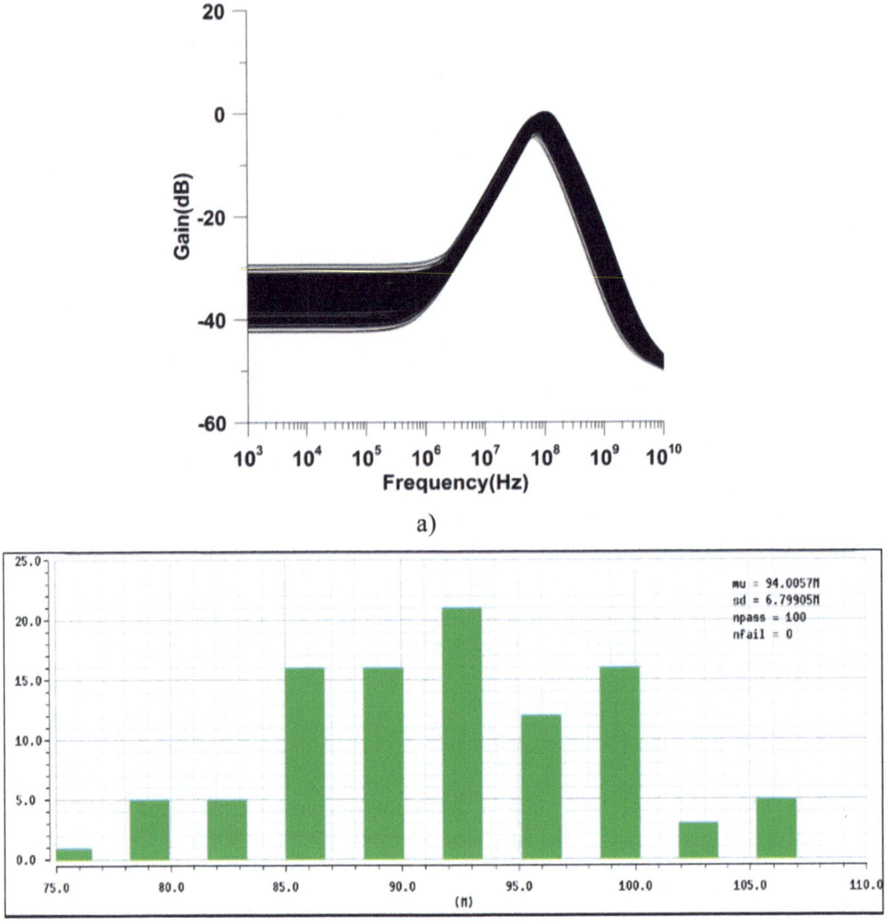

a)

b)

Fig. 9.18 Monte-Carlo analysis of the proposed structure: (**a**) BP frequency response variation over extended temperature range from −40 °C to 120 °C with process variations and mismatches. (**b**) Monte-Carlo analysis at corner of 27 °C. (Ozenli et al. 2018a, b)

Exemplary Circuit III

The core structure of the proposed single-input multi-output current-mode filter employing four transistors is given in Fig. 9.19 and its AC small signal equivalent circuit is shown in Fig. 9.20 with the assumption that all transistors operate in the saturation region. It is important to note that external resistors and capacitors have not been used in the proposed filter. For NMOS transistors, their bulk terminals are connected to ground, whereas for PMOS transistors, their bulk terminals are connected to their source terminals. From Eqs. (9.19), (9.20), and (9.21), it can be

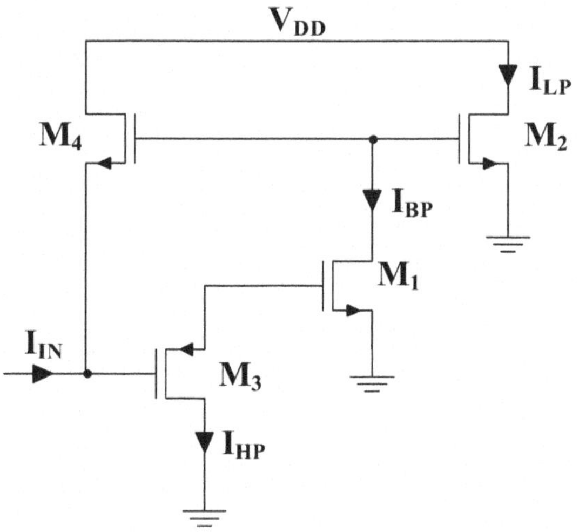

Fig. 9.19 The core structure of the proposed CM single-input multi-output filter. (Yesil et al. 2017)

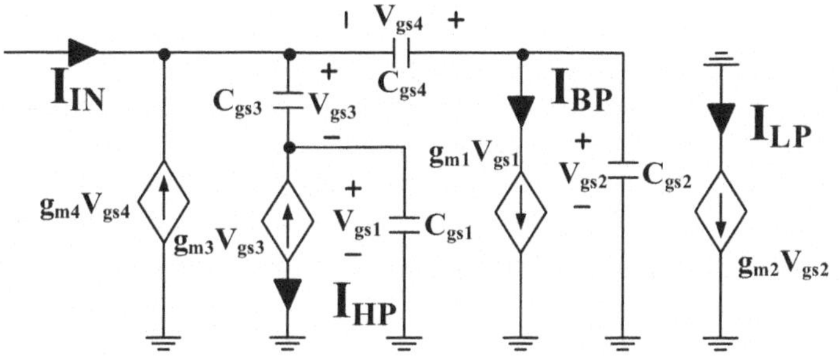

Fig. 9.20 AC small signal equivalent circuit of CM filter. (Yesil et al. 2017)

observed that C_{gs1} and C_{gs2} are essential gates to source capacitances of AC small signal equivalent circuit and these capacitances are grounded. In other words, both of the grounded MOS capacitances (MOS-cap) described in the literature (Allen and Holberg 2002; Razavi 2017) can be replaced by physical capacitances in order to get a high capacitance value so that the proposed CM filter can operate at lower frequencies.

Considering the AC small signal equivalent circuit and neglecting the output transconductance values, g_{ds}, the current transfer functions of the presented CM filter can be obtained as follows:

$$\frac{I_{LP}}{I_{IN}} = -\frac{\frac{g_{m1}g_{m2}g_{m3}}{g_{m4}}}{s^2 C_{gs1}C_{gs2} + sC_{gs2}g_{m3} + g_{m1}g_{m3}} \tag{9.19}$$

$$\frac{I_{BP}}{I_{IN}} = \frac{s\frac{C_{gs2}g_{m1}g_{m3}}{g_{m4}}}{s^2C_{gs1}C_{gs2} + sC_{gs2}g_{m3} + g_{m1}g_{m3}} \tag{9.20}$$

$$\frac{I_{HP}}{I_{IN}} = -\frac{s^2\frac{C_{gs1}C_{gs2}g_{m3}}{g_{m4}}}{s^2C_{gs1}C_{gs2} + sC_{gs2}g_{m3} + g_{m1}g_{m3}} \tag{9.21}$$

The pole frequency, quality factor, and gain with respect to the output of CM filter can be expressed as

$$\omega_0 = \sqrt{\frac{g_{m1}g_{m3}}{C_{gs1}C_{gs2}}} \tag{9.22}$$

$$Q = \sqrt{\frac{C_{gs1}g_{m1}}{C_{gs2}g_{m3}}} \tag{9.23}$$

$$G_{LP} = \frac{g_{m2}}{g_{m4}}, \quad G_{BP} = \frac{g_{m1}}{g_{m4}}, \quad G_{HP} = \frac{g_{m3}}{g_{m4}} \tag{9.24}$$

It is clearly observed that gains of the proposed filter can be adjusted electronically by g_m value of M_4 without changing the pole frequency and quality factor of the presented CM filter. Besides, the transconductance gain of M_2 can change the gain of the LP filter without affecting its pole frequency and quality factor. On the other hand, BS and AP filter outputs can be easily obtained by the interconnection of relevant outputs of the presented filter, without using additional circuit element, as the following:

$$\frac{I_{BS}}{I_{IN}} = \frac{I_{HP} + I_{LP}}{I_{IN}} = -\frac{g_{m3}}{g_{m4}}\frac{\left(s^2C_{gs1}C_{gs2} + g_{m1}g_{m2}\right)}{s^2C_{gs1}C_{gs2} + sC_{gs2}g_{m3} + g_{m1}g_{m3}} \tag{9.25}$$

$$\frac{I_{AP}}{I_{IN}} = \frac{I_{HP} + I_{BP} + I_{LP}}{I_{IN}} = -\frac{g_{m3}}{g_{m4}}\frac{\left(s^2C_{gs1}C_{gs2} - sC_{gs2}g_{m1} + g_{m1}g_{m2}\right)}{s^2C_{gs1}C_{gs2} + sC_{gs2}g_{m3} + g_{m1}g_{m3}} \tag{9.26}$$

Note that AP filter operates as the ideal one when the matching condition gm1 = gm3 is fulfilled (Figs. 9.21 and 9.22).

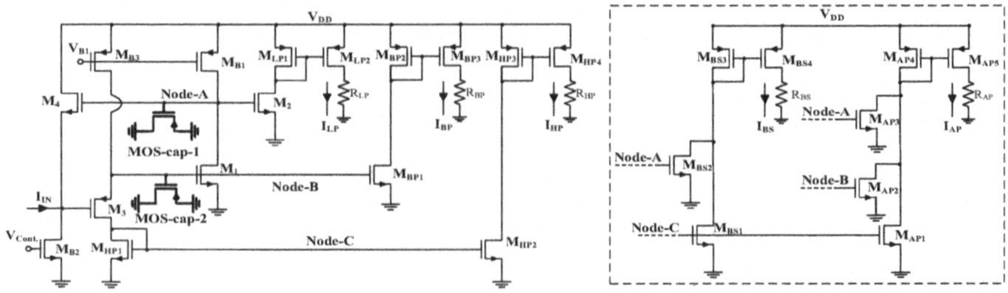

Fig. 9.21 The overall proposed CM filter. (Yesil et al. 2017)

Fig. 9.22 Layout of the proposed structure with 1278 μm^2 of chip area. (Yesil et al. 2017)

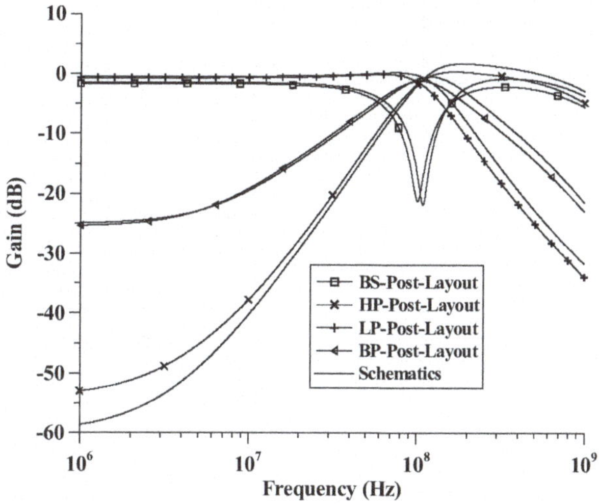

Fig. 9.23 Frequency response of the proposed filter structure. (Yesil et al. 2017)

Figure 9.23 depicts the frequency responses of the presented filter given in Fig. 9.21. The center frequency and quality factor of the proposed structure in post-layout are measured as 104 MHz and 0.88, respectively. Gain and phase responses of the all-pass filter are given in Fig. 9.24. It can be easily inferred that a small discrepancy between post-layout and schematic responses comes from the extracted parasitics of the interconnections between transistors. On the other hand, gains of LP, BP, and HP filters can be tuned electronically by changing the biasing voltage of $V_{Cont.}$. The relationship between their gains and $V_{Cont.}$ can be observed in Figs. 9.25, 9.26, and 9.27.

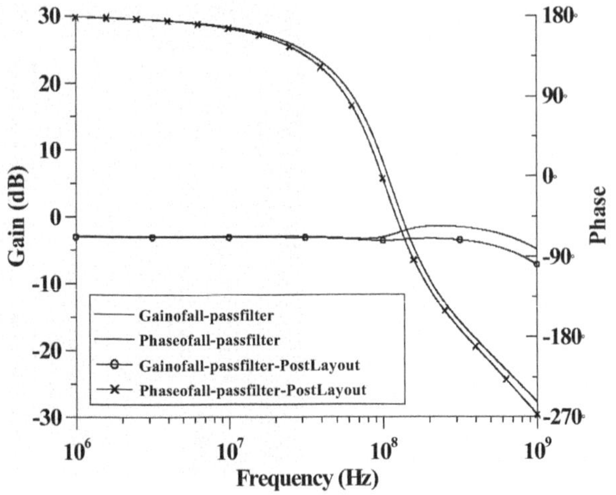

Fig. 9.24 Frequency response of the all-pass configuration. (Yesil et al. 2017)

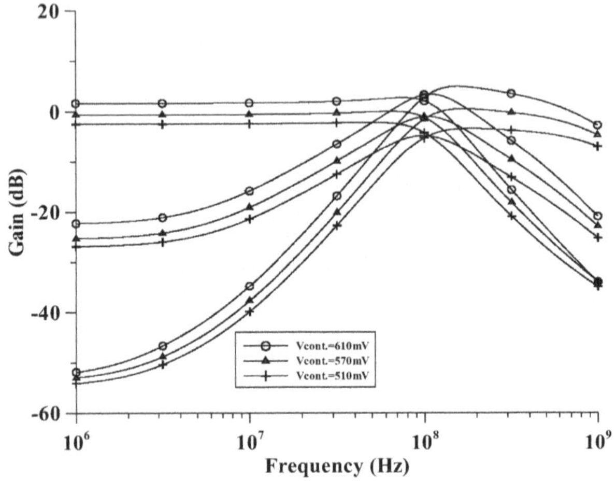

Fig. 9.25 Variation of the filter gains at LP, BP, and HP outputs with $V_{\mathrm{Cont.}}$ voltage. (Yesil et al. 2017)

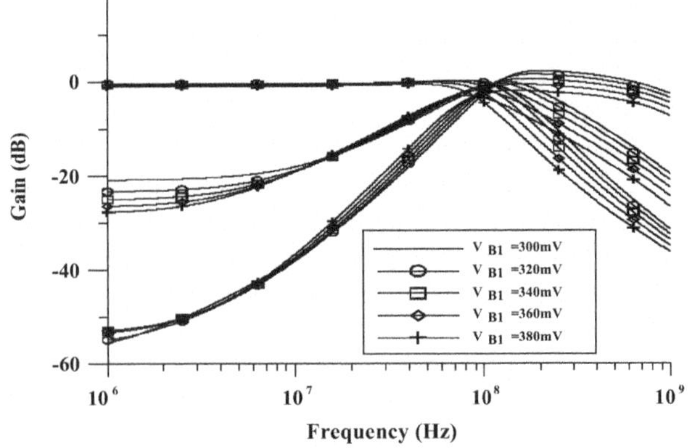

Fig. 9.26 Tunability of the operating frequencies at LP, BP, and HP outputs with V_{B1}. (Yesil et al. 2017)

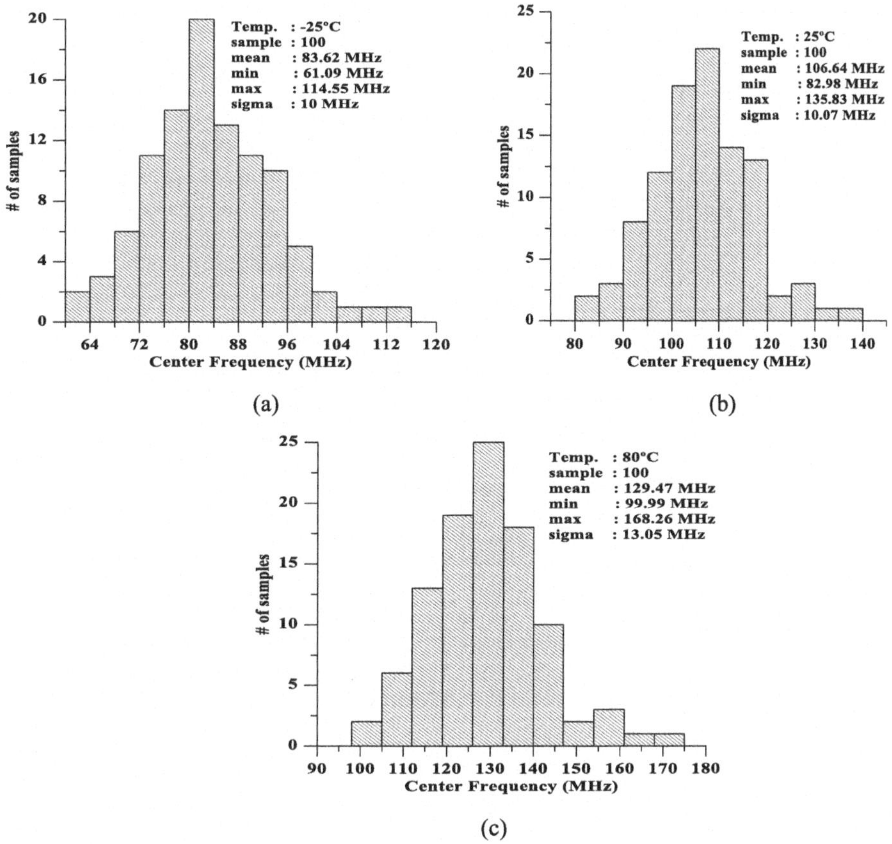

Fig. 9.27 Monte-Carlo simulations of the BP's center frequency at different temperatures: (**a**) −25 °C, (**b**) 25 °C, and (**c**) 80 °C. (Yesil et al. 2017)

Exemplary Circuit IV

At the end of the aforementioned automation cycle, the proposed NMOS and PMOS first-order current-mode phase shifters are given in Fig. 9.28a, b, respectively. The core structures consist of three transistors and a capacitor. It is assumed that all transistors are operated in the saturation region.

$$\frac{I_{\text{AP-NMOS}}}{I_{\text{IN}}} = -\frac{g_{\text{m3}}(g_{\text{m3}} - sC)}{g_{\text{m2}}(g_{\text{m1}} + sC)}, \quad \frac{I_{\text{AP-PMOS}}}{I_{\text{IN}}} = \frac{g_{\text{m3}}(g_{\text{m2}} - sC)}{g_{\text{m2}}(g_{\text{m1}} + sC)} \tag{9.27}$$

If we assume that $g_{m1} = g_{m2} = g_m$, NMOS phase shifter gives inverting all-pass function and the PMOS one produces noninverting all-pass output. In this regard, the

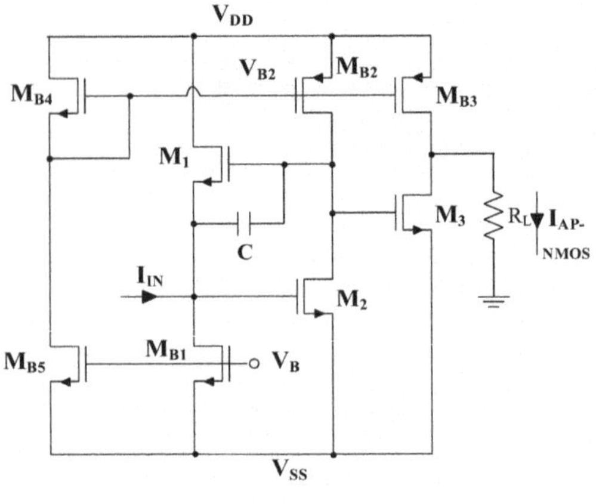

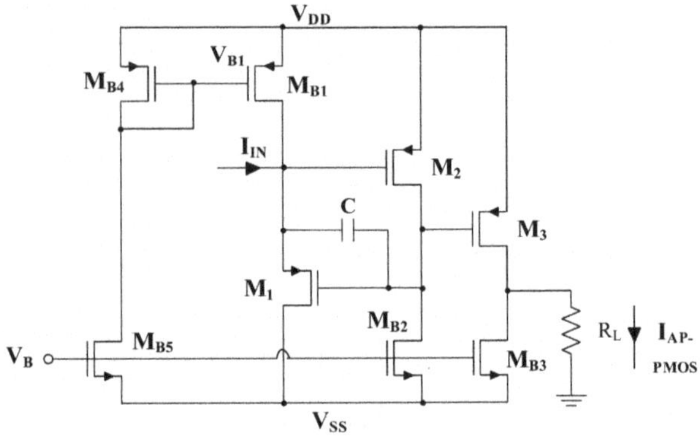

Fig. 9.28 Proposed first-order phase shifters with bias transistors. (**a**) NMOS based. (**b**) PMOS based. (Ozenli et al. 2018a, b)

pole frequency and gain expressions are the same for both of them and given in Eq. (9.28), respectively:

$$\omega_0 = \frac{g_{\rm m}}{C}, H_0 = \frac{g_{\rm m3}}{g_{\rm m}} \tag{9.28}$$

The filters' gain can be adjusted by g_{m3}. Taking Eqs. (9.27) and (9.28) into consideration, the phase response of the filters can be calculated as follows, respectively:

$$\Phi(\omega) = \pi - 2\tan^{-1}\left(\omega \frac{C}{g_m}\right) \tag{9.29}$$

Table 9.1 MOS-only automation results (geometric programming) with 4.5 pF floating capacitor

Designs	NMOS based		PMOS based	
Transistors	W(μm)	L(μm)	W(μm)	L(μm)
M_1, M_2	3	0.5	8	0.2
M_3	2	0.9	2	0.25
M_{B1}	3	0.4	8	0.3
M_{B2}	10	0.6	5	0.6
M_{B3}	30	0.2	6	0.4
M_{B4}	10	0.2	16	0.3
M_{B5}	3	0.2	8	0.3

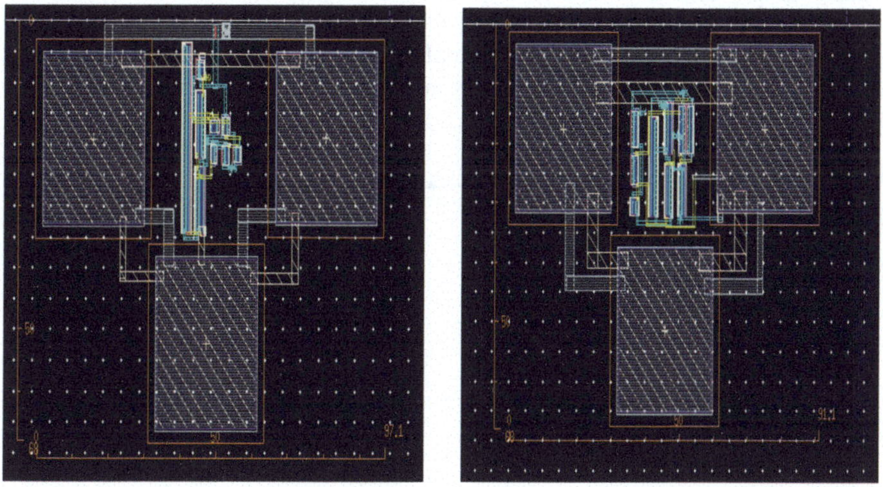

a) b)

Fig. 9.29 Proposed first-order AP architectures' layouts. (**a**) NMOS based (68 μm × 91.1 μm). (**b**) PMOS based (68 μm × 91.1 μm). (Ozenli et al. 2018a, b)

$$\Phi(\omega) = -2\tan^{-1}\left(\omega\frac{C}{g_m}\right) \tag{9.30}$$

When the core structures in Fig. 9.28 are examined carefully and the condition (C_{gs1}>>C_{gs2}, C_{gs3}) is satisfied, these can be employed without external capacitors. However, it is not a matter anymore so that poly1–poly2 layers allow us to use floating capacitance on the chip. Hence, a floating capacitor shown in Fig. 9.28 can be integrated into the MOS transistors. In Table 9.1, transistor dimensions are given. By taking into account these dimensions, overall NMOS and PMOS phase shifters are laid out respectively as illustrated in Fig. 9.29.

Concluding Remarks

The presented MOS-only and MOS-C circuits based on 0.18 μm TSMC environment in this chapter are easily implementable for encrypted communication or FM band applications, where low power consumption and small chip area are very crucial. Besides, most of the proposed circuits are very compact structures, which can be controlled with only a single voltage port for a wide range of operating regions. Noise, THD results, and Monte-Carlo analyses contribute to their promising characteristics as well.

The given reformulations of the small signal parameters are used to directly connect design variables with design specifications. After removing the dependency on technology parameters using surface fitting by using SPICE simulation-based data, these techniques can be easily implementable for different processes. To alleviate mean errors between simulators and proposed reformulations, simulation

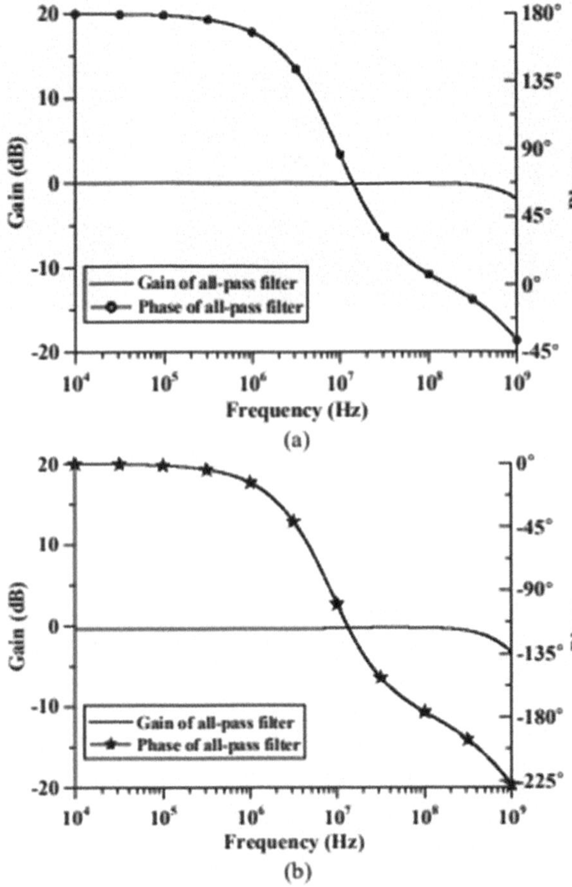

Fig. 9.30 Proposed first-order phase shifters' frequency responses: (**a**) NMOS based and (**b**) PMOS based

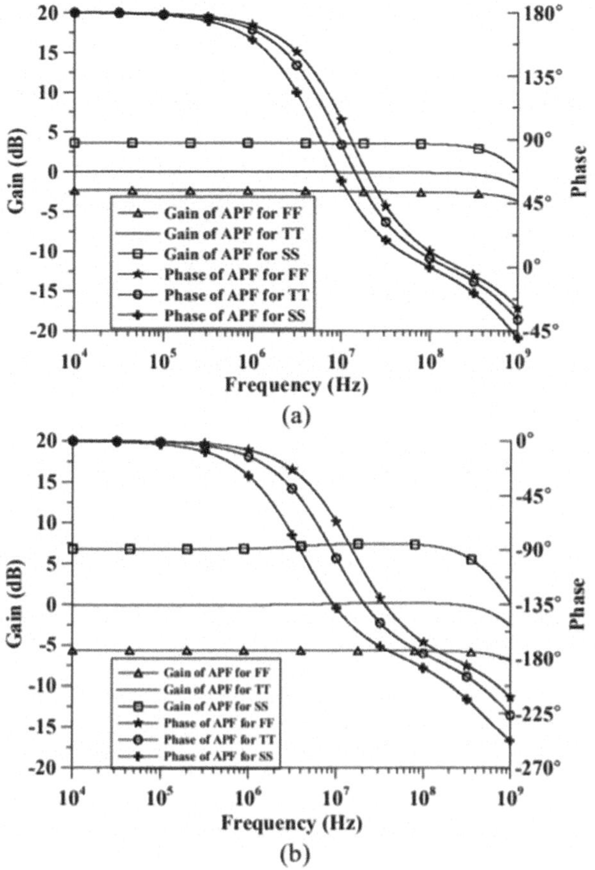

Fig. 9.31 Corner analyses of the proposed phase shifters with regard to FF, SS, and TT transistors: (**a**) NMOS based and (**b**) PMOS based

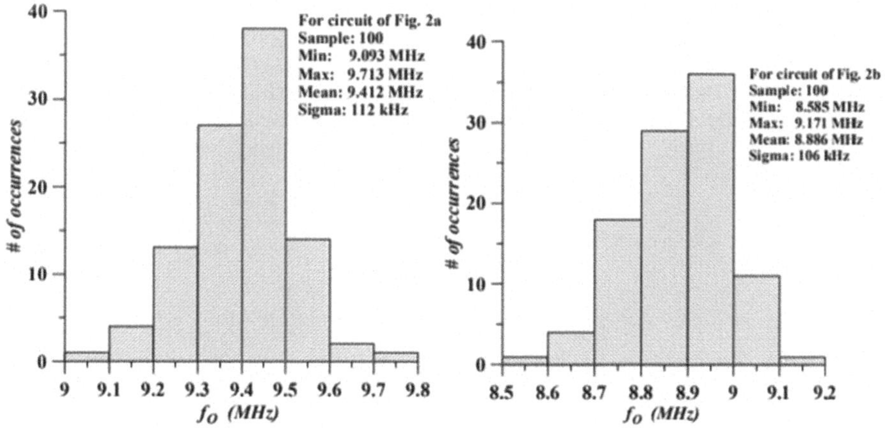

Fig. 9.32 Monte-Carlo simulation for the proposed circuits. (**a**) NMOS based. (**b**) PMOS based. (Ozenli et al. 2018a, b)

data are collected from both of CADENCE-SPECTRE and SPICE. In this regard, the proposed automation techniques offer new opportunities to the analog designers not only filter-based applications but also main analog circuit blocks such as op-amps and OTAs extending the given reformulations.

Moreover, it is shown that MOSFET small signal parameters can be modeled by using polynomial regression in the subthreshold region. Therefore, the proposed design automation algorithm is applied to circuits targeting biomedical applications. Generated circuits perform for a wide range of low frequencies with pretty low power consumption. Also, they can give promising results as real biomedical signals are injected to the filter inputs (Figs. 9.30, 9.31, and 9.32).

References

Allen, P. E., & Holberg, D. R. (2002). CMOS analog circuit design, Oxford Univ. Press.

Daems, W., Gielen, G. and Sansen, W. (2003). Simulation-based generation of posynomial performance models for the sizing of analog integrated circuits. IEEE Transactions on Computer-Aided Design of Integrated Circuits and Systems, 22(5), 517–534.

Lakys, Y. and Fabre, A. (2013). Multistandard transceivers: state of the art and a new versatile implementation for fully active frequency agile filters. Analog Integrated Circuits and Signal Processing, 74(1), 63–78.

Ozenli, D. (2018), "New opportunities in MOS-only filter design," Ph.D. dissertation, Dept. Electronics and Communications Eng., Istanbul Technical Univ., Istanbul, Turkey.

Özenli, D. and Kuntman, H. (2017). Mos-only automation technique with an exemplary mos-only BP/LP agile filter design. Istanbul University-Journal of Electrical & Electronics Engineering, 17(2), 3361–3367.

Ozenli, D. and Kuntman, H. H. (2018). A novel low power MOSFET-C band pass filter for low frequency applications with subthreshold models based on polynomial regression. Analog Integrated Circuits and Signal Processing, 97(1), 97–105.

Ozenli, D., Alaybeyoglu, E., Kuntman, H. and Cicekoglu, O. (2018a). MOSFET-Only filter design automation based on polynomial regression with exemplary circuits. AEU-International Journal of Electronics and Communications, 84, 342–354.

Ozenli, D., Yesil, A., Arslan, E. and Kuntman, H. H. (2018b). Novel First Order Current Mode MOS-C Phase Shifters. Elektronika ir Elektrotechnika, 24(1), 31–35.

Razavi, B. (2017). Design of analog CMOS integrated circuits, McGraw Hill. Press.

Yesil, A., Ozenli, D., Arslan, E. and Kuntman, H. (2017). Current mode single-input multi-output MOSFET-only filter. AEU-International Journal of Electronics and Communications, 80, 157–164.

Chapter 10
Versatile Active Elements, Structure, and Applications

Basic Concept

As an active building block, operational amplifier played a predominant role in the last two decades and an enormous number of publications exist in the literature on various circuit examples so that the design engineer can make choice of them. However, op-amp-based circuits exhibit several drawbacks in their performance arising from the limited bandwidth and slew rate of these active elements. Therefore, the current-mode approach has been increasingly recognized as a way to overcome the op-amp drawbacks and to realize high-speed systems. In the last decade and especially in recent years, new current-mode active building blocks like second-generation current conveyors (CCII+ and CCII−) and current feedback op-amps (CFOA) received considerable attention due to their larger dynamic range and wider bandwidth. In addition, different types of active elements like electronically controlled current conveyor (ECCII), differential voltage current conveyor (DVCC), differential difference current conveyor (DDCC), third-generation current conveyor (CCIII), dual-output operational transconductance amplifier (DO-OTA), and four-terminal floating nullor (FTFN) are presented in the literature.

In a recent work, a new active building block, CDBA (current differencing buffered amplifier), is introduced by Acar and Özoguz to provide further possibilities in the circuit synthesis and to simplify the implementation (Acar and Özoğuz 1999). The basic aim of this paper is to present a CDBA-based multipurpose filter topology which realizes LP, BP, and HP transfer functions simultaneously. In this study a single CDBA is used as the starting point. Most cascadable current-mode filters permit cascadability due to their high output impedances. But most of them do not exhibit low input impedance. The proposed circuit however permits low input impedance due to unconditionally grounded input pin of the CDBA (Özcan et al. 2000, 2002). This will bring further advantages, for example, there will be no loading problem for the current-mode signal source.

H. Kuntman, D. Özenli, *Trends in Circuit Design for Analog Signal Processing*, Analog Circuits and Signal Processing,
https://doi.org/10.1007/978-3-030-96836-6_10

In the last two decades, several works on active devices were performed and a lot of works appeared in the available literature. One important work is performed by Biolek et al. (2008) where the main problem is described and several new active devices are introduced:

> *"The reason consists in the variety of specific requirements, imposed on the analog subsystems working in various operating modes and under their interaction with digital circuits: high dynamic range, noise immunity, low supply voltage and power consumption, high linearity and high bandwidth, low nonlinear distortion, specific impedance levels, etc. The development of new circuit elements over the last ten years, for example CDBA, CDTA, ICCII, FDCCII, MCCIII, CCOA, DDOFA, FBFTFN, CFA-OTA, CC-CFA, CCTA, CCCDTA, CCCCTA, CTTA, CCCDBA, DC-CDBA, CMI, DCCF, etc. shows that the procedure of finding new circuit elements is up-to-date. The work suggests a methodology for prospective continuation of this process, which is based on the concurrence of simple building blocks such as CDU, CF, CI, OTA, voltage buffer and OpAmp."*

Biolek et al. introduced complex active devices combining sub-blocks such as CDTA (current differencing transconductance amplifier), VDBA (voltage differencing buffered amplifier), VDTA (voltage differencing transconductance amplifier) etc., composed of a current differencing circuit followed by voltage buffer or transconductance amplifier, etc. The authors give a survey of current conveyors, followed by a figure reflecting operational amplifiers and hybrid elements, and described other active elements, composed of sub-circuits. These circuits provide flexibility in the design of analog circuits.

In this section we describe CMOS realization of some versatile active devices such as CDBA, CDTA, and VDTA, giving CMOS realization topologies with application examples.

Current Differencing Buffered Amplifier (CDBA)

The circuit symbol of the CDBA is shown in Fig. 10.1a, where p and n are input and w and z are output terminals. The equivalent circuit of the CDBA is given in Fig. 10.1b. The current differencing buffered amplifier is characterized by Eq. (10.1).

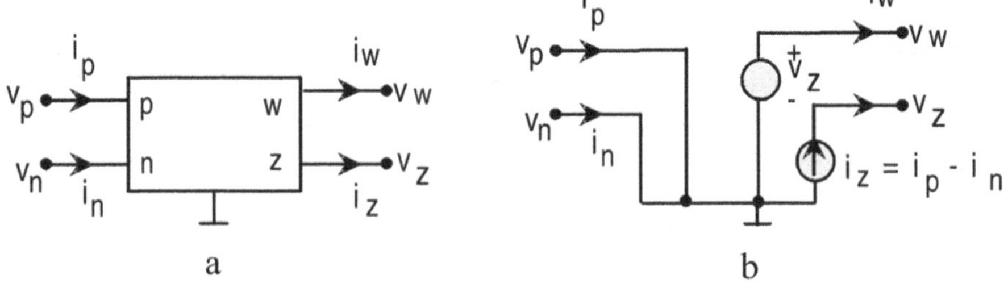

Fig. 10.1 (**a**) Circuit symbol of the CDBA. (**b**) Equivalent circuit of the CDBA. (Özcan et al. 2002)

$$\begin{bmatrix} i_z \\ v_w \\ v_p \\ v_n \end{bmatrix} = \begin{bmatrix} 0 & 0 & 1 & -1 \\ 1 & 0 & 0 & 0 \\ 0 & 0 & 0 & 0 \\ 0 & 0 & 0 & 0 \end{bmatrix} \begin{bmatrix} v_z \\ i_w \\ i_p \\ i_n \end{bmatrix} \tag{10.1}$$

According to the above matrix equation and equivalent circuit of Fig. 10.1b, the current through z terminal is the difference of the currents through p terminal and n terminal; hence, the z terminal is called current output; p and n terminals are noninverting and inverting input terminals, respectively. Since the voltage at the w terminal follows the voltage of z terminal, it is called voltage output. Note that the input terminals, through which i_p and i_n flows, are internally grounded. The CDBA can easily be implemented with bipolar and CMOS technologies (Özcan et al. 2002).

Filter Topology Employing CDBA

The current-mode filter topology proposed is illustrated in Fig. 10.2.

Transfer functions from the input to i_{o1},i_{o2} and i_{o3} outputs are respectively given by

$$\begin{aligned} \frac{i_{o1}}{i_{\text{in}}} &= \frac{\frac{G_2}{C_2}s}{s^2 + \frac{G_1 + G_2 + G_3}{C_1}s + \frac{G_1 G_2}{C_1 C_2}} \\ \frac{i_{o2}}{i_{\text{in}}} &= \frac{\frac{G_2 G_3}{C_2}s}{s^2 + \frac{G_1 + G_2 + G_3}{C_1}s + \frac{G_1 G_2}{C_1 C_2}} \\ \frac{i_{o3}}{i_{\text{in}}} &= \frac{s^2 + \frac{G_1 + G_2 + G_3}{C_1}s}{s^2 + \frac{G_1 + G_2 + G_3}{C_1}s + \frac{G_1 G_2}{C_1 C_2}} \end{aligned} \tag{10.2}$$

where the pole frequency and the pole quality factor are defined as

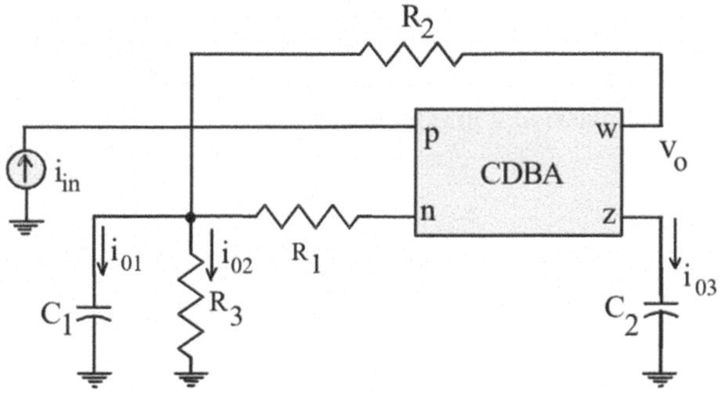

Fig. 10.2 Filter topology employing CDBA. (Özcan et al. 2002)

$$\omega_p = \sqrt{\frac{G_1 G_2}{C_1 C_2}}$$
$$Q_p = \frac{1}{G_1 + G_2 + G_3}\sqrt{\frac{G_1 G_2 C_1}{C_2}} \tag{10.3}$$

The passive ω_p and Q_p sensitivities are determined as

$$\begin{aligned}
&S_{G_1}^{\omega_p} = S_{G_2}^{\omega_p} = 1/2 \; S_{G_3}^{\omega_p} = 0\\
&S_{C_1}^{\omega_p} = S_{C_2}^{\omega_p} = -1/2\\
&S_{G_1}^{Q_p} = \frac{G_2 - G_1 + G_3}{2(G_1 + G_2 + G_3)}\\
&S_{G_2}^{Q_p} = \frac{G_1 - G_2 + G_3}{2(G_1 + G_2 + G_3)}\\
&S_{G_3}^{Q_p} = \frac{G_3}{G_1 + G_2 + G_3}\\
&S_{C_1}^{Q_p} = 1/2\\
&S_{C_2}^{Q_p} = -1/2
\end{aligned} \tag{10.4}$$

In this form the circuit represents a dual-function filter where Q_p can be adjusted independently. Furthermore, the circuit is easily cascadable and no additional active element is needed to pick up the signals on the passive elements, since the input of the next stage is grounded.

In this circuit (Fig. 10.2) the high-pass filter function can be obtained by means of the i_{o3} output. But as it is observed from Eq. (10.4), the transfer characteristic represents the sum of high-pass and band-pass functions; therefore, the HP filter characteristic must be extracted from Eq. (10.4), which will be described next.

Realization of High-Pass filter To obtain HP characteristic two modifications must be made on the circuit proposed as described below.

The first realization circuit of the HP filter is obtained by subtracting Eq. (10.2) from Eq. (10.4) which yields

$$\frac{i_{o3} - i_{o1}}{i_{\text{in}}} = \frac{s^2 + \frac{G_1+G_2+G_3}{C_1} s - \frac{G_2}{C_2} s}{s^2 + \frac{G_1+G_2+G_3}{C_1} s + \frac{G_1 G_2}{C_1 C_2}} \tag{10.5}$$

As it can be clearly seen from Eq. (10.7), we get only HP filter characteristic if the coefficients of the s-terms are set equal to each other, which results in the relation between passive components as

$$\frac{C_2}{C_1} = \frac{G_2}{G_1 + G_2 + G_3} \tag{10.6}$$

This can be easily achieved by introducing an additional CDBA and applying the currents i_{o3} and i_{o1} to its p and n inputs, respectively. The resulting circuit is illustrated in Fig. 10.3a. On the other hand, Eq. (10.6) gives the relation between the quality factor Q_p and the element values as

$$\frac{C_2}{C_1} = \frac{G_1G_2}{Q_p^2(G_1 + G_2 + G_3)^2} \tag{10.7}$$

Combination of Eqs. (10.8) and (10.9) yields the condition which must be satisfied by the values of conductances $G1$, $G2$, and $G3$. This condition is given by

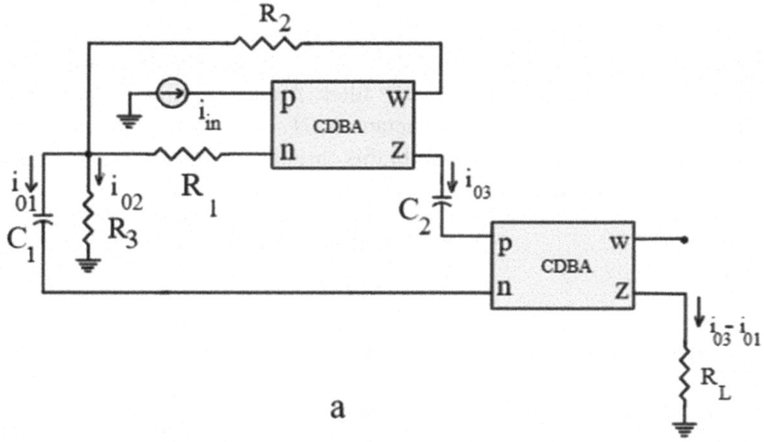

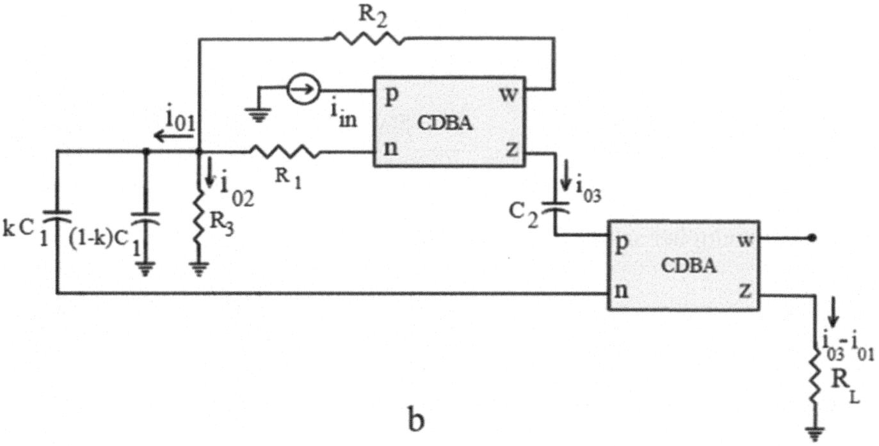

Fig. 10.3 Realization of HP circuits. (**a**) First realization, Eqs. (10.7, 10.8, 10.9, 10.10, 10.11, and 10.12). (**b**) Second realization, Eqs. (10.13) and (10.14). (Özcan et al. 2002)

$$Q_p^2 = \frac{G_1}{G_1 + G_2 + G_3} \tag{10.8}$$

From Eq. (10.10) it is easily observed that this realization of the HP filter is only suitable for low Q applications. Note that there are many parameters and element values to be determined which cause difficulties in the circuit design. To overcome these difficulties, the conditions to be satisfied can be simplified by choosing $G_3 = 0$ which reduces to

$$\begin{aligned} Q_p^2 &= \frac{G_1}{G_1 + G_2} \\ \frac{C_2}{C_1} &= \frac{G_1 G_2}{Q_p^2 (G_1 + G_2)^2} \end{aligned} \tag{10.9}$$

With adequate choice of values, the desired filter function is obtained.

The second realization circuit of the HP filter: Another simple realization of HP filter is also possible by setting the conductances $G1$, $G2$, and $G3$ equal to each other and adding extra capacitor to the circuit. In this case the quality factor which is given in Eq. (10.6) is modified to

$$Q_p = \frac{1}{3}\sqrt{\frac{C_1}{C_2}} \tag{10.10}$$

which gives the relation between capacitor values as

$$\frac{C_2}{C_1} = \frac{1}{9Q_p^2} \tag{10.11}$$

On the other hand, similar to Eq. (10.7), for obtaining HP characteristic subtraction of the BP characteristic from Eq. (10.4) is necessary which results in

$$\frac{i_{o3} - k i_{o1}}{i_{in}} = \frac{s^2 + \frac{3G}{C_1} s - k \frac{G}{C_2} s}{s^2 + \frac{G_1 + G_2 + G_3}{C_1} s + \frac{G_1 + G_2}{C_1 C_2}} \tag{10.12}$$

where k is a multiplier satisfying the condition

$$k = \frac{3C_2}{C_1} = \frac{1}{3Q_p^2} \tag{10.13}$$

This can be provided by splitting the capacitor into two components such as

$$
\begin{aligned}
C_{1A} &= kC_1 \\
C_{1B} &= (1-k)C_1
\end{aligned}
\tag{10.14}
$$

and injecting the current through the C_{1A} to the n input of the second CBDA in Fig. 10.3a while the capacitor C_{1B} is a grounded capacitor. Note that $C_1 = C_{1A} + C_{1B}$. The resulting circuit is illustrated in Fig. 10.3b. The circuit is similar to the circuit of Fig. 10.3a. Note that the HP realization circuit of Fig. 10.3b is more suitable than Fig. 10.3a for high-Q applications at the expense of large component spread due to the flexibility provided by the multiplier *k*.

Influence of the Non-idealities

In the case of non-ideal CDBA, voltage and current tracking errors arising from the active element can seriously influence the circuit behavior. It is necessary therefore to give the effect of the non-idealities on the circuit performance. Including the voltage and current tracking errors, the current differencing buffered amplifier is characterized by Eq. (10.1).

$$
\begin{bmatrix} i_z \\ v_w \\ v_p \\ v_n \end{bmatrix} = \begin{bmatrix} 0 & 0 & a_p & -a_n \\ b & 0 & 0 & 0 \\ 0 & 0 & 0 & 0 \\ 0 & 0 & 0 & 0 \end{bmatrix} \begin{bmatrix} v_z \\ i_w \\ i_p \\ i_n \end{bmatrix}
\tag{10.15}
$$

where α_p and α_n are the tracking coefficients for i_p and i_n, respectively; *b* represents the voltage tracking coefficient. Including these non-idealities into the transfer function, the pole frequency and the pole quality factor are modified as follows:

$$
\begin{aligned}
\omega_p &= \sqrt{\frac{a_n b G_1 G_2}{C_1 C_2}} \\
Q_p &= \frac{1}{G_1 + G_2 + G_3}\sqrt{\frac{a_n b G_1 G_2 C_1}{C_2}}
\end{aligned}
\tag{10.16}
$$

The active sensitivities are calculated as

$$
\begin{aligned}
S_{a_n}^{\omega_p} &= S_{b_n}^{\omega_p} = 1/2 \\
S_{a_n}^{Q_p} &= S_{b_n}^{Q_p} = 1/2
\end{aligned}
\tag{10.17}
$$

Experimental Verification

The performance of the circuit is tested experimentally by using the bipolar realization of the CDBA shown in Fig. 10.4 consisting of two CFOAs, namely, AD844 of analog devices. The supply voltages were taken as $V_{CC} = +12$ V and $V_{EE} = -12$ V. The element values were chosen as $R_1 = 1$ k, $R_2 = 1$ k, $R_3 = 1$ k, $C_1 = 4.5$ nF, and $C_2 = 1$ nF, which results in a universal filter with a pole frequency of $f_p = 100$ kHz and with a quality factor of $Q_P = 0.707$. To obtain HP filter characteristic, the k coefficient is calculated as $k = 1/3.(Q_P)^2$. For $Q_P = 0.707$, k is obtained as $k = 2/3$. As a result, the capacitance values of C_{1A} andC_{1B} are obtained as $C_{1A} = k.C_1 = 2/3 \times 4.5$ nF $= 3$ nF and $C_{1B} = (1 - k)\ C_1 = 1/3 \times 4.5$ nF $= 1.5$ nF.

The resulting frequency characteristics obtained from experiments are illustrated with ideal frequency responses in Fig. 10.5. It can be easily observed from Fig. 10.5 that the frequency response of the proposed filter is in good agreement with theoretical ideal results in the operating region. The deviation in the actual band-pass and high-pass characteristics at low frequencies is caused by the finite output impedances of the active elements. The limited bandwidths of the CDBA result in the deviations from the ideal filter frequency responses at high frequencies. From Fig. 10.3 it is obvious that the HP output can be also obtained from the low-impedance w terminal of the second CDBA. Note that the currents i_{o1} and i_{o2} realizing BP and LP filters can be similarly picked up and obtained at a low output impedance by the use of an additional CDBA. The dependence of the output signal level on the load resistance R_L is determined experimentally and illustrated in Fig. 10.6. Note that signal amplitude increases linearly with increasing load resistance and signal amplitudes up to 10 V are possible at the output for appropriate resistance values.

The large signal behavior of the circuit is demonstrated by applying an input signal amplitude of 1000 μA and picking up the current io1 realizing BP filter with a second CDBA. The resulting output waveform obtained at z and w outputs for a load resistance of $R_L = 10$ kΩ is shown in Fig. 10.7. The results prove that the circuit operates even at larger signal levels of the order of 1000 μA properly. Note that w terminal provides a low-impedance BP output which is a further advantage provided by the circuit. From Fig. 10.7 it is obvious that a maximum output swing of 20 V_{PP} is possible even at a frequency of 100 kHz. The maximum value of the

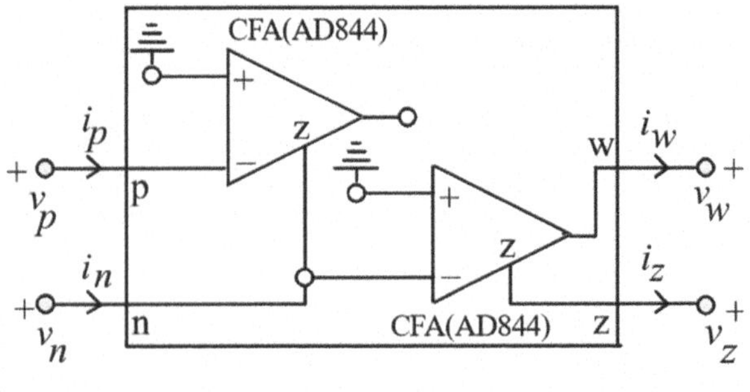

Fig. 10.4 Realization of the CDBA with CFOAs of AD844. (Özcan et al. 2002)

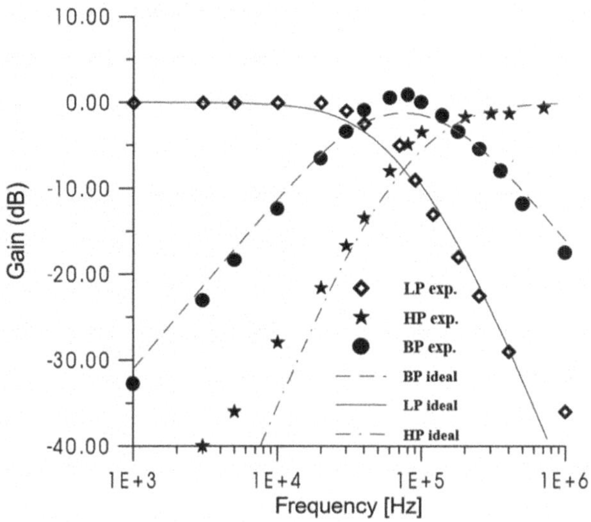

Fig. 10.5 Theoretical and experimental frequency response of the proposed filter topology. (Özcan et al. 2002)

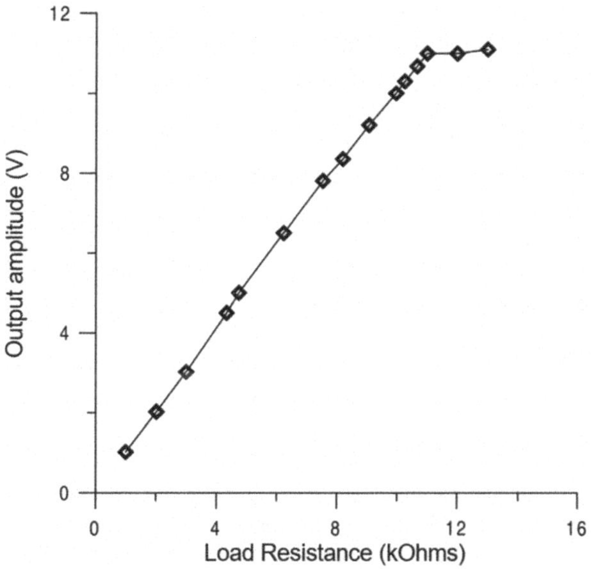

Fig. 10.6 Dependence of the output voltage at z terminal of CDBA2 on the load resistance RL for a sinusoidal 100 kHz input signal of 1000 μA. (Özcan et al. 2002)

input signal I_{in} depends on the realization topology of the CDBA and can be increased by appropriate design of the active element which results in an improved large signal behavior. The CFOA-based realization of the CDBA used for the experiments exhibits an input resistance value of 64.65 Ohm and an input capacitance value of 4.6 pF for p and n terminals; the output resistance and capacitance of the z terminal are specified as $R_z = 2.2$ MOhms and $C_z = 5.5$ pF. The output resistance of the w terminal is determined as $R_W = 15$Ohm. The related capacitance to the w terminal exhibits very low values and can be easily neglected. To reduce the

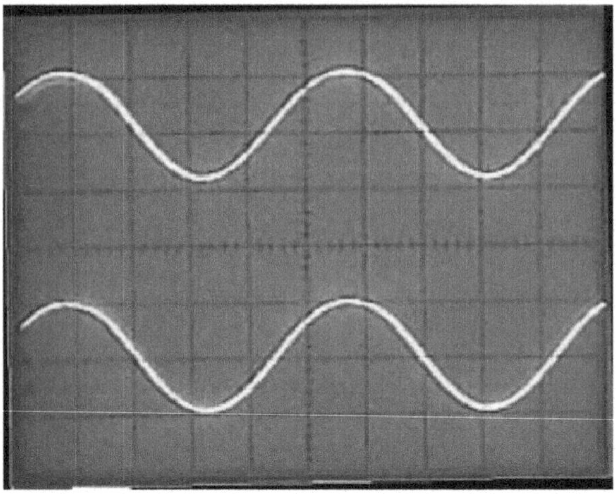

Fig. 10.7 Output waveforms for a sinusoidal 100 kHz input signal of 1000 μA and for a load resistance of $R_L = 10$ kOhm. Lower trace w-output and upper trace z-output (voltages) of CDBA2. Vert.: 10 V/div. Hor.: 2 μs/div. (Özcan et al. 2002)

influence of the device parameters and the load impedance on the circuit behavior, wideband and high impedance building blocks for the realization of the z terminal and low-impedance structures for the realization of the input stages can be used in the design of high-performance CDBA topologies. A design example of a fourth-order Butterworth BP filter with a center frequency of $f = 100$ kHz is chosen to demonstrate that the proposed filter topology can be easily cascaded by connecting the terminal of the corresponding element to the input of the next stage. The filter is realized by cascading two similar topologies illustrated in Fig. 10.2. The pole frequencies and the pole quality factors of the first and second stages are specified as $f_{p1} = f_{p2} = 100$ kHz, $Q_{p1} = 1.315$, and $Q_{p2} = 0.543$, respectively. All of the resistance values are taken as 1 kΩ. For the specified pole frequencies and pole quality factors, the capacitance values are calculated as $C_{11} = 6.3$ nF, $C_{21} = 403$ pF for the first stage and $C_{12} = 2.59$ nF, $C_{22} = 977$ pF for the second stage. The cascade connection is provided simply by connecting the lower terminal of the capacitor C_{11} of the first stage to the input of the second stage. The CDBAs are realized again with CFOAs (AD844 of analog devices) as shown in Fig. 10.4 and supplied with symmetrical voltages of ±12 V. SPICE simulation results are given in Fig. 10.8 with the ideal frequency response of the fourth-order Butterworth filter. It can be observed from Fig. 10.8 that the performance of the actual circuit is in good agreement with the ideal response which demonstrates clearly that two similar stages can be easily cascaded to achieve a high-order filter. The deviations in the passband gain and in the high-frequency region arise from the non-idealities of the active elements and can be overcome by the adequate design of the CDBAs.

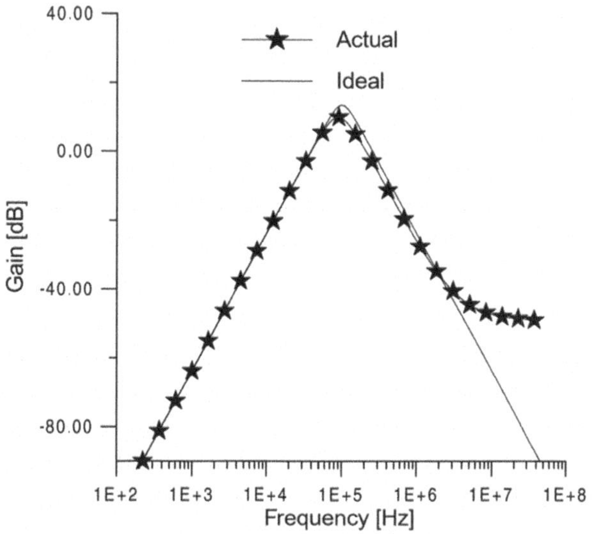

Fig. 10.8 Frequency responses of ideal and actual fourth-order Butterworth BP filters with a center frequency of f = 100 kHz; filters obtained by cascading two biquad sections. (Özcan et al. 2002)

Concluding Remarks

In this section a circuit example for CDBA-based multipurpose filter is presented. The basic filter cell is a dual-function filter realizing LP and BP transfer functions. For this realization no element matching is necessary. The filter permits independent adjustment of Q exhibiting low input impedance and a good large signal behavior. Due to the low input impedance provided by the active element, CDBA, the proposed filter topology can be easily cascaded by connecting the terminal of the corresponding element to the input of the next filter stage. The passive ω_o and Q sensitivities are shown to be low. Using an additional CDBA and matching conditions, the realization of an HP filter is also possible. Due to current-mode operation large output swings even at high frequencies are possible. Furthermore, the LP and BP filter functions can be picked up and obtained at low output impedance by the use of an additional CDBA, similar to HP filter.

Current Differencing Transconductance Amplifier (CDTA): A New, Improved CMOS Realization of the CDTA and Its Filter Applications

Circuits based on the current-mode approach, such as current conveyors, current amplifiers, and operational transconductance amplifiers, seem to have a better signal dynamic range and closed-loop bandwidth performance than conventional voltage amplifiers. Moreover, they are particularly suitable whenever the input sources and/or the outputs are current signals. Great interest has been devoted to the analysis and design of the current differencing transconductance amplifier proposed by Biolek, mainly because it exhibits better performance, particularly higher speed and better bandwidth, than classic voltage-mode operational amplifiers, which are limited by a constant gain-bandwidth product. This device, with two current inputs and two kinds of current output, provides an easy implementation of current-mode active filters. A CDTA is a synthesis of the well-known advantages of the current differencing amplifier and a transconductance amplifier to facilitate the implementation of current-mode active filters. The main features of a CDTA are low gain errors (high accuracy), high linearity, high transconductance, and wide frequency response. In addition, high output resistance at terminal X of the CDTA is required to enable easy cascading without the need for additional active elements in applications. The conventional CDTA structure available in the literature employs a well-known current differencing amplifier followed by a standard transconductor. As a result, the overall performance of this circuit differs strongly from the expected active element characteristics. The circuit exhibits low output resistance and limited linearity range, which causes deviations in the expected performance of the application circuits employing the CDTA. There are several works available in the literature on extending the performance of current-mode active elements, including operational transconductance amplifiers (OTAs) and second-generation current conveyors (CCIIs), where input stages with a wide linearity range and output stages providing high output impedance are necessary. However, good linearity, high accuracy, and high output resistance can be obtained by using high-performance current mirrors and adequate input stages in the structure of the CDTA. The aim of this paper is to introduce a new, improved CMOS configuration of a CDTA providing low input impedances at ports p and n, very high output impedances at ports z and x, an extended linearity range, and high transconductance. To provide high-output impedance values, cascode current mirrors and regulated cascode current mirrors were used in the CDTA structure. Furthermore, the input linearity range of the transconductor section was increased by introducing a cross-coupled quad configuration. To demonstrate the performance of the CDTA circuit, a biquad filter was chosen. The next sections include the PSPICE simulations of the CDTA device characteristics and the filter characteristics. The simulations showed that the proposed CDTA circuit exhibits a very good performance and that the results obtained for the filter are in good agreement with the theory (Kacar and Kuntman 2011).

A CDTA consists of an input current difference and a dual-output transconductance stage. The input stage takes the difference of input signals and transfers this difference current to the intermediate z terminal, where this current is converted to voltage via an external impedance. A CDTA is a five-terminal current-mode active building block. It can be also considered as a current operational amplifier, as in Fig. 10.1, and the defining equation matrix is given in Eq. (10.18).

$$\begin{bmatrix} V_p \\ V_n \\ I_z \\ I_x \end{bmatrix} = \begin{bmatrix} 0 & 0 & 0 & 0 \\ 0 & 0 & 0 & 0 \\ 1 & -1 & 0 & 0 \\ 0 & 0 & 0 & \pm g_m \end{bmatrix} \begin{bmatrix} I_p \\ I_n \\ V_x \\ V_z \end{bmatrix} \tag{10.18}$$

Considering the deviation of the voltage and current gains from their ideal values, the defining equation of the CDTA in Fig. 10.1b becomes

$$\begin{bmatrix} V_p \\ V_n \\ I_z \\ I_x \end{bmatrix} = \begin{bmatrix} 0 & 0 & 0 & 0 \\ 0 & 0 & 0 & 0 \\ \alpha_p & -\alpha_n & 0 & 0 \\ 0 & 0 & 0 & \pm g_m \end{bmatrix} \begin{bmatrix} I_p \\ I_n \\ V_x \\ V_z \end{bmatrix}, \tag{10.19}$$

where αp and αn are current gains, $\alpha_p = 1 - \varepsilon_p$, and $\alpha_n = 1 - \varepsilon_n$. Here, ε_p and ε_n are the current tracking errors, and their absolute values are much less than the unit value (Fig. 10.9).

The differential input current flows over the z terminal. An external impedance is usually connected to this node and the voltage over this impedance is converted to the output currents by the output transconductors with transconductance gm for the positive output and – gm for the negative output. According to above equation and the circuit of Fig. 10.1, the current through terminal z follows the difference of the currents through the terminals p and n ($i_p - i_n$) and flows from terminal z into impedance Zz. The voltage drop at terminal z is transferred to a current at terminal x (i_x) by a transconductance gain (g_m), which is electronically controllable by an external bias current.

CMOS Realization of the CDTA and Simulation Results

A CMOS realization of the CDTA available in the literature is shown in Fig. 10.10a. The proposed CMOS realization of the CDTA element is shown in Fig. 10.10b. Transistors M1 to M24 form the input differential current-controlled current source (DCCCS) stage, which is used for transforming the differential input current to the intermediate voltage, or the voltage at the z terminal. Transistors M25 to M60 form the high-performance dual-output transconductor stage. In the literature, some linear transconductance elements have been presented. The cross-coupled quad configuration proposed, shown in Fig. 10.11, has superior linearity and input voltage range compared with conventional source-coupled differential pairs. In this circuit, the large-signal i–v transfer characteristic will be given by

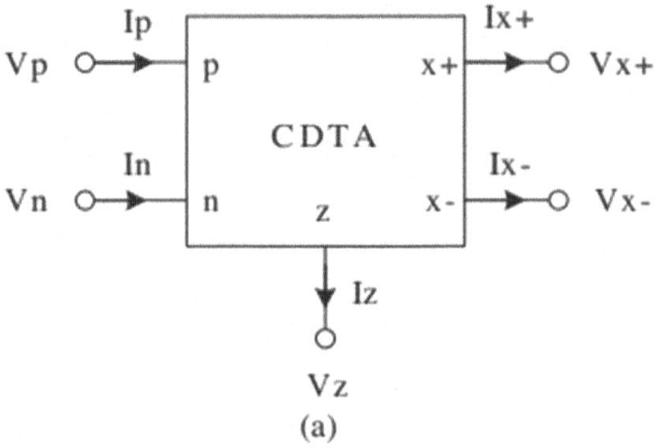

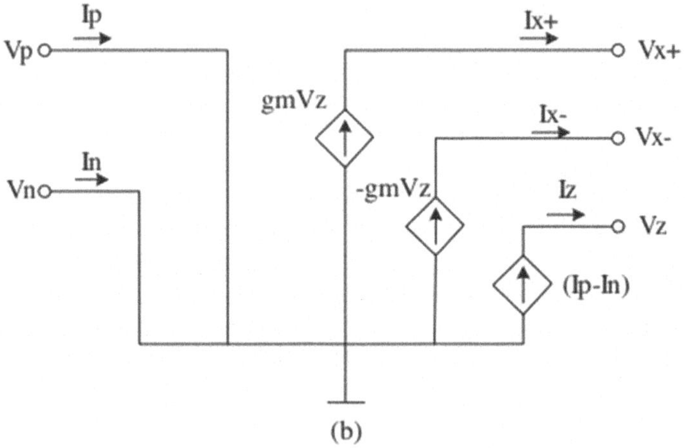

Fig. 10.9 The (**a**) circuit symbol and (**b**) equivalent circuit of the CDTA. (Kacar and Kuntman 2011)

$$\begin{aligned} i &= (n+1)kIv_1 - (kv_2)4I, \\ v &\leq (n+1)Ik \end{aligned} \tag{10.20}$$

This means that the linear region of the input differential voltage v increases by the amount of (n + 1)/2. For this purpose, transistors M_1, M_2, M_3, and M_4 should be identical in channel length, but M_3 and M_4 should be made n times wider than M_1 and M_2. The high-performance output stage, shown in Fig. 10.12, was proposed in Zeki and Kuntman (1997). The two complementary current mirrors are composed of transistors M_{1P}–M_{7P} and M_{1N}–M_{7N}. In the regulated-gate-cascode (RGC) parts of the output stage (transistors M_{2i}, M_{3i}, and M_{4i} for i = P,N), high output impedance is achieved by the negative active-feedback loop along M4i and the source follower M_{3i}. The current mirror M_{6i}–M_{7i} employs the RGC stage to achieve very high output impedance.

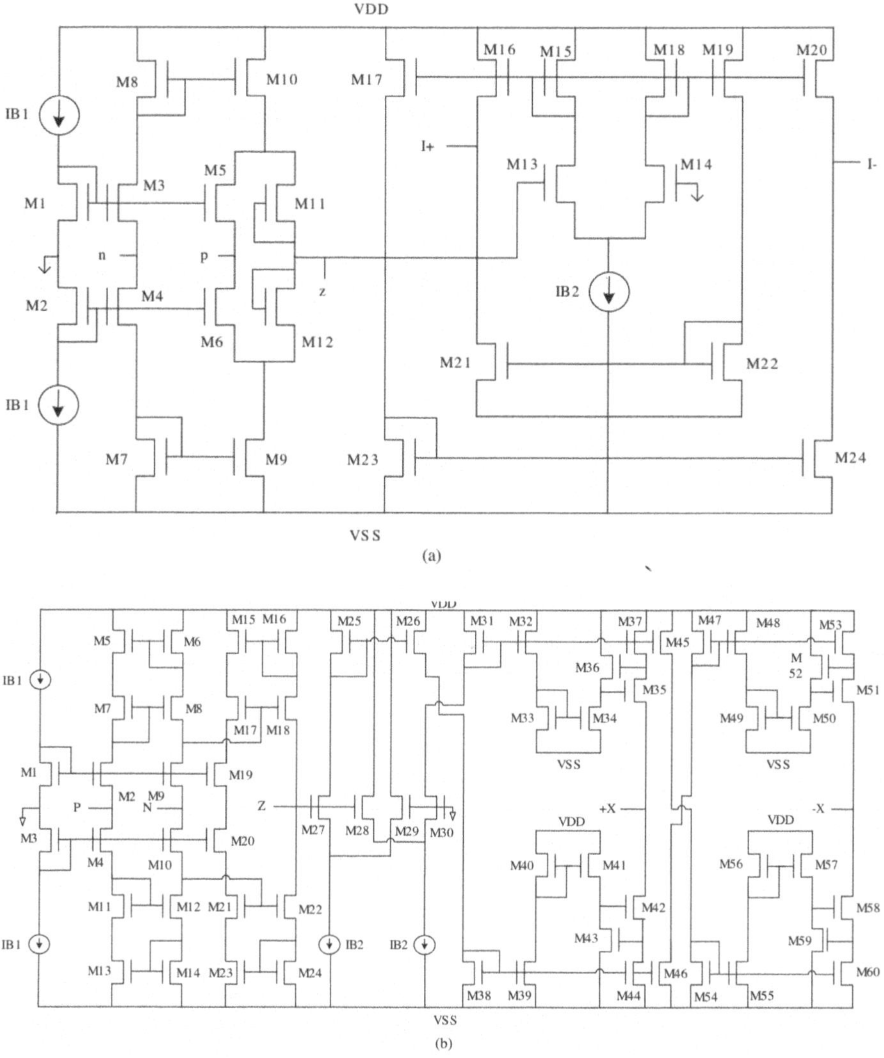

Fig. 10.10 The (**a**) classical (Keskin and Biolek 2006) and (**b**) proposed CMOS implementation of a CDTA. (Kacar and Kuntman 2011)

Input resistances of the p and n terminals and output resistance of z, x+, and x− can be calculated approximately using the following equations:

$$
\begin{aligned}
R_p &\cong \frac{1}{(g_{m2} + g_{m4}) + (g_{ds2} + g_{ds4})} \\
R_n &\cong (r_{ds9} r_{ds8} g_{m9}) // (r_{ds10} r_{ds12} g_{m10}) \\
R_Z &\cong \frac{1}{(g_{m18} + g_{m22}) + (r_{ds18} // r_{ds22})}
\end{aligned}
\qquad (10.21)
$$

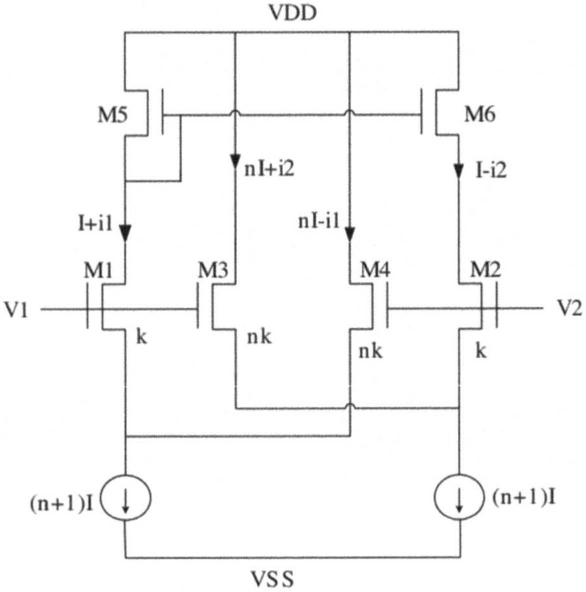

Fig. 10.11 Circuit realization of the cross-coupled quad configuration proposed in Nedungadi and Viswanathan (1984)

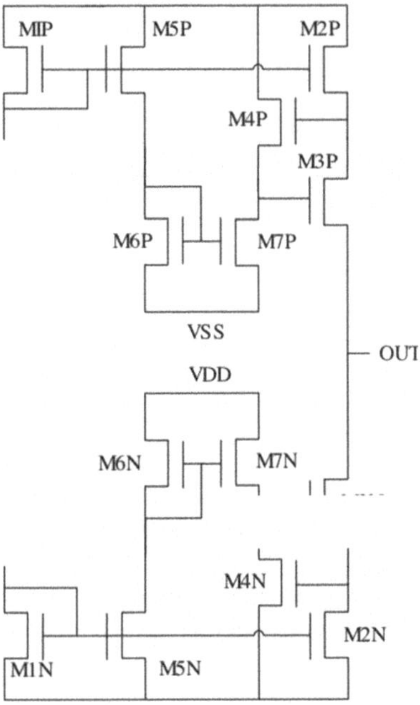

Fig. 10.12 Circuit realization of the current output stage proposed in Zeki and Kuntman (1997)

Table 10.1 Transistor aspect ratios for the proposed circuit

Transistors	$W(\mu m)$	$L(\mu m)$
M_1, M_2, M_9, M_{39}, M_{43}, M_{55}, M_{59}	0.7	0.7
M_3, M_4, M_{10}	2.8	0.7
$M_5 - M_8$, $M_{15} - M_{18}$	28	0.7
$M_{11} - M_{14}$, $M_{21} - M_{24}$	14	0.7
M_{20}, M_{28}, M_{29}, M_{38}, M_{42}, M_{44}, M_{46}, M_{54}, M_{58}, M_{60}	7	0.7
M_{25}, M_{26}	22.4	0.7
M_{27}, M_{30}	2.8	2.8
M_{31}, M_{35}, M_{37}, M_{45}, M_{47}, M_{51}, M_{53}	21	0.7
M_{32}, M_{36}, M_{48}, M_{52}	2.1	0.7
M_{33}, M_{34}, M_{40}, M_{41}, M_{49}, M_{50}, M_{56}, M_{57}	1.75	0.7

$$\begin{aligned} R_x &\cong [g_{m58}g_{m59}r_{ds57}r_{ds60}(r_{ds57}//r_{ds58})]//[g_{m51}g_{m52}r_{ds50}r_{ds53}(r_{ds50}//r_{ds52})] \\ R_{x+} &\cong [g_{m43}g_{m42}r_{ds41}r_{ds44}(r_{ds41}//r_{ds43})]//[g_{m35}g_{m36}r_{ds34}r_{ds37}(r_{ds34}//r_{ds36})] \end{aligned} \tag{10.22}$$

The performance of the proposed CDTA was verified using the PSPICE simulation program. The MOS transistors were simulated using TSMC CMOS 0.35 μm process model parameters. The aspect ratios of the transistors are given in Table 10.1. The supply voltages and biasing currents are given by $V_{DD} = -V_{SS} = 1.5$ V, $I_{B1} = 100$ μA, and $I_{B2} = 50$ μA, respectively.

The input terminals' current transfer characteristics are given in Fig. 10.13, obtained when one input is open-circuited. The input stage transfers the difference of the input currents to the z terminal with good accuracy, as demonstrated in the figures. Since few internal nodes exist over the signal path from the input to the z terminal of the input stage, a high frequency operation is satisfied, exploiting the high-frequency capability of current-mode signal processing. Figure 10.14 shows AC responses of the variation of the current at the z terminal with respect to input currents. Variation of the dual-output transconductance stage's output currents with respect to the input current is given in Fig. 10.15. The lower and upper boundaries of current Ix for the proposed CDTA were determined as $I_{xmax} = 530$ μA and $I_{xmin} = 530$ μA for the X terminal.

The transconductance of the proposed CDTA has a large bandwidth, which makes it suitable for high frequency operations. Transconductance is a very important feature of CDTAs because it directly affects circuit equations. The transconductance for the proposed CDTA is higher than that of classical CDTAs. The frequency responses of the input impedances at terminals n and p and the output impedances z and x are given in Figs. 10.9, 10.10, and 10.11, respectively. The output resistances for the proposed CDTA at terminals Z and X ± were found to be 1.08 GΩ, 178 GΩ, and 167 GΩ, respectively. It can be seen that at Z and X±, the output resistances for the proposed CDTA are much higher than those of the classical CDTA. The simulation results confirm the high linearity and high performance of the

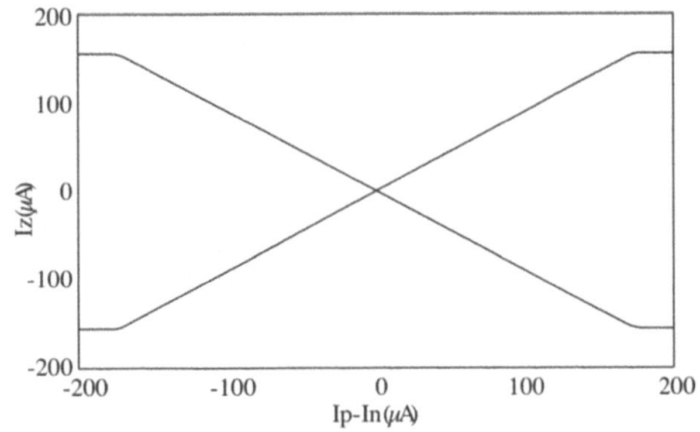

Fig. 10.13 Current transfer from p and n to z. (Kacar and Kuntman 2011)

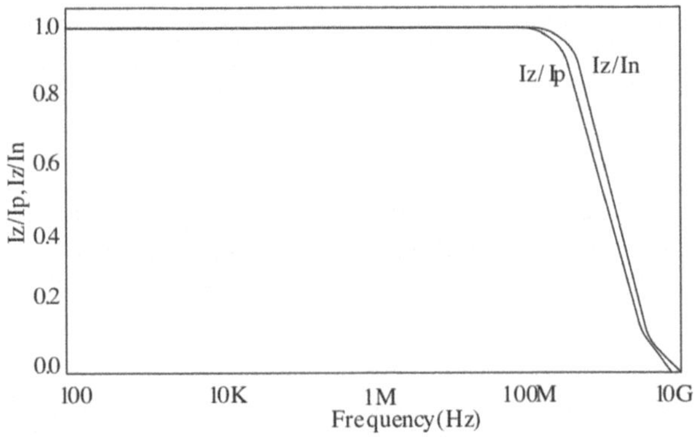

Fig. 10.14 Frequency response of I z /I p and I z /I n. (Kacar and Kuntman 2011)

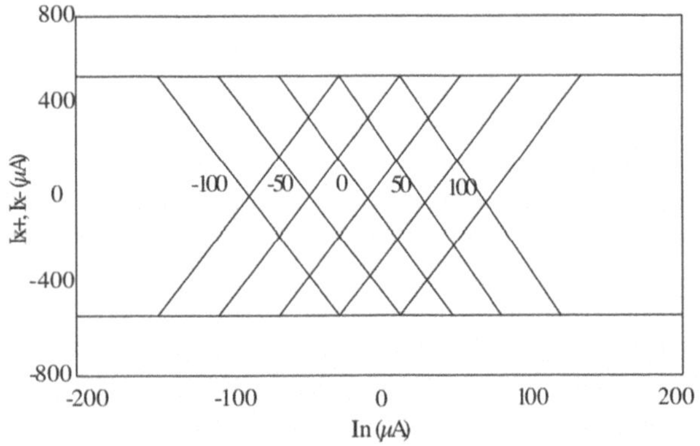

Fig. 10.15 Variation of the z terminal current with respect to input currents (I_P: parametric). (Kacar and Kuntman 2011)

circuit in terms of good input and output resistances and wide bandwidths for both the voltage and current operations. The main performances of the proposed and classical CDTAs are summarized in Table 10.2, where the improvements provided by the proposed circuit can be easily observed. The p and n terminal input impedances were lower than those of the conventional circuit, the linearity range of the output currents I_{X+} and I_{X-} was extended to twice the amount, and the linearity

Table 10.2 Circuit performances of the classical and proposed CDTAs

	Proposed	Classical
Supply voltages	±1.5 V	±2.5 V
Output current range	$I_{B1} = 100\,\mu A$, $I_{B2} = 50\,\mu A$	$I_{B1} = 85\ \mu A$, $I_{B2} = 200\,\mu A$
$I_Z/I_P(-3\,\text{dB})$ bandwidth	609 MHz	632 MHz
$I_Z/I_n(-3\,\text{dB})$ bandwidth	462 MHz	483 MHz
Linearity	$-424\,\mu A \le I_{X+}, I_{X-} \le 412\,\mu A$	$-196\,\mu A \le I_{X+}, I_{X-} \le 198\,\mu A$
	$-240\,mV \le V_{Z+}$, $V_{Z-} \le 240\,mV$ mVA	$-160\,mV \le V_{Z+}$, $V_{Z-} \le 160\,mV$
p input impedance	812 Ω	3.9 k Ω
n input impedance	348 Ω	1.59 kΩ
z output impedance	1.08 MΩ	303 kΩ
X+ output impedance	178 GΩ	211 kΩ
x- output impedance	167 GΩ	268 kΩ
Power consumption	3.61 mW	3.45 mW

range of V_Z was extended by the same amount. The output impedance of the X+ and X− terminals increased from the level of several hundred kΩ to the order of the GΩ level. The Z and x terminals' impedances were close to ideal at low frequencies. Although impedances were seen to decrease at high frequencies, they were high enough for these terminals. As a result, the proposed circuit can be assumed to operate with high performance in a wide frequency range. The power consumption is nearly the same for both CDTA circuits.

Biquad Filter and Simulation Results

A biquad filter is a very important filter for analog signal processing. The proposed universal filter, employing current differencing transconductance amplifiers, is shown in Fig. 10.16. Each of the proposed circuits is composed of two CDTAs. The configuration uses only two capacitors, without any resistors. The proposed current-mode universal biquad is shown in Fig. 10.16. Circuit analysis yields the following filter current transfer functions:

$$\begin{aligned}\frac{I_{HP}}{I_{\text{IN}}} &= \frac{C_1C_2s^2}{g_{m1}g_{m2} + C_2g_{m1}s + C_1C_2s^2},\\ \frac{I_{BP}}{I_{\text{IN}}} &= \frac{C_2g_{m1}s}{g_{m1}g_{m2} + C_2g_{m1}s + C_1C_2s^2},\\ \frac{I_{LP}}{I_{\text{IN}}} &= \frac{g_{m1}g_{m2}}{g_{m1}g_{m2} + C_2g_{m1}s + C_1C_2s^2},\end{aligned} \tag{10.23}$$

The pole angular frequency ωo and quality factor Q are given by:

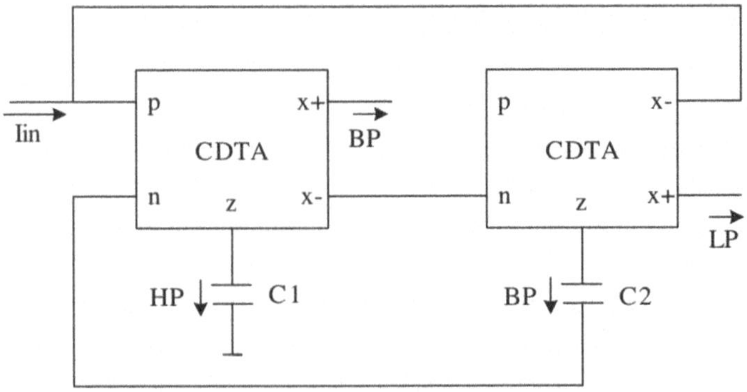

Fig. 10.16 Proposed biquad filters employing CDTAs. (Kacar and Kuntman 2011)

$$\omega_0 = \sqrt{\frac{g_{m1} g_{m2}}{C_1 C_2}},$$
$$Q = \sqrt{\frac{g_{m2} C_1}{g_{m1} C_2}}. \tag{10.24}$$

Finally, to verify the theoretical prediction of the proposed biquad filter, a PSPICE simulation with a 0.35 μm TSMC process was performed. The CMOS implementation of a CDTA is shown in Fig. 10.10b. The aspect ratios of the MOS transistors are given in Table 10.1. The supply voltages were taken as ±1.5 V.

In Fig. 10.12, the current-mode biquad was designed for f0 = 10.7 MHz by choosing gm1 = gm2 = 673 uA/V and C1 = C2 = 1 pF. Figure 10.17 shows the simulated and theoretical response of the low-pass, band-pass, and high-pass filters.

Application Example and Simulation Results

The aim was to construct a double-tuned amplifier with band-pass characteristics employing circuits constructed with CDTAs and passive elements. The center frequency f_0 and the bandwidth B were given as $f_0 = 10.7$ MHz and B = 2 MHz, respectively. There are many applications that require narrow band-pass tuned amplifiers such as video signal processing, TV receivers, and wireless communication stages. By using the filter topology shown in Fig. 10.16, a fourth-order band-pass filter was implemented which is shown in Fig. 10.18. To obtain a filter with a center frequency of 10.7 MHz and a bandwidth of 4.2 MHz, the pole frequencies and the quality factors of the resonant circuits were chosen as $f_{P1} = 7.59$ MHz, $f_{P2} = 15.9$ MHz, $Q_{P1} = 1$, and $Q_{P2} = 1$, respectively. The simulation result of the filter response is given in Fig. 10.18; it was in very good agreement with the predicted theory (Fig. 10.19).

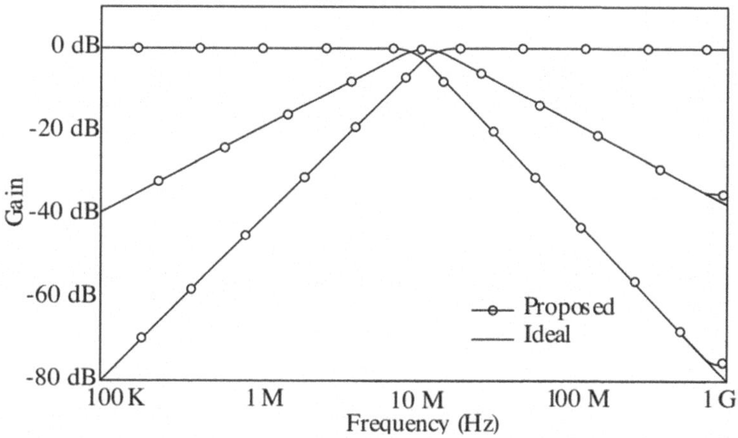

Fig. 10.17 Frequency response of the proposed biquad filter. (Kacar and Kuntman 2011)

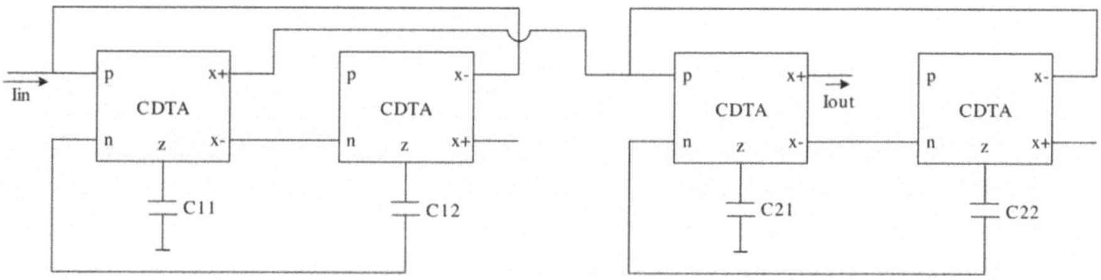

Fig. 10.18 A fourth-order band-pass filter realized with CDTAs. (Kacar and Kuntman 2011)

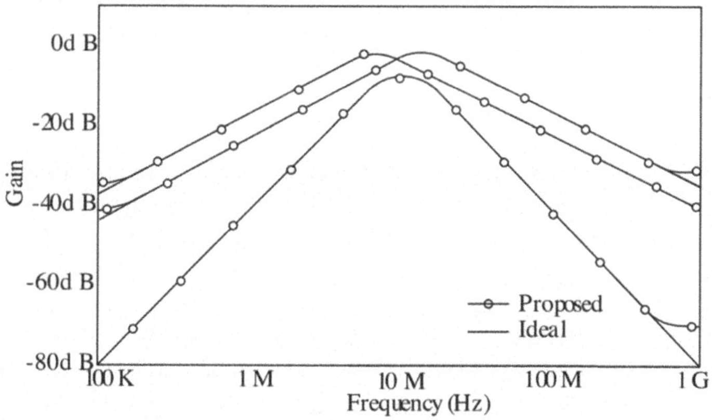

Fig. 10.19 PSPICE simulation results of frequency response of the fourth-order band-pass filter. (Kacar and Kuntman 2011)

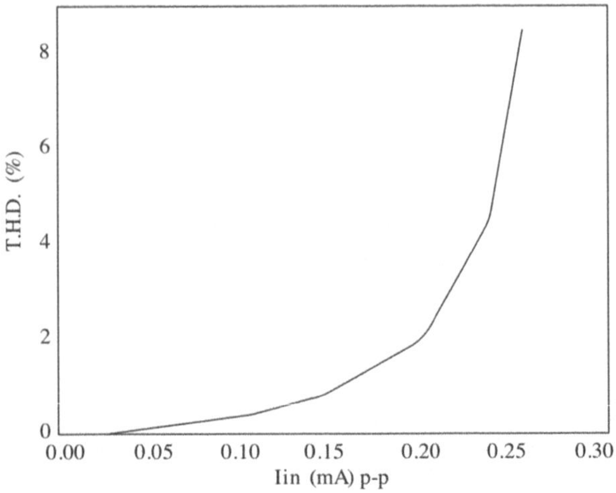

Fig. 10.20 THD of the fourth-order band-pass filter for an input signal of 100 μA at 10 MHz. (Kacar and Kuntman 2011)

The total harmonic distortion (THD) results for the fourth-order BP filter from Fig. 10.18 are given in Fig. 10.20, which clearly shows that for an input signal lower than 200 μA, the THD remains within acceptable limits, thus confirming the practical utility of the proposed circuit. In Fig. 10.20, for an input signal of 100 μA at 10 MHz, the variation of the THD value versus output load resistance is shown. For a load resistance lower than 20 k, the THD is low, less than 3%. Output amplitude change with load resistance for constant input signal is given in Fig. 10.21; it is seen that at 10 MHz, a large swing of $V_O = 1.43$ V is obtained at the output.

Concluding Remarks

In this section, a high-performance CMOS implementation of current differencing transconductance was presented. Simulated device characteristics showed that the proposed circuit exhibits a very good performance. The simulation results confirmed the high performance provided by the circuit in terms of low input, high output resistances, and a wide linearity range for both the voltage and current operations. The performance of the proposed circuit was tested with an application example of a current-mode universal biquad filter. A fourth-order band-pass filter was realized to illustrate the practical use of the proposed topologies as an application example. PSPICE simulation results were given to verify the theory. It was seen that the simulation results verified the theory. The circuit can be considered another good alternative for the circuit designer besides the other topologies in the literature realized with CMOS and BJT technologies.

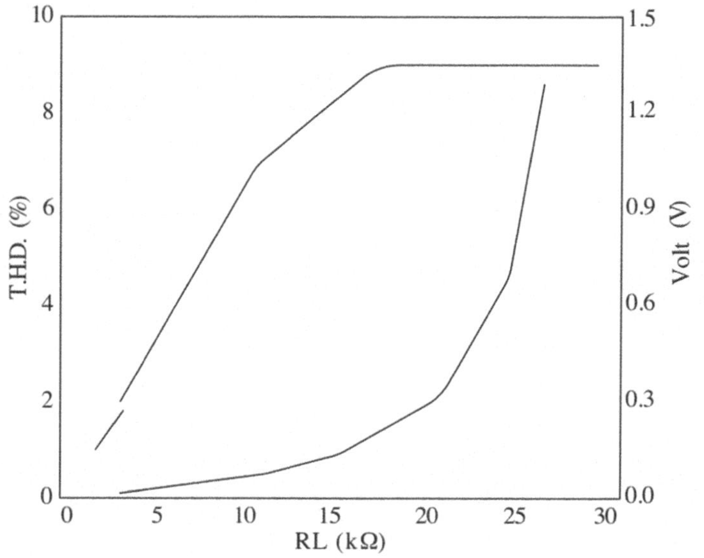

Fig. 10.21 THD and dependence of the output amplitude versus load resistance for an input signal of 100 μA at 10 MHz. (Kacar and Kuntman 2011)

The Voltage Differencing Transconductance Amplifier (VDTA)

The voltage differencing transconductance amplifier (VDTA) is a recently introduced active element for analog signal processing. Differential input voltage (*VP–VN) is* transferred to current at terminal Z by first transconductance gain, and the voltage drop at the terminal Z is transferred to current at terminals X+ and X− (negative of X+) by second transconductance gain. The VDTA is a building block for analog signal processing with two different transconductance values. Transconductance values are electronically controllable by external biasing currents. The proposed filter can operate properly from 90 MHz to 800 MHz (Figs. 10.22 and 10.23).

$$\begin{bmatrix} I_Z \\ I_{X+} \\ I_{X-} \end{bmatrix} = \begin{bmatrix} g_{m1} & -g_{m1} & 0 \\ 0 & 0 & g_{m2} \\ 0 & 0 & -g_{m2} \end{bmatrix} \begin{bmatrix} V_{VP} \\ V_{VN} \\ V_Z \end{bmatrix} \tag{10.25}$$

The VDTA is very flexible in different mode filter realizations. The advantages obtained are that the values of transconductances are adjusted by bias currents. The circuits can be realized without any requirement of external resistors.

Fig. 10.22 VDTA block diagram

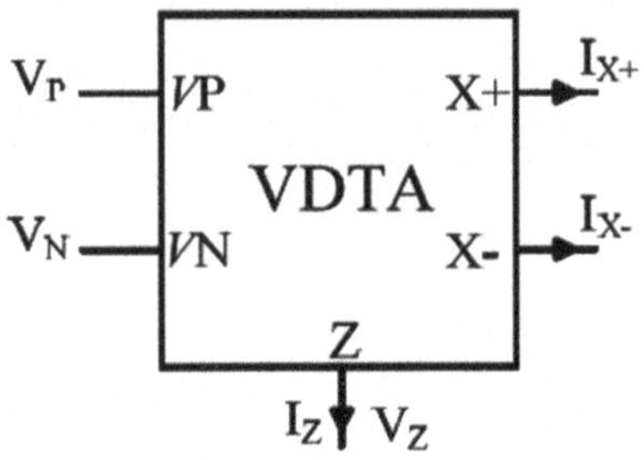

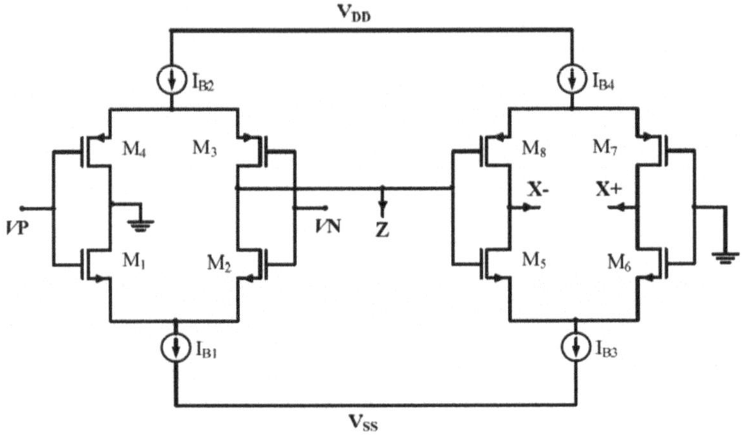

Fig. 10.23 Simple CMOS implementation of the VDTA; *gi is the transconductance value of ith transistor*

Application Example of the VDTA-Based Filter (Fig. 10.24)

The realization of voltage-mode filter with dual input and dual outputs is given. This proposed circuit employing a single VDTA and only two capacitors realizes *LP*, *BP*, and *HP* filter.

$$
\begin{aligned}
&\text{If } V_1 = V_{\text{IN}} \text{ and } V_2 = 0, \text{ then}\\
&BP \rightarrow \frac{V_{\text{OUT1}}}{V_{\text{IN}}} = \frac{sC_1 g_{m1}}{s^2 C_1 C_{2T} + sC_1 g_{m2} + g_{m1} g_{m2}},\\
&LP \rightarrow \frac{V_{\text{OUT2}}}{V_{\text{IN}}} = \frac{g_{m1} g_{m2}}{s^2 C_1 C_{2T} + sC_1 g_{m2} + g_{m1} g_{m2}}.\\
&\text{If } V_2 = V_{\text{IN}} \text{ and } V_1 = 0, \text{ then}\\
&HP \rightarrow \frac{V_{\text{OUT1}}}{V_{\text{IN}}} = \frac{s^2 C_1 C_2}{s^2 C_1 C_{2T} + sC_1 g_{m2} + g_{m1} g_{m2}},\\
&BP \rightarrow \frac{V_{\text{OUT2}}}{V_{\text{IN}}} = \frac{sC_2 g_{m2}}{s^2 C_1 C_{2T} + sC_1 g_{m2} + g_{m1} g_{m2}}
\end{aligned}
\tag{10.26}
$$

The natural frequency ω0 and quality factor *Q* can be given as follows:

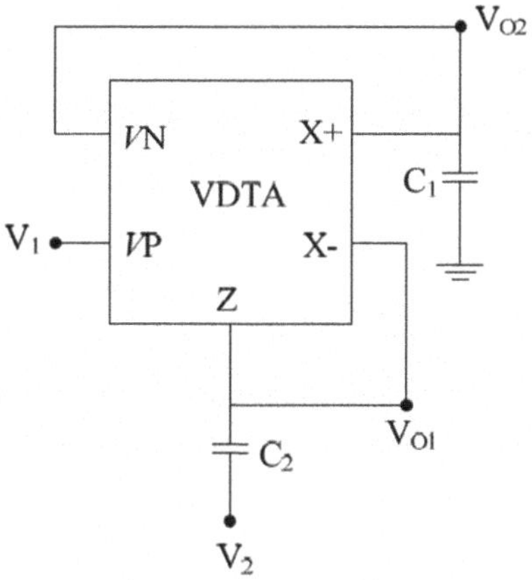

Fig. 10.24 The realization of voltage-mode filter with dual inputs and dual outputs

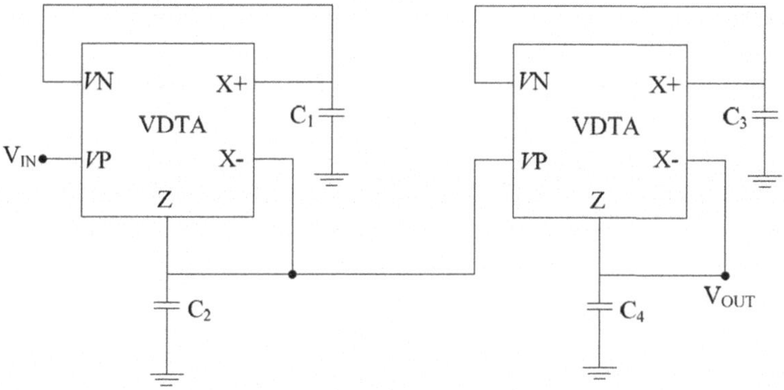

Fig. 10.25 Double tuned amplifier realization

$$\omega_0 = \sqrt{\frac{g_{m1}g_{m2}}{C_1C_{2T}}}, Q = \sqrt{\frac{C_{2T}g_{m1}}{C_1g_{m2}}} \tag{10.27}$$

The passive sensitivities of natural frequency and quality factor for the proposed circuit are given as

$$S_{g_{m1}}^{\omega_0} = S_{g_{m2}}^{\omega_0} = -S_{C_1}^{\omega_0} = -S_{C_{2T}}^{\omega_0} = \frac{1}{2}, S_{g_{m1}}^{Q} = -S_{g_{m2}}^{Q} = S_{C_{2T}}^{Q} = -S_{C_1}^{Q} = \frac{1}{2} \tag{10.28}$$

Double tuned amplifier realization (Fig. 10.25).

$$\frac{V_{OUT}}{V_{IN}} = K_0 \frac{\frac{\omega_{p1}}{Q_{p1}}s}{s^2 + \frac{\omega_{p1}}{Q_{p1}}s + \omega_{p1}^2} \cdot \frac{\frac{\omega_{p2}}{Q_{p2}}s}{s^2 + \frac{\omega_{p2}}{Q_{p2}}s + \omega_{p2}^2} \tag{10.29}$$

The bandwidth *of this band-pass filter is equal* to the diameter of the pole circle

$$B = \Delta\omega/2\pi \tag{10.30}$$

The pole frequencies are determined to slightly different frequencies:

$$\begin{aligned} f_{p1} &= \frac{\omega_{p1}}{2\pi} = f_0 + \frac{B}{2}\sin(45^\circ), \\ f_{p2} &= \frac{\omega_{p2}}{2\pi} = f_0 - \frac{B}{2}\sin(45^\circ) \end{aligned} \tag{10.31}$$

To obtain a filter with a center frequency 100 MHz and bandwidth of 80 MHz, the pole frequencies and quality factors of the proposed circuits are chosen as

$$fp1 = 71.716 \ \text{MHz}, fp2 = 128.284 \ \text{MHz and} \, Qp1 = Qp2 = 1$$

The transconductance and capacitance values providing these properties are determined as

$$\begin{aligned} &g_{m1,2,3,4} = 636.3 \ \mu\text{A/V} \, \text{and} \, C_1 = 0.789 \, \text{pF}, C_2 = 0.639 \, \text{pF}, \\ &C_3 = 1.412 \, \text{pF}, C_4 = 1.262 \, \text{pF} \end{aligned}$$

In this application of the VDTA, the disadvantage caused by external resistors was avoided. The proposed filter is constructed employing only a single active element, VDTA, and two capacitors; the value of transconductances can be adjusted by biasing currents (Figs. 10.26 and 10.27).

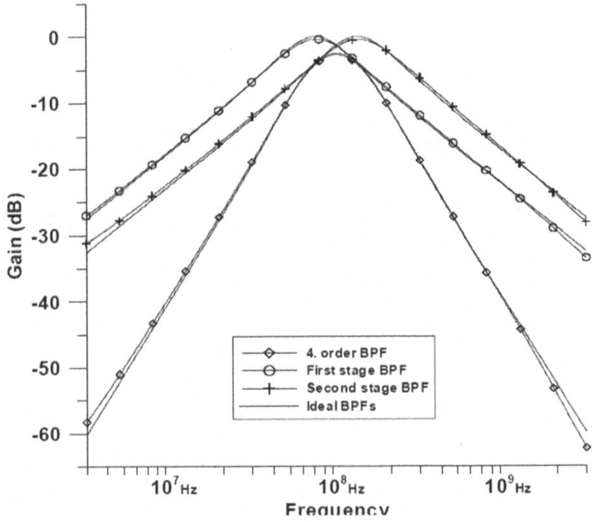

Fig. 10.26 Ideal and simulated frequency responses of the designed double tuned amplifier

Fig. 10.27 Dependence of the center frequency on biasing currents (IBs)

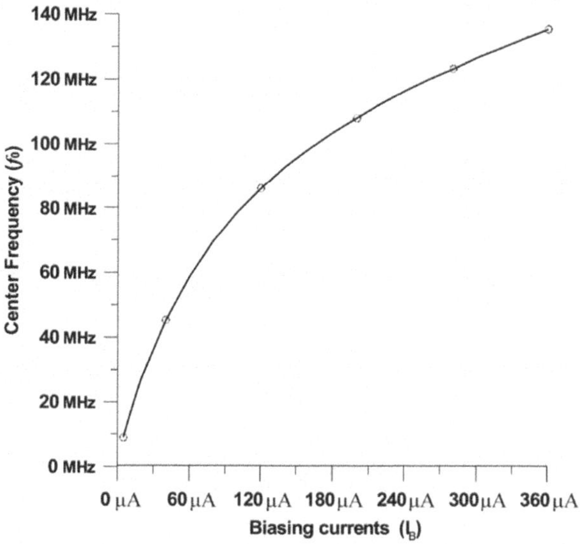

Application Example: Realization of Frequency Agile Filters for Encrypted Communications and Multi-standard Transceivers

General Information

The reconfigurable filter called as frequency agile filter is one of the background circuit multi-standard transceivers. The frequency agile filter allows to sense different protocols with only one circuit. Such a circuit can be a signal processing part of different positioning system bands (GPS, GLONASS, Beidou, GNSS, and Galileo) to process different positioning system protocols in the same circuit. Also, the design of reconfigurable low-pass filter is one of the most critical parts of the Zero-IF receivers (Alaybeyoglu, 2017; Alaybeyoglu et al. 2015; Alaybeyoğlu and Kuntman 2017, 2018).

Furthermore, frequency agile filter can be considered as a part of frequency-hopping system used in military applications for secure communication. Frequency-hopping spread spectrum (FHSS) and direct-sequence spread spectrum (DSSS) are two types of encoding system. In FHSS transmission technology, the data signal is modulated with a narrowband carrier signal hopping arbitrarily frequency to frequency. The random modulation can be predictable by only the desired receivers with the aid of frequency agile filters.

In this section, new recent CMOS realization methods for agile filters suitable for encrypted communication will be presented including the concept, topologies, IC design, layouts, and simulation results. All data is taken from the recent research works performed in Istanbul Technical University. Two different technologies (TSMC 0.18 μm and AMS 0.18 μm) are used to design frequency agile filters.

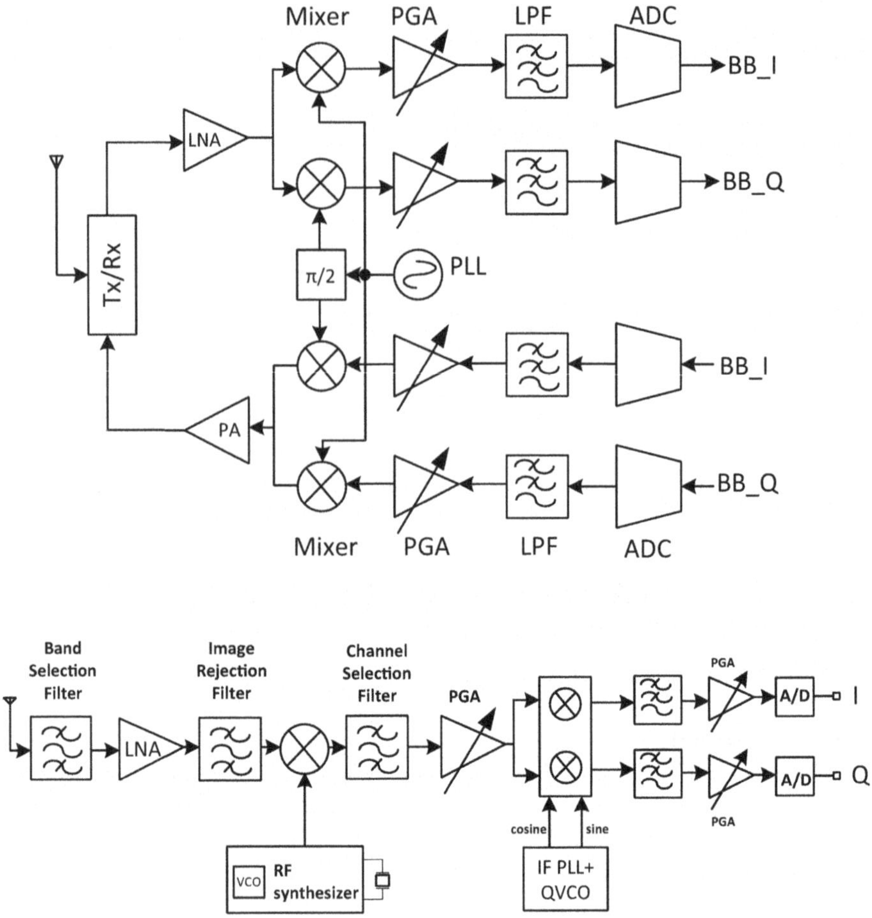

Fig. 10.28 The general aspects of the transceiver and receiver architecture. (Alaybeyoğlu and Kuntman 2017)

To design multi-standard transceiver, LNA, local oscillators, mixers, filters, and such architectures must be reconfigurable. The general aspects of the conventional transceiver and receiver architecture are given in Fig. 10.28.

To make the encrypted communications or secure communications not exploitable by an unauthorized receiver, various techniques of encoding are used. The encoding can be done either on the level of information or on the level of frequency.

Encoding on the level of information consists of applying coding algorithms with the addition of keys to the binary information, to make it indecipherable when the suitable key of deciphering is absent.

The second technique of coding is at the frequency level. This technique is called "spread spectrum": it proceeds by spreading out of the spectrum. The spreading out of the spectrum is currently widely used for wireless communication (the 802.11 family of standards), military, industrial avionics, scientific, and civil usages. This technique makes the emitted signal resistant to the interferences of other signals and the reflections due to multi-path. The transmitted signals can thus share bands with other types of transmission, which makes it possible to use the bandwidth more effectively.

The spreading out of spectrum can be carried out in two different ways: by direct sequence (DSSS) or frequency hopping (FHSS). In a DSSS system, the spreading function is a code word used to modulate the RF carrier. In FHSS the spreading function controls the specific frequency. So for a signal to be audible by a receiver, it is necessary that its frequency-hopping pattern be the same as the frequency-hopping pattern of the transmitter. It will be inaudible for other receivers. Note that for FHSS systems, reconfigurable filters can be used.

This technique was first introduced with an aim of making military communications safe. It consists of emitting on different carriers. Thus, a receiver which does not have the code to synchronize with transmissions will not be able to receive the transmitted signals. In that transceiver system the frequency of the carrier is periodically modified (hopped) following a specific sequence of frequencies as, for example, shown in Fig. 10.29.

For civil use, some walkie-talkies that employ FHSS technology have been developed for unlicensed use on the 900 MHz band.

Typical applications also include cellular deployments for fixed broadband wireless access (BWA). These applications in the unlicensed spectrum use the frequency band known as "2.4 GHz." On the other hand, this technique is still widely used in many military equipment. Additional equipment of encoding of the voice like KY-57 and of information are added to overcome limitations. In the state of the art of current military equipment which uses frequency hopping, one can quote

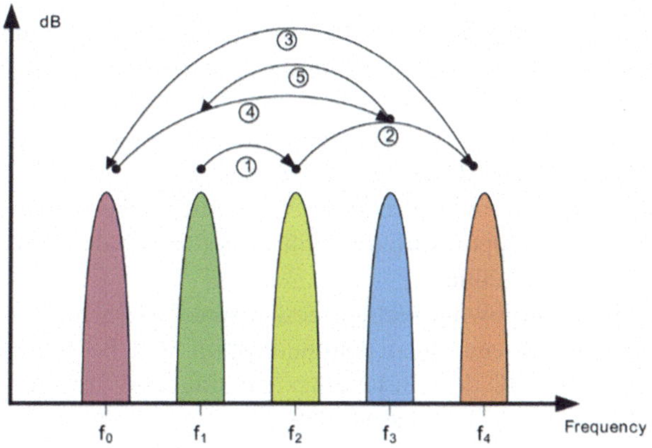

Fig. 10.29 FHSS, example of frequency-hopping pattern (1, 2, 3, 4, 5 ...) of a transmitter. (Alaybeyoğlu and Kuntman 2017)

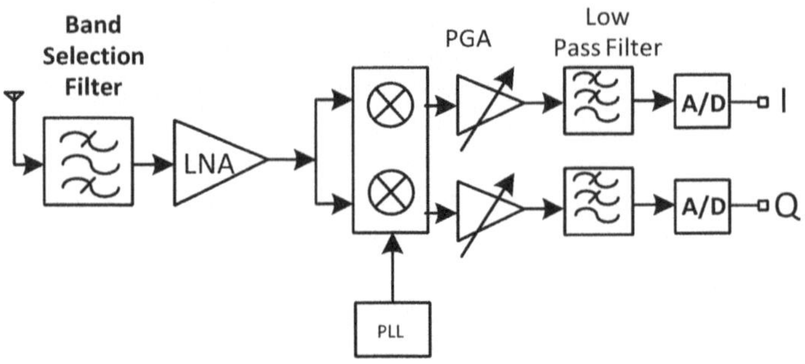

Fig. 10.30 The general structure of Zero-IF receiver. (Alaybeyoğlu and Kuntman 2016b, 2017)

TRANSEC, HAVEQUICK, and SINCGARS, for which the frequency hopping or agility range is up to 500 MHz.

HAVE QUICK (HQ) is a frequency-hopping system used to protect military UHF radio traffic. Progress in electronics in the 1970s reached a point where anyone with an inexpensive radio frequency scanner or receiver set could intercept military communications. HQ program was a response to this problem.

HAVE QUICK is not an encryption system, though many HAVE QUICK radios can be used with encryption, e.g., the KY-58 VINSON system. Single Channel Ground and Airborne Radio System (SINCGARS) is a combat-net radio (CNR) currently used by the USA and allied military forces.

Furthermore, the complexity of the multi-standard transceivers requires new design techniques. For example, multi-standard transceivers must suit for more than one protocol as GSM, GPS, WCDMA, Wi-Fi (IEEE 802.11a/b/g/n), WiMAX, Bluetooth, Zee Bee, and ultra-wideband (UWB). Despite the cost-effectiveness of the multi-standard transceivers, the background of the multi-standard transceivers is more and more complicated than the conventional transceivers. The signal coming from the antenna of any receiver is passing through band-pass filter and low-noise amplifier (LNA), respectively. The signal needs one or more down-conversion to achieve baseband. New trend to design multi-standard receivers is direct conversion receivers to minimize the complexity and chip occupation of the circuit. In this case, the incoming signal to the receiver is passing through directly baseband without channel selection. The block diagram of the direct conversion receiver is given in Fig. 10.30. Direct conversion receiver is also called as Zero-IF receivers. The reconfigurable low-pass filter has great importance in the design of Zero-IF receiver's structures.

At the same time, frequency agile filter is the critical part in the design of receiver circuits to support different protocols for global positioning systems (GPS, Beidou, GNSS, and Galileo).

In the following section, new recent CMOS realization methods for agile filters suitable for encrypted communication will be presented including the concept, topologies, IC design, layouts, and simulation results. The examples are based on CDTA- and VDTA-based filters.

Method Designed with Feedback Technique

Frequency agile filter designed by Fabre and Lakys is constructed with a classical second-order filter which has low-pass and band-pass filter given in Fig. 10.31 (Alaybeyoğlu and Kuntman 2017). This method is constituted with this basic element as low-pass filter output and band-pass filter output. Low-pass output is fed to the input by passing through an amplifier as given in Fig. 10.32.

The general equations for classical voltage-mode band-pass and low-pass filter are given in (Eqs. 10.31) and (10.32), respectively.

$$F_{BP}(s) = \frac{V_{BP}}{V_{\mathrm{IN}}}(s) = \frac{a's}{1 + as + bs^2} \tag{10.31}$$

$$F_{LP}(s) = \frac{V_{LP}}{V_{\mathrm{IN}}}(s) = \frac{d'}{1 + as + bs^2} \tag{10.32}$$

The center frequency of the band-pass filter is $f_0 = 1/2\pi\sqrt{b}$. This value also corresponds to -3 dB cutoff frequency of low-pass filter. Quality factor, the gain at f_0 for the band pass, the band pass -3 dB bandwidth, and the gain of the low-pass filter at low frequencies are given as $Q = \sqrt{b}/a$, $G_{BP} = a'/a$, $\Delta f = a/2\pi b$, and $G_{LP} = d'$, respectively.

The new voltage-mode circuit is given in Fig. 10.5, and the input signal of the system is changed as given in (3). The further calculation for band-pass output proves that band-pass function is unchanged as given in Eq. (10.34). Equation 10.35 also proves the low-pass filter output remains low pass. All these equations are the same for the current-mode approach.

$$V_E = V_{\mathrm{IN}} - A.V_{LP} \tag{10.33}$$

$$F_{BP}(s) = \frac{V_{BP}}{V_{\mathrm{IN}}}(s) = \frac{\frac{a's}{(1-Ad')}}{1 + \frac{as}{(1-Ad')} + \frac{bs^2}{(1-Ad')}} \tag{10.34}$$

$$F_{LP}(s) = \frac{I_{LP}}{I_{\mathrm{IN}}}(s) = \frac{\frac{d'}{(1-Ad')}}{1 + \frac{as}{(1-Ad')} + \frac{bs^2}{(1-Ad')}} \tag{10.35}$$

Fig. 10.31 The basic second-order filter structure. (Lakys et al. 2009, 2011)

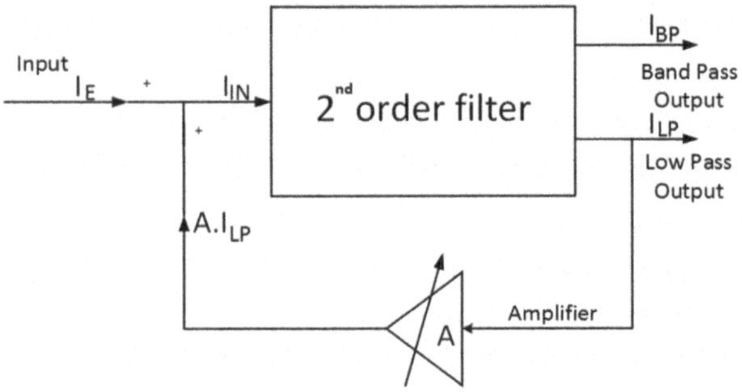

Fig. 10.32 Second-order current-mode frequency agile filter. (Lakys et al. 2009, 2011)

This feedback method is realized with the CDTA's (current differencing transconductance amplifier) second-order biquadratic filter given in Fig. 10.16.

$$\frac{I_{HP}}{I_{\text{IN}}} = \frac{s^2 C_1 C_2}{s^2 C_1 C_2 + s g_{m1} C_2 + g_{m1} g_{m2}} \tag{10.36}$$

$$\frac{I_{BP}}{I_{\text{IN}}} = \frac{s g_{m1} C_2}{s^2 C_1 C_2 + s g_{m1} C_2 + g_{m1} g_{m2}} \tag{10.37}$$

$$\frac{I_{LP}}{I_{\text{IN}}} = \frac{g_{m1} g_{m2}}{s^2 C_1 C_2 + s g_{m1} C_2 + g_{m1} g_{m2}} \tag{10.38}$$

$$\omega_0 = \sqrt{\frac{g_{m1} g_{m2}}{C_1 C_2}} \tag{10.39}$$

$$Q = \sqrt{\frac{g_{m2} C_1}{g_{m1} C_2}} \tag{10.10}$$

$$S_{g_{m1}}^{\omega_0} = S_{g_{m2}}^{\omega_0} = -S_{C_1}^{\omega_0} = -S_{C_2}^{\omega_0} = 0.5 \tag{10.11}$$

$$S_{g_{m1}}^{Q} = S_{g_{m2}}^{Q} = S_{C_1}^{Q} = -S_{C_2}^{Q} = 0.5 \tag{10.12}$$

The filter characteristics of the post-layout simulations are shown in Fig. 10.33. The capacitance values of C_1 and C_2 are 5 pF. C_1, C_2 are especially determined for appropriate center frequency range to support different protocols for global positioning systems (GPS, Beidou, GNSS, and Galileo) and SINCGARS uses 25 kHz channels in the very-high-frequency (VHF) FM band, from 30 to 87.975 MHz.

The complete implementation of the frequency agile band-pass filter operating in current mode that has been directly deduced from the scheme in Fig. 10.34 for gain A is shown in Fig. 10.8. The output of the frequency agile filter is given in Fig. 10.35.

The configurability of the agile filter is realized with switches. All feedback CDTA's g_m is designed according to the positioning systems and SINCGARS.

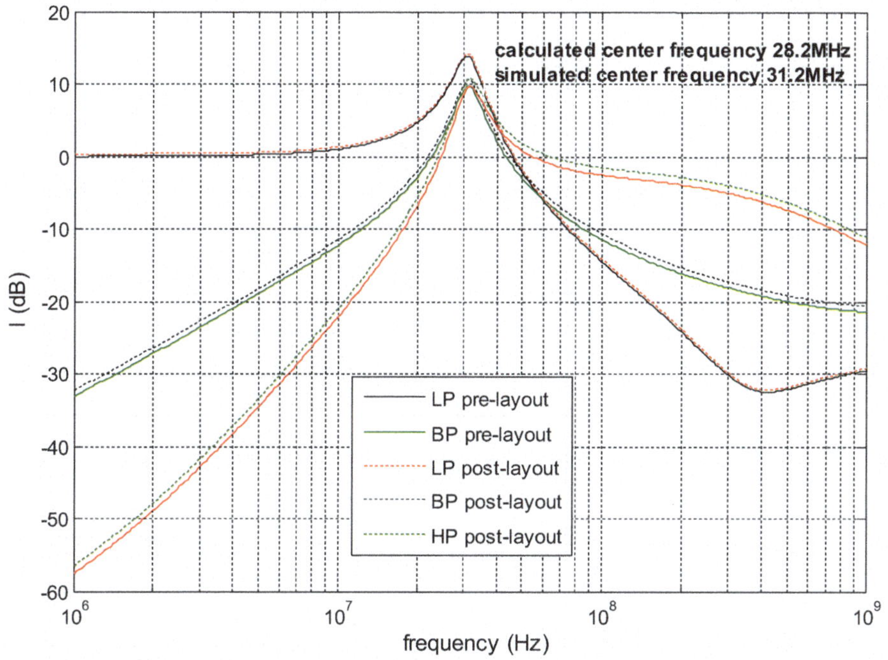

Fig. 10.33 CDTA-based biquadratic filter structure. (Alaybeyoğlu and Kuntman 2016b, 2017)

The transconductance of the CDTAs used in frequency agile filter structure is given in Fig. 10.36. It is obviously seen that the higher g_{m_fb} is used to achieve higher frequencies. The transient analysis of the designed filter at 54.9 MHz is given in Fig. 10.37. The distortion for 100 μA pp. input current is 3.8%. The layout of the overall design is given in Fig. 10.38.

The total chip area occupies 132 μm × 73 μm. Further completion of layouts is required to obtain clean Design Rule Check (DRC), and ElectroStatic Discharge (ESD) protected IC from fabrication. The maximum power consumption of the frequency agile filter working at 88 MHz is 8.53 mW.

The new band-pass function is given in Eq. (10.13). The center frequency and quality factor of the designed filter are changed according to Eqs. (10.14) and (10.15), respectively.

$$\frac{I_{BP}}{I_E} = \frac{\frac{C_2}{g_{m2}\left(1-g_{m\_fb}R_z\right)}s}{1+\frac{C_2}{g_{m2}\left(1-g_{m\_fb}R_z\right)}s+\frac{C_1C_2}{g_{m1}g_{m2}\left(1-g_{m\_fb}R_z\right)}s^2} \tag{10.13}$$

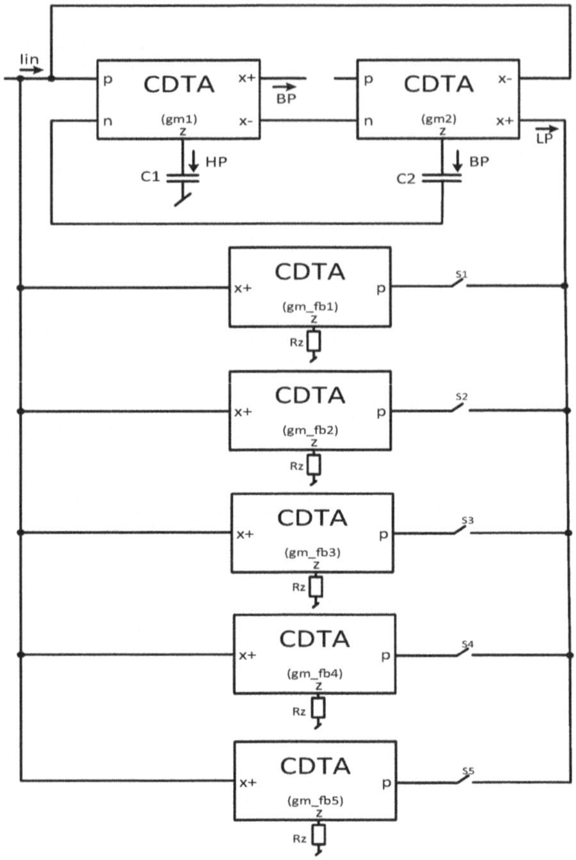

Fig. 10.34 CDTA-based biquadratic filter structure. (Alaybeyoğlu and Kuntman 2016b, 2017)

$$\omega_0 = \sqrt{1 - g_{m\_fb} R_z} \sqrt{\frac{g_{m1} g_{m2}}{C_1 C_2}} \tag{10.14}$$

$$Q = \sqrt{1 - g_{m\_fb} R_z} \sqrt{\frac{C_1 g_{m1}}{g_{m2} C_2}} \tag{10.15}$$

The Butterworth filter is designed for military application of SINCGARS. Figure 10.9 proves that the proposed filter can operate between the range of 30 and 87.975 MHz. There is no more information about SINCGARS due to security reasons. Some criteria as quality factor, stop-band attenuation, passband gain, and characteristics of the designed filter can be adjusted for the application frequency range of 30–87.975 MHz.

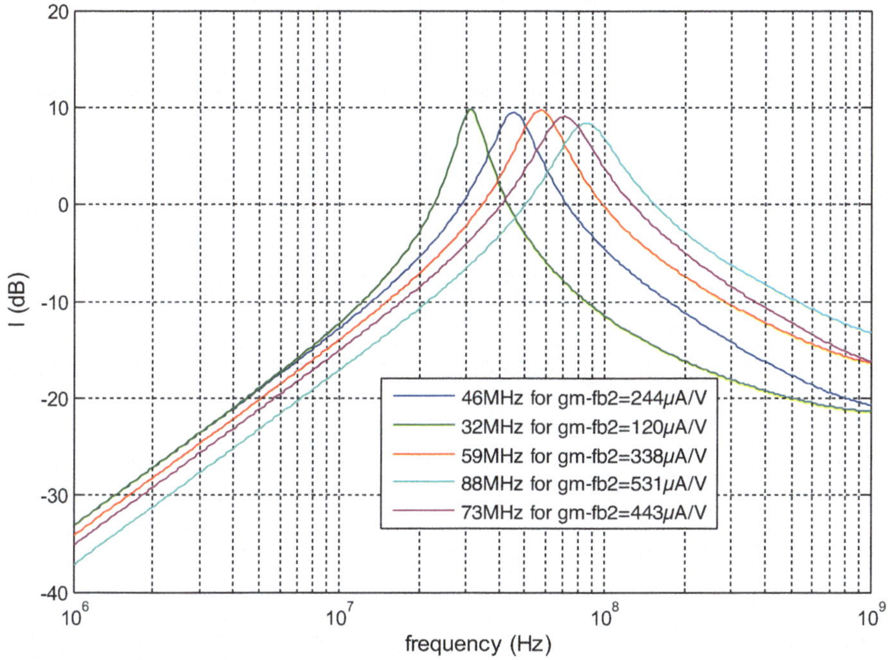

Fig. 10.35 The output of the frequency agile filter designed with feedback technique. (Alaybeyoğlu and Kuntman 2016b, 2017)

VDTA-Based Design

The CMOS realization of the VDTA can be implemented by using two floating current sources proposed by Arbel and Goldminz in 1992. But the improved version of the floating current source is more flexible to obtain high-quality factor Q for analog filter design. The CMOS realization of the voltage differencing transconductance amplifier is given in Fig. 10.39.

The layout of the voltage differencing transconductance amplifier is given in Fig. 10.40. There are nine big pieces in the layout. These pieces are bipolar transistors of temperature-independent current source. Two of the bipolar transistors are used as dummy for good matching.

The layout occupies 248 μm × 90 μm area (0.022 mm^2) with biasing circuit and temperature-independent current generator. The voltage differencing transconductance amplifier occupies only 35 μm × 54 μm area on the chip. The capacitor sizes are selected as 4 μm × 6.5 μm to realize the value of 50 fF.

The CMOS implementation is a significant advantage of the designed VDTA-based voltage-mode filter. In contrary to BiCMOS implementation, CMOS production is cheaper. Thanks to improved floating current source's very basic CMOS

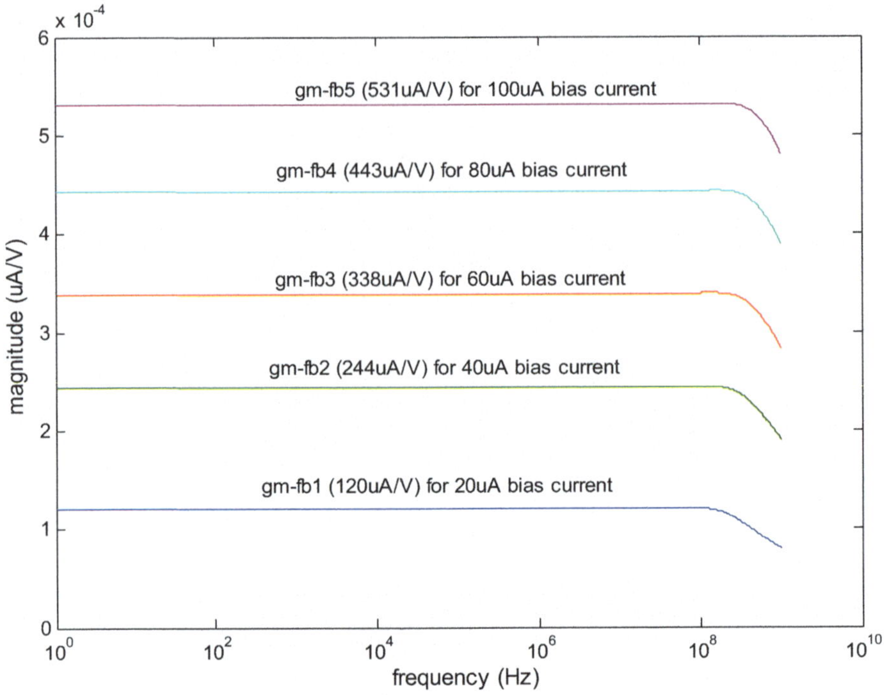

Fig. 10.36 The transconductance of the CDTAs. (Alaybeyoğlu and Kuntman 2016b, 2017)

implementation, the chip size occupies very less chip area. Furthermore, the same technique is also considered to be used in passive radar receivers which are operating up to 1 GHz frequencies.

Single-block VDTA second-order frequency agile filter structure is given in Fig. 10.41 (Alaybeyoğlu and Kuntman 2016a, b, c). The transfer functions of the low-pass and band-pass output are given in Eqs. (10.24) and (10.25), respectively. Equation 10.26 shows the pole angular frequency. Equation 10.27 gives the quality factor of the filter.

$$\frac{V_{BP}}{V_{\text{in}}} = \frac{sC_1 g_{m1}}{s^2 C_1 C_2 + sC_1 g_{m1} + g_{m1} g_{m2}}$$

$$\frac{V_{LP}}{V_{\text{in}}} = \frac{g_{m1} g_{m2}}{s^2 C_1 C_2 + sC_1 g_{m1} + g_{m1} g_{m2}}$$

$$\omega_0 = \sqrt{\frac{g_{m1} g_{m2}}{C_1 C_2}}$$

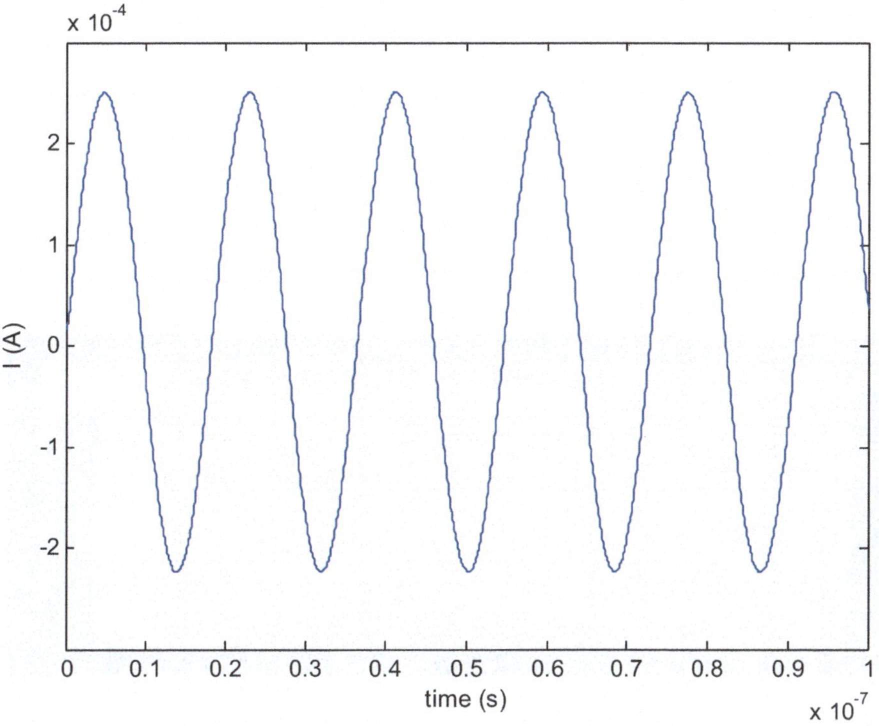

Fig. 10.37 Transient analysis with respect to input current at 54.9 MHz. (Alaybeyoğlu and Kuntman 2016b, 2017)

Fig. 10.38 The layout of the overall design. (Alaybeyoğlu and Kuntman 2016b, 2017)

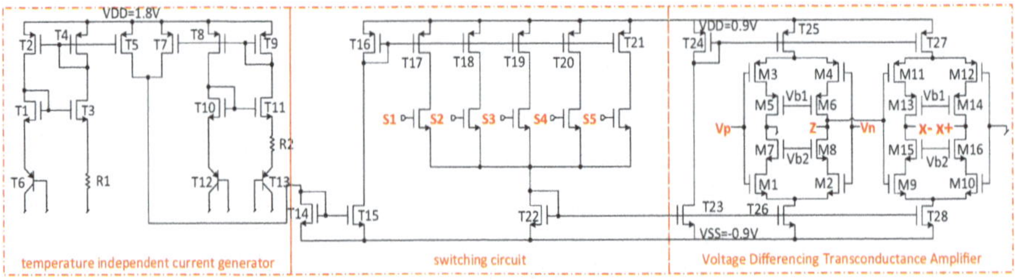

Fig. 10.39 CMOS realization of the switched VDTA. (Alaybeyoğlu and Kuntman 2016c, 2017)

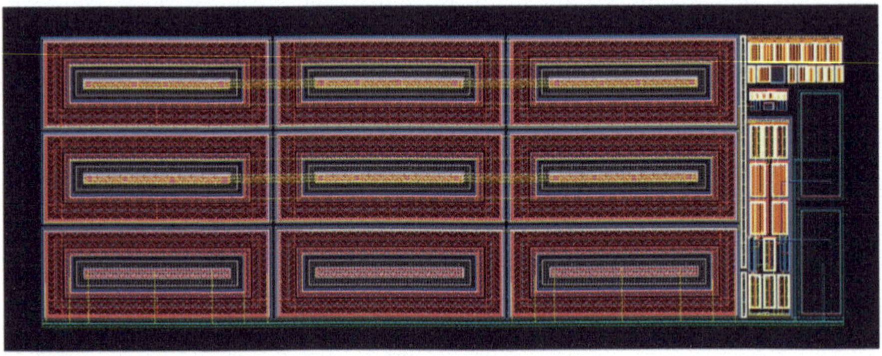

Fig. 10.40 The layout of the frequency agile filter enhanced with g_m-C. (Alaybeyoğlu and Kuntman 2016c, 2017)

$$Q = \sqrt{\frac{C_2 g_{m1}}{C_1 g_{m2}}}$$

Figure 10.42 shows the output of the designed frequency agile filter. The proposed filter can operate properly from 90 MHz to 800 MHz. Table 10.3 gives the on/off positions to obtain the different band-pass outputs given in Fig. 10.42. The tuning ratio of the designed filter can improve with different biasing currents. The transient responses of the filter according to the switch on/off position are given in Figs. 10.43 and 10.44, respectively. The maximum on/off time of the designed filter is 5 ns.

The maximum noise of the designed filter with g_m-C filter is 152 nV/sqrt(Hz). Worst-case analysis is performed for the maximum frequency of the designed circuit to check its behavior under various conditions of temperature and supply voltages. The fabrication conditions are selected with six parameters (wp, ws, wo, wz, fff, ssf). The temperature conditions are selected (−50°, 120°). The supply voltage conditions are selected (±1.1 V, ±0.9 V). The band-pass characteristics are preserved under these worst-case conditions.

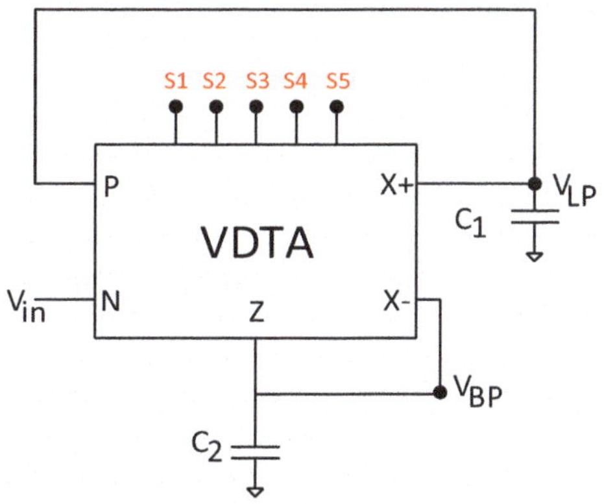

Fig. 10.41 VDTA second-order frequency agile filter structure. (Alaybeyoğlu and Kuntman 2016c, 2017)

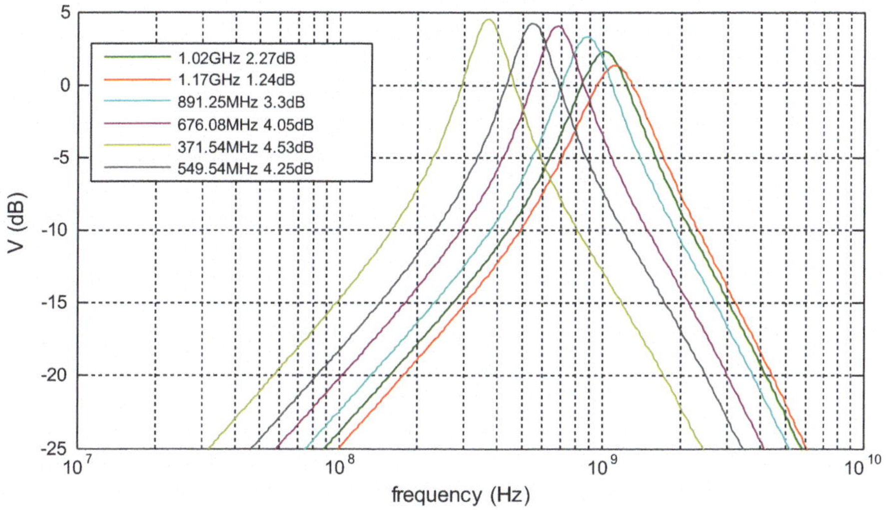

Fig. 10.42 The output of the proposed frequency agile filter enhanced with g_m-C. (Alaybeyoğlu and Kuntman 2016c, 2017)

The standard deviation of the center frequency for 200 random cases is found to be 49.417 MHz. All these solutions are acceptable for the analog filter design procedure. The variation of process, temperature, and supply voltage (PVT) can be eliminated with automatic tuning circuits.

Table 10.3 Switch on/off positions

Center frequency	Switches on
1.17 GHz	S_1, S_2, S_3, S_4, S_5
1.02 GHz	S_2, S_3, S_4, S_5
891.25 MHz	S_3, S_4, S_5
676.08 MHz	S_4, S_5
549.54 MHz	S_1
371.54 MHz	S_5

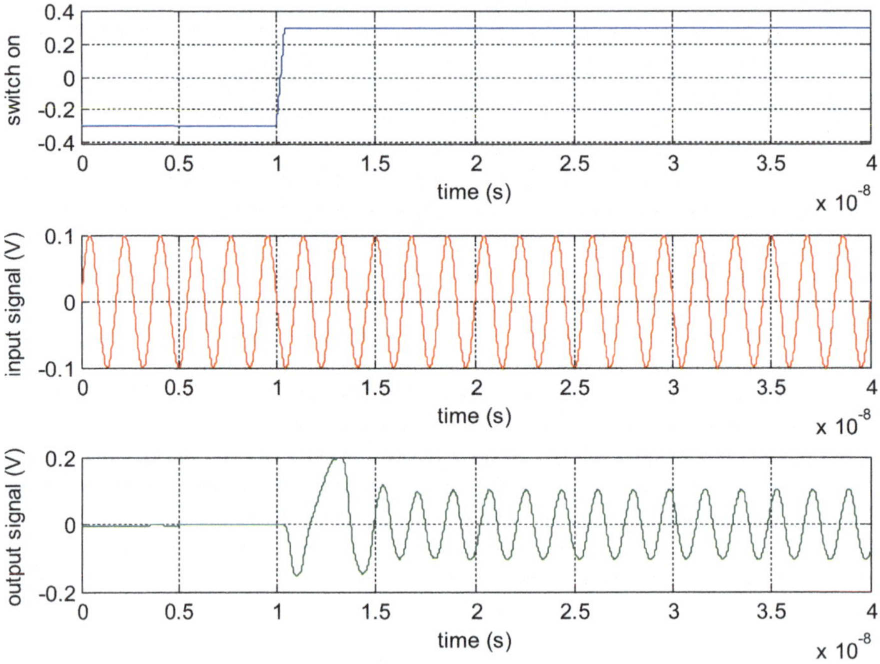

Fig. 10.43 Transient response of the frequency agile filter off/on time. (Alaybeyoğlu and Kuntman 2016c, 2017)

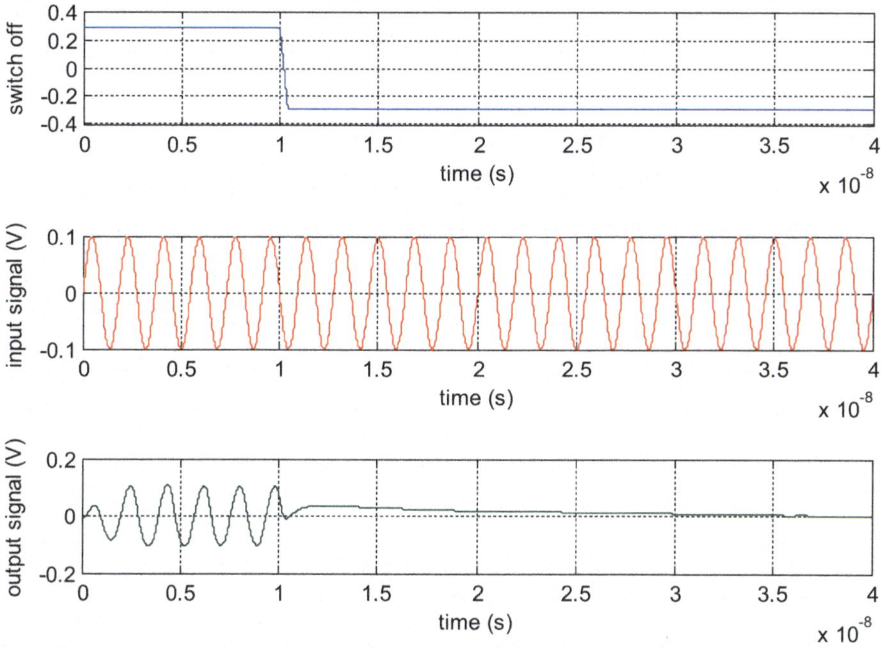

Fig. 10.44 Transient response of the frequency agile filter on/off time. (Alaybeyoğlu and Kuntman 2016c, 2017)

References

A. D. Inc.: AD844 current feedback op-amp. Data Sheet.

Acar, C., & Ozoguz, S. (1999). A new versatile building block: current differencing buffered amplifier suitable for analog signal-processing filters. Microelectronics journal, 30(2), 157–160.

Alaybeyoglu, E. (2017). New Possibilities in the Design of CMOS Frequency Agile Filters. Ph. D. thesis.

Alaybeyoglu, E., Atasoyu, M., & Kuntman, H. (2015, July). Frequency agile filter structure improved by MOS-only technique. In 2015 38th International Conference on Telecommunications and Signal Processing (TSP) (pp. 1–5). IEEE.

Alaybeyoğlu, E., & Kuntman, H. (2015). Sadece MOS Tekniği ile Geliştirilmiş Yüksek Performanslı Atik Süzgeç Tasarımı High Performance Frequency Agile Filter Structure Designed with MOS-Only.

Alaybeyoğlu, E., & Kuntman, H. (2016a). A new CMOS ZC-CDTA realization and its filter applications. Turkish Journal of Electrical Engineering & Computer Sciences, 24(3), 746–761.

Alaybeyoğlu, E., & Kuntman, H. (2016b). A new frequency agile filter structure employing CDTA for positioning systems and secure communications. Analog Integrated Circuits and Signal Processing, 1–11.

Alaybeyoğlu, E., & Kuntman, H. (2016c). CMOS implementations of VDTA based frequency agile filters for encrypted communications. Analog Integrated Circuits and Signal Processing, 1–10.

Alaybeyoğlu, E. and Kuntman H.(2017). New Realization Methods of Frequency Agile Filters for Encrypted Communications and Multi-Standard Transceivers (Invited Talk), Proc. of

ICTUS'2017: International Conference on Infocom Technologies and Unmanned Systems, pp. 99–112, Amity University, Dubai, 18–20 December 2017.

Alaybeyoğlu, E. and Kuntman H., (2018). A new implementation of the reconfigurable analog baseband low pass filter with cell-based variable transconductance amplifier, Analog Integrated Circuits and Signal Processing, 97, 87–96.

Arbel, A., & Goldminz, L. (1992). Output stage for current-mode feedback amplifiers, theory and applications. Analog Integrated Circuits and Signal Processing, 2(3), 243–255.

Biolek, D., Senani, R., Biolkova, V., & Kolka, Z. (2008). Active elements for analog signal processing: classification, review, and new proposals. Radioengineering, 17(4), 15–32.

Kacar, F., & Kuntman, H. H. (2011). A new, improved CMOS realization of CDTA and its filter applications. Turkish Journal of Electrical Engineering & Computer Sciences, 19(4), 631–642.

Keskin, A. Ü., & Biolek, D. (2006). Current mode quadrature oscillator using current differencing transconductance amplifiers (CDTA). IEE Proceedings-Circuits, Devices and Systems, 153(3), 214–218.

Lakys, Y., Godara, B., & Fabre, A. (2009, November). Cognitive and encrypted communications, part 2: A new approach to active frequency-agile filters and validation results for an agile bandpass yopology in SiGe-BiCMOS. In 2009 International Conference on Electrical and Electronics Engineering-ELECO 2009 (pp. II–16). IEEE.

Lakys, Y., Godara, B., & Fabre, A. (2011). Cognitive and encrypted communications: state of the art and a new approach for frequency-agile filters. Turkish Journal of Electrical Engineering & Computer Sciences, 19(2), 251–273.

Nedungadi, A. & Viswanathan, T. R. (1984), Design of Linear CMOS Transconductance Elements IEEE Trans. Circuits System, vol CAS31, pp 891–889.

Özcan, S., Kuntman, H., & Çiçekoğlu, O. (2002). Cascadable current mode multipurpose filters employing current differencing buffered amplifier (CDBA). AEU-International Journal of Electronics and Communications, 56(2), 67–72.

Özcan, S., Toker, A., Acar, C., Kuntman, H., & Çiçekoğlu, O. (2000). Single resistance-controlled sinusoidal oscillators employing current differencing buffered amplifier. Microelectronics Journal, 31(3), 169–174.

Zeki, A. & Kuntman, H. (1997), A CMOS CCII suitable for continuous-time active networks Proc. Int. Syrnp. On circuits and Systems (SCS '97), Iasi, Romania, 2, 393–396.

Chapter 11
DTMOS-Based Ultra-Low-Voltage Low-Power Circuit Design

Basic Concept

The demand for portable applications has continuously increased during the last decade. Smart phones, tablet computers, netbooks, and several wireless devices are everywhere. However, this trend of ever-decreasing supply voltage levels of circuits for low power consumption has raised the necessity to design analog circuits that are capable of operating under very low supply voltage levels. That severely limits the performance of analog circuits which share supply voltages with digital circuits. There is a strong need for new ideas and perspectives to analog circuit design to meet the requirements of modern electronic devices. Until now, especially in digital circuits, decreasing supply voltage levels for transistors having thinner gate oxides has been utilized as a solution to lower the power consumption and increase chip density by laying out more transistors on a smaller chip area. However, further decreasing the channel lengths has created leakage current problems. Moreover, threshold voltages cannot be kept below certain limits to minimize the leakages in the chip. Conventional analog circuits suffer from very low supply voltages of digital circuits and relatively high threshold voltage levels to prevent large leakage currents in standard CMOS process technology. As a solution to the problems of conventional circuits, DTMOS, dynamic threshold voltage metal oxide transistor, was presented by Assederaghi et al. (1994) which operates as a low-leakage as well as low-voltage, low-power device.

H. Kuntman, D. Özenli, *Trends in Circuit Design for Analog Signal Processing*,
Analog Circuits and Signal Processing,
https://doi.org/10.1007/978-3-030-96836-6_11

DTMOS Transistor

DTMOS transistor shows high threshold characteristic when it is off to minimize the leakage, and at the same, it behaves as a low threshold device for high current drivability under low voltage supplies. This feature makes it as a promising candidate for modern ultra-low-voltage analog circuits. As shown in Fig. 11.1a, a PMOS transistor's gate and body terminals are connected without requiring any additional processing steps and it can be produced in standard CMOS technology. DTMOS transistor and its commonly used circuit symbol are shown in Fig. 11.1. This configuration however goes back to earlier dates. Assederaghi et al. extensively describe the device and express the underlying reasons of its operation. The idea is to connect the gate and body of the device to dynamically change the threshold voltage of the transistor by utilizing the relation

$$V_{\mathrm{TH}} = V_{\mathrm{TO}} + \gamma\left(\sqrt{|2\Phi_{\mathrm{F}}| + V_{\mathrm{SB}}} - \sqrt{|2\Phi_{\mathrm{F}}|}\right). \tag{11.1}$$

DTMOS transistor, under the same V_{GS} voltage, behaves as a high-transconductance MOSFET. As it is seen in Fig. 11.2, it conducts more current than a regular MOSFET. In the figure, V_{DS} is kept at −0.1 V constant voltage, while V_{GS} is swept from −0.4 V to 0 V. The reason behind this mechanism is the threshold voltage reduction due to the positive source body voltage (Uygur and Kuntman 2013a, b, 2014).

DTMOS transistor, under the same V_{GS} voltage, behaves as a high transconductance MOSFET. As it is seen in Fig. 11.2, it conducts more current than a regular MOSFET. In the figure, V_{DS} is kept at −0.1 V constant voltage, while V_{GS} is swept from −0.4 V to 0 V. The reason behind this mechanism is the threshold voltage reduction due to the positive source body voltage. DTMOS with its plain structure is not usable for supply voltages exceeding 0.7 V. Although it is possible to use it with an addition of an extra limiter transistor (Assederaghi et al. 1997), this would almost double the chip area for digital circuits, and additionally, it increases parasitic effects. Furthermore, the operation of all chip components strongly depends

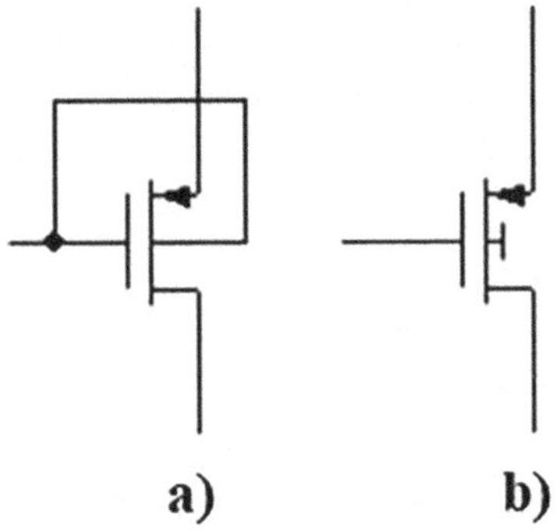

Fig. 11.1 DTMOS transistor and its circuit symbol. (**a**) Bulk-gate connection, (**b**) Circuit symbol

Fig. 11.2 DTMOS and standard MOS drain currents

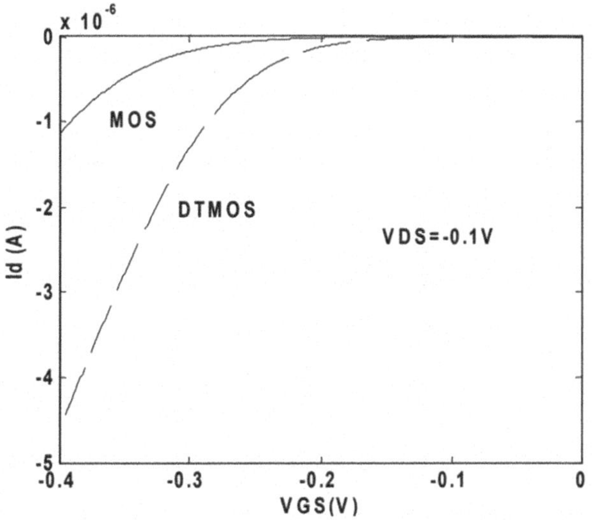

on those limiter transistors which decreased robust operation performance because any high on diode current totally disrupts transistor operation. For these reasons, supply voltages are chosen low enough to limit any forward-biased diode currents in this study. In the literature, transistors with source body junctions forward biased close to 0.4–0.5 V voltage levels are used without transistor operation being affected by the parasitic elements (Zhang et al. 2006, Chatterjee et al. 2005). The main reason is that the mobile carrier concentrations caused by parasitic elements for supply voltages less than 0.5 V do not reach high levels in modern highly doped bodies to affect the overall transistor operation. Another question might arise on whether the conventional MOSFET models are sufficient for modeling properly the operation of the DTMOS transistor. Actually, mostly used MOSFET models such as BSIM and EKV are developed under the assumption that the channel is free from mobile carriers, which is the total depletion approximation. However, for a DTMOS transistor, this is not the case because there are mobile carriers and the total depletion approximation is not valid now. Additionally, vertical drain body and source body junction currents add another dimension and this might require two-dimensional device models. However, these are not necessary if the supply voltage is kept below 0.4–0.5 V voltage levels and the device channel length is not chosen very small to prevent short channel effects. Conventional models are still applicable to DTMOS transistor to model the device and the related circuits provided that the mentioned specifications exist. Therefore, we have used 0.4 V supply voltage and transistors with minimum channel lengths 2 μm in our designs to be in agreement with results and compact models experimentally proven in the related literature.

The advantages provided by DTMOS transistors are demonstrated on two examples, namely, on a DTMOS-based 0.4 V ultra-low-voltage low-power VDTA design and on a 0.4 V DTMOS CCII circuit design (Uygur and Kuntman 2013a, b, 2014).

DTMOS-Based 0.4 V Ultra-Low-Voltage Low-Power VDTA Design and Its Application to EEG Data Processing

In this section, we have used DTMOS transistors in our VDTA design to enable ultra-low-power, ultra-low-voltage operation under ±0.2 V symmetric supply voltages. Then, this design was applied to a fourth-order double-tuned band-pass filter for EEG data processing (Berg and Mirmotahari, 2011). There are other low-power filter designs in the literature. For example, the filter circuit in the paper by Lee et al. (Lee et al. 2004, 2008) uses switched op-amp, switched capacitor technique. Although it has 0.9 V supply voltage, the power consumption is 262 μW which can be considered quite large for ultra-low-power designs. Another work uses gm-C technique and consumes 230 nW, but its supply voltage is 2.8 V and the THD values are as high as 5% (Salthouse and Sarpeshkar, 2003). More promising alternative was proposed by Young et al. (Young et al. 2010). However, in this circuit, 1 V supply voltage was used. Our VDTA-based circuit, including DTMOS transistors, uses significantly lower supply voltage of ±0.2 V with a power consumption of only 12.7 nW (Uygur and Kuntman 2013a, b).

Voltage differencing transconductance amplifier (VDTA) has voltage inputs, an alternative of current differencing transconductance amplifier (CDTA) where the inputs are currents (Biolek 2003; Biolek et al. 2008). These voltage inputs result in new propositions to conventional circuit solutions for analog signal processing (Yesil et al. 2011). The circuit symbol is given in Fig. 11.3 with its definition relations shown in matrix form.

The voltage difference of input terminals is multiplied by a transconductance of g_{m1} which becomes the I_Z current that is flowing over the impedance at the Z terminal forming the voltage at Z terminal. This voltage is then multiplied by positive and negative $\pm g_{m2}$ transconductances to form the output $\pm I_X$ currents. VDTA element can be generated by connecting two OTA circuits in a cascaded fashion.

DTMOS-based ultra-low-voltage, ultra-low-power proposed VDTA circuit is illustrated in Fig. 11.4. The transistors from M1 to M9 constitute the first OTA and the transistors from M10 to M18 constitute the second OTA where the transistors from M1 to M5 are DTMOS and similarly in the second part the transistors from M10 to M14 are DTMOS. These DTMOS transistors efficiently use available voltage headroom under the ultra-low supply voltage of ±0.2 V. In an n-well standard CMOS process, PMOS transistors can be connected as DTMOS transistors,

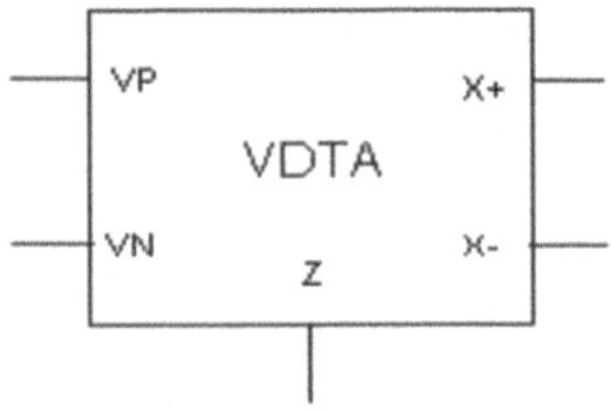

Fig. 11.3 VDTA element's circuit symbol with its definition relations shown in matrix form

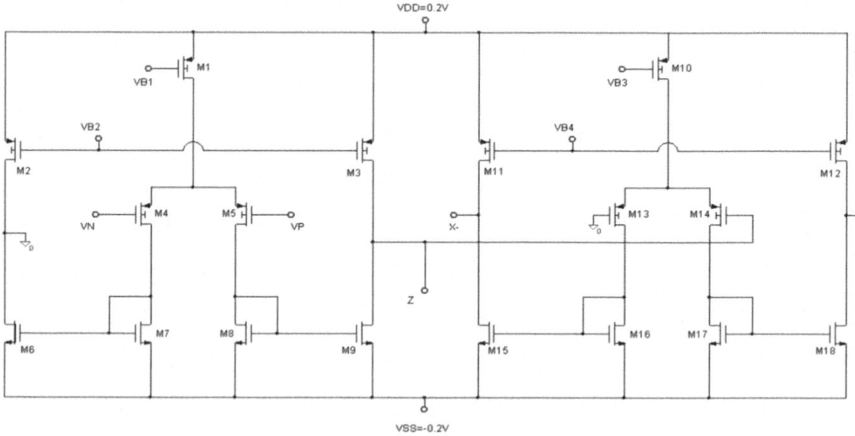

Fig. 11.4 The proposed DTMOS-based ultra-low-voltage ultra-low-power VDTA circuit. (Uygur and Kuntman 2013a, b)

whereas NMOS transistors share a common well in an n-well process and their body terminals cannot be connected to their gates to generate NMOS DTMOS transistors. That requires expensive triple-well processes using deep n-wells to produce NMOS transistors with their own wells. This restriction limits the overall performance of the proposed circuit where most of the voltage headroom is consumed over the NMOS transistors. As a result, strong inversion operation and high frequency applications are not possible using this circuit. Therefore, the transistors in this circuit are biased in weak inversion where a MOS transistor's drain current is given by

$$I_D = I_s\left(\frac{W}{L}\right) \exp\left(q\frac{V_{\rm GS} - V_{\rm TH}}{nkT}\right)\left[1 - \exp\left(-q\frac{V_{\rm DS}}{kT}\right)\right] \tag{11.2}$$

If $V_{\rm DS} \geq 3$ kT/q, the transistor will saturate in weak inversion. The transconductance g_m is described by

$$g_m = q\frac{I_D}{nkT} \tag{11.3}$$

where the parameters in Eqs. (11.3) and (11.4) have their usual meanings.

Using TSMC 0.18 μm BSIM3v3.2 process parameters, PSPICE gives $g_{m1} = g_{m2} = 54$ nA/V for the proposed VDTA circuit when the transistor dimensions in Table 11.1 are used (Table 11.2).

VDTA-based EEG filter EEG (electroencephalography) is a commonly used way of recording brain electrical activity by connecting electrodes to scalp. It has a wide usage area in biomedical applications and plays an important role in diagnosing several brain disorders (İşcan et al. 2011). There is a need for low-voltage and low-power portable EEG processing circuits. EEG data obtained by connecting

Table 11.1 The proposed VDTA circuit transistor dimensions

Transistors	Width	Length
M1, M2, M3, M10, M11, M12	5 μm	2 μm
M4, M5, M13, M14	300 μm	2 μm
M7, M8, M16, M17	50 μm	5 μm
M6, M9, M15, M18	100 μm	5 μm

Table 11.2 Performance summary of the proposed VDTA

Technology	TSMC 0.18 μm
Supply Voltage	±0.2 V
Input Voltage Range	−170 mV – 60 mv
Power Consumption	5.96 nW
Transconductances ($g_{m1} = g_{m2}$)	54 nA/V
Transconductance 3dB Frequency	3.3 kHz
Input Resistancce @300Hz	628 MΩ
Output Resistance@300Hz	1.85 GΩ

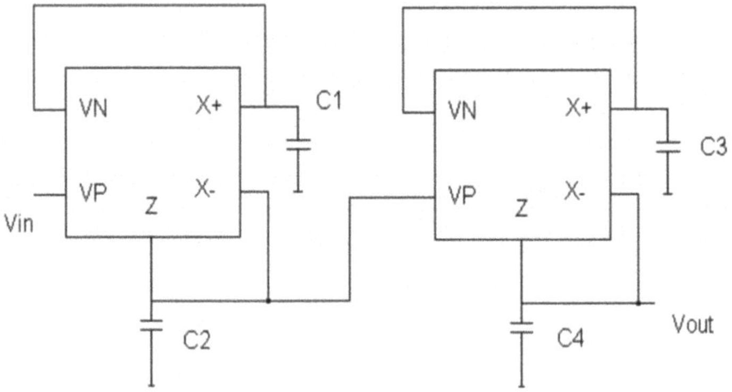

Fig. 11.5 VDTA-based double-tuned band-pass filter. (Uygur and Kuntman 2013a, b)

electrodes to human scalp has very low-amplitude voltage signals in microvolt range which are amplified by an instrumentation amplifier and then unwanted frequency components are filtered out according to the application. In our study, we assume that the signals coming from scalp electrodes are applied to a low-noise instrumentation amplifier which sufficiently amplifies the signals by achieving the required noise performance of EEG signals when the succeeding stages in the system have high noise characteristics. The VDTA-based band-pass filters are used in this study. The fourth-order band-pass filter circuit is shown in Fig. 11.5.

The double-tuned circuit is comprised of two band-pass filters which are tuned using two different pole frequencies. The transfer function of the filter in Fig. 11.5 is given in Eq. 11.4

$$\frac{V_{\text{out}}}{V_{\text{in}}} = H_0 \frac{\frac{\omega_{p1}}{Q_{p1}} s}{s^2 + \frac{\omega_{p1}}{Q_{p1}} s + \omega_{p1}^2} \frac{\frac{\omega_{p2}}{Q_{p2}} s}{s^2 + \frac{\omega_{p2}}{Q_{p2}} s + \omega_{p2}^2} \tag{11.4}$$

where H_0 is the gain factor. Natural frequencies $\omega_{p1,2}$ and quality factors $Q_{p1,2}$ of the filter are determined according to the relations in the following equations:

$$\begin{aligned} \omega_{p1,2} &= \sqrt{\frac{g_{m1,3} g_{m2,4}}{C_{1,3} C_{2,4}}}, \\ Q_{p1,2} &= \sqrt{\frac{C_{2,4} g_{m1,3}}{C_{1,3} g_{m2,4}}} \end{aligned} \tag{11.5}$$

The filter circuit was used in processing real EEG data measurements which will be explained in the next subsection. For our EEG application, the requirement was a fourth-order band-pass filter with a band-pass between 4 Hz and 35 Hz. For the double-tuned filter, we have used the pole frequency relations in where B is the bandwidth and f_0 is the center frequency.

$$\begin{aligned} f_{p1} &= \frac{\omega_{p1}}{2\pi} = f_0 + \frac{B}{2} \sin(45°), \\ f_{p2} &= \frac{\omega_{p2}}{2\pi} = f_0 - \frac{B}{2} \sin(45°) \end{aligned} \tag{11.6}$$

The EEG filter parameters, $B = 31$ Hz, $f_0 = 19.5$ Hz, and $Q_{p1} = Q_{p2} = 1$, were chosen. The pole frequencies are $f_{p1} = 30.45$ Hz and $f_{p2} = 8.54$ Hz. To realize these pole frequencies, VDTA transconductances and capacitor values were determined according to the relations in Eqs. (11.5) and (11.7) which give $C_1 = C_2 = 1.006$ nF, $C_3 = C_4 = 0.282$ nF when VDTA transconductances $g_{m1} = g_{m2} = g_{m3} = g_{m4} =$ 54 nA/V are chosen. Relatively large capacitors should be connected to the filtering circuit externally. Double-tuned band-pass filter using the circuit in Fig. 11.5 was simulated using PSPICE program with the above-calculated passive element values. Ideal and simulated frequency responses of the filter are shown in Fig. 11.6. The deviation from the ideal one, after a few kHz frequencies, is caused by the proposed VDTA's bandwidth which is limited to a few kHz range because the transistors in the active circuit operate in weak inversion. Actually, there is a tradeoff between bandwidth and power consumption. High biasing currents bring higher bandwidth at the expense of high power consumption. Fortunately, for our EEG data filtering application, our active block's bandwidth was quite sufficient and did not lead to any problems.

In order to investigate the time domain response of the double-tuned VDTA-based band-pass filter, an input sine wave with a frequency of 20 Hz and peak-to-peak amplitude of 100 mV is applied and the corresponding response in Fig. 11.7 is obtained. The decrease in the amplitude is actually the characteristic of the double-tuned filter design because two different intersections of the two band-pass filters to

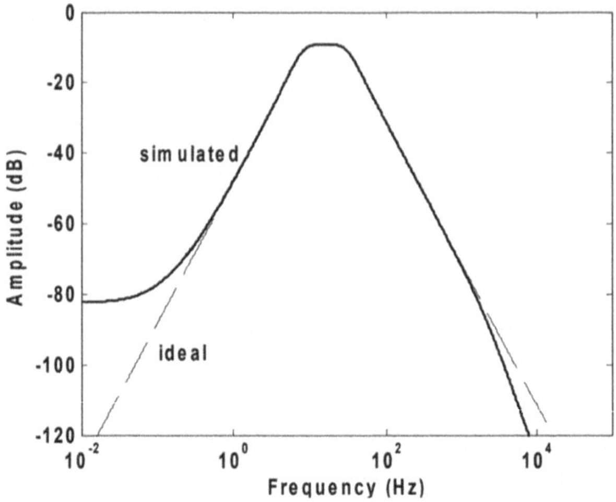

Fig. 11.6 Ideal and simulated frequency responses of the filter. (Uygur and Kuntman 2013a, b)

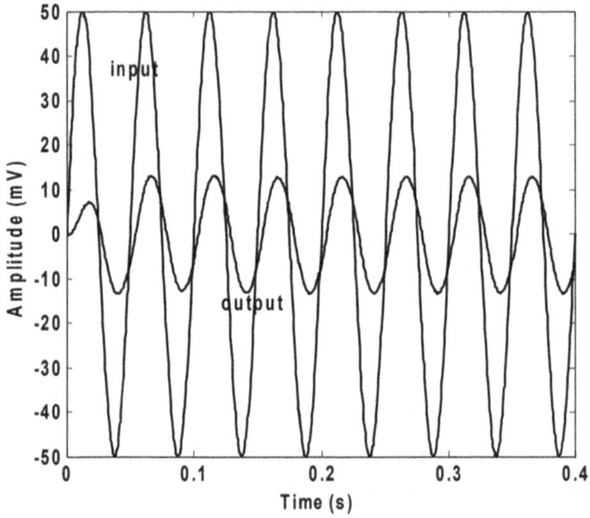

Fig. 11.7 Time response of the filter for a 20 Hz, 100 mV (p-p) input sine wave signal. (Uygur and Kuntman 2013a, b)

obtain double-tuned response cause an expected decrease in the amplitude which can be increased by an additional circuitry if necessary.

Processing of the Measured EEG Data

In order to further investigate the characteristics of the proposed band-pass filter, real measurements from an SSVEP (steady-state visually evoked potential)-based BCI (brain-computer interface) EEG experiment were used with the help of MATLAB program. SSVEP dominantly appears in the visual cortex of the brain and it is the result of a person's attention on flickering lights. It is measured by using EEG methods. Constant frequency signals visually stimulate a person, and this affects the person's EEG signal at the same frequency which can be used as a means to brain–computer interaction which is currently an active research topic.

EEG measurement setup and the data were used for applying input signal data to our filter. The experimental setup is shown in Fig. 11.8. EEG signal is sensed via electrodes connected to the scalp. These signals are then fed into an amplifier system specialized on EEG data recordings. Amplified signal data are then transferred to computer for further processing. The part of the experiment we used for filtering is comprised of computer recordings of the EEG signals of the subjects while they are looking at four red circles at the four corners of the computer monitor flickering at four different constant frequencies (4.60 Hz, 6.43 Hz, 8.03 Hz, and 10.70 Hz). Applied input data is taken from the OZ channel of the connected 16 electrodes. This channel is more sensitive to the visual stimulations.

In our filter application, we have just used the recordings for 2 s with 1001 data points from OZ channel recordings of the subject's visual attention on the left red circle that is flickering on the monitor with a constant frequency of 10.70 Hz. Filter input and output signals are illustrated in Fig. 11.8. For figure clarity, only the data of the first 0.4 s is shown in the figure and the output signal is multiplied by a factor of filter gain loss to compensate for the decrease in amplitude. Input data signal amplitudes are also multiplied by a factor to manage and to use them properly as inputs to the filter.

In Figs. 11.9 and 11.10, pre-filter and post-filter frequency spectrums of our EEG signal are depicted respectively. As shown in the figures, unwanted frequency harmonics are successfully filtered out from the signal by the VDTA filter and the

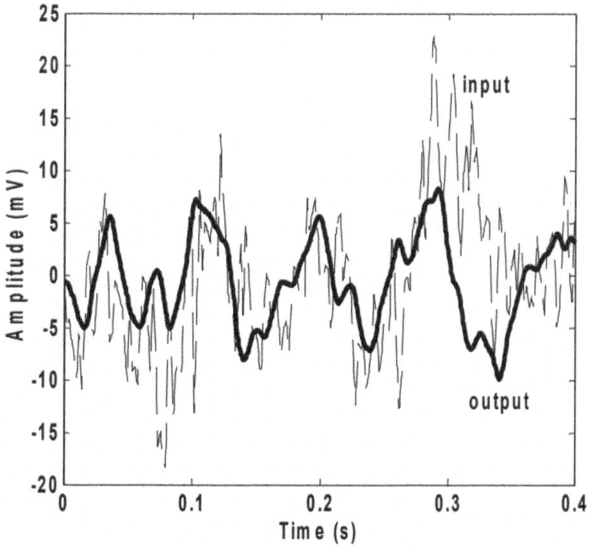

Fig. 11.8 Time response of the filter output for the measurement EEG data for 0.4 s. (Uygur and Kuntman 2013a, b)

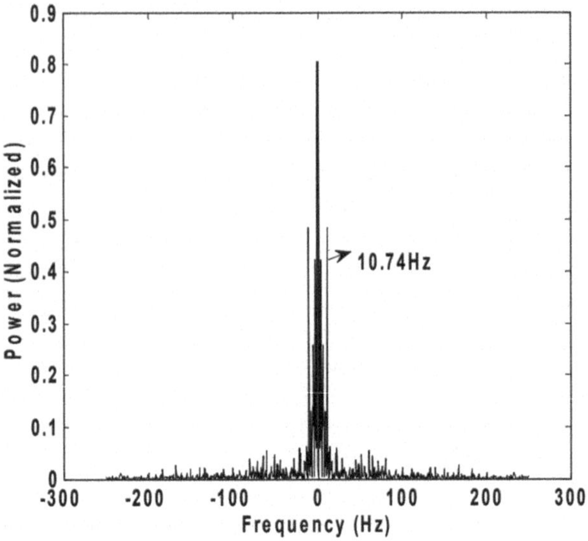

Fig. 11.9 Pre-filter frequency spectrum of the EEG data. (Uygur and Kuntman 2013a, b)

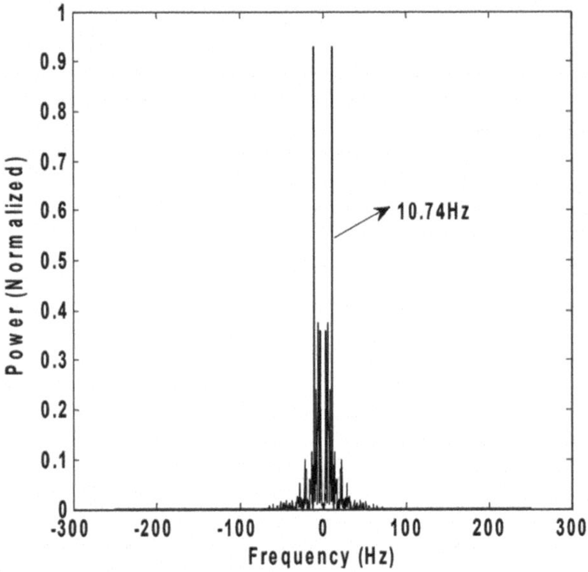

Fig. 11.10 Post-filter frequency spectrum of the EEG data. (Uygur and Kuntman 2013a, b)

main frequency component of 10.74 Hz has become clearer after the filter. That is almost the same frequency of the applied visual stimulation signal to the subject which shows the validity of SSVEP study where the subject's brain reacts to the applied flickering light by producing EEG signal at the same frequency as the flickering. That phenomenon is used in brain–computer interface applications.

In this study, DTMOS-based VDTA circuit was proposed. The circuit is capable of working under an ultra-low supply voltage ±0.2 V and only consuming 5.96

nW. DTMOS transistors are used to efficiently exploit shrank voltage headroom. For very low power consumption, the transistors were used in weak inversion where DTMOS transistors are very suitable to this mode of operation due to their well subthreshold slope characteristic. Using the VDTA, a band-pass filter was designed for EEG data processing. The proposed circuit was used in a fourth-order double-tuned band-pass filter. The filter consists of two VDTA cells and two externally connected capacitors. According to PSPICE simulations, both VDTA and the double-tuned filter have performed well. The filter was successfully used to filter out unwanted frequency components of an SSVEP-based BCI system's amplifier outputs. Although, in measurements, there was amplifying and filtering integrated in the measurement hardware, there was still a need for additional filtering which is usually done by digital filtering via software. Instead of digital filtering approach, the proposed filter was utilized to investigate its performance in a practical application. It is found that both PSPICE and MATLAB results are in close agreement with the theory. The proposed DTMOS-based VDTA circuit is suitable for ultra-low-power, ultra-low-voltage analog signal processing applications.

A Very Compact, 0.4 V DTMOS CCII Employed in an Audio-Frequency Filter

This section presents an ultra-low-voltage, ultra-low-power, very compact, dynamic threshold voltage MOS transistor (DTMOS)-based CCII circuit (Uygur and Kuntman 2014). The proposed circuit is capable of operating under ±0.2 V symmetric supply voltages. The circuit topology is very compact and consists of only four DTMOS transistors and four ordinary NMOS transistors. The total power consumption of the circuit is found as only 214 nW, while all transistors are working in the subthreshold region. The current conveyor has 570 kHz 3 dB bandwidth from X to Y terminal for the voltage gain and has low, 0.2% following error between these terminals for inputs not exceeding ±60 mV. TSMC 0.18 μm process technology parameters are used in the design of the proposed CCII block which is then employed in an audio-frequency, second-order, band-pass filter configuration where real speech signals are fed to the input of the filter to further investigate its characteristics. A close agreement is found between theoretical study and simulated response.

In this study, a dynamic threshold voltage MOS transistor (DTMOS)-based CCII circuit is proposed. The circuit is the CCII circuit from the related publications where it is preliminarily described. Additional simulation results for the current conveyor and filter application, more detailed explanations, and further mathematical derivations are given in this extended version.

DTMOS-based ultra-low-voltage, ultra-low-power, subthreshold CCII circuit has been designed using DTMOS transistors. The proposed circuit is depicted in Fig. 11.11. The circuit structure is very compact consisting of only eight transistors,

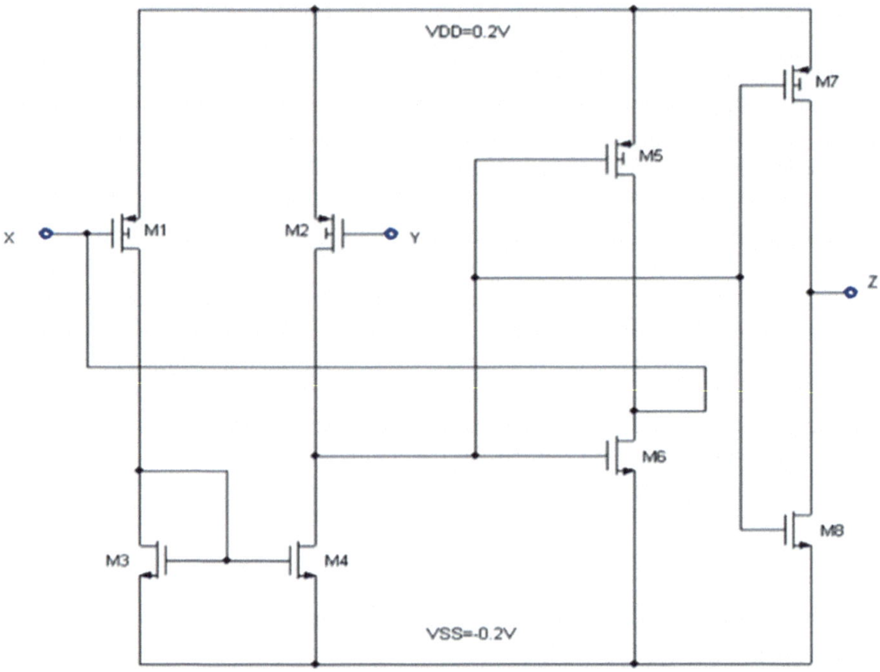

Fig. 11.11 DTMOS-based ultra-low-voltage, ultra-low-power subthreshold CCII circuit. (Uygur and Kuntman 2014)

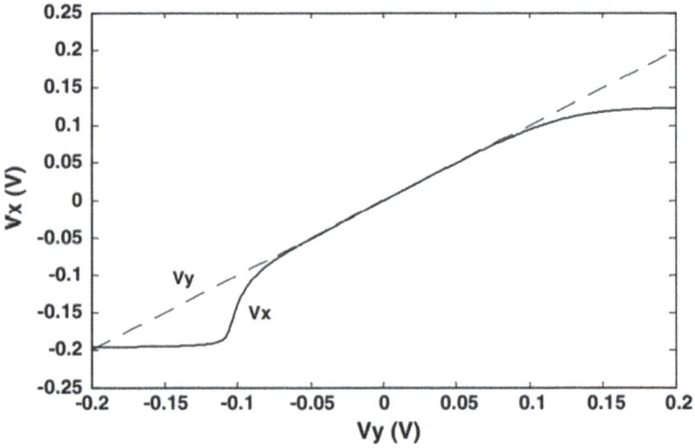

Fig. 11.12 The change of V_X voltage versus V_Y voltage. (Uygur and Kuntman 2014)

all operating in the subthreshold region. The usage of only eight transistors increases the power consumption efficiency of the design and minimizes the parasitics as well. In the proposed CCII circuit in Fig. 11.11, supply voltages are chosen as ±0.2 V and

Table 11.3 Transistor dimensions of the proposed CCII

Transistor	Width	Length
M1, M2, M5, M7	300 μm	2 μm
M3, M4	50 μm	2 μm
M6, M8	320 μm	0.4 μm

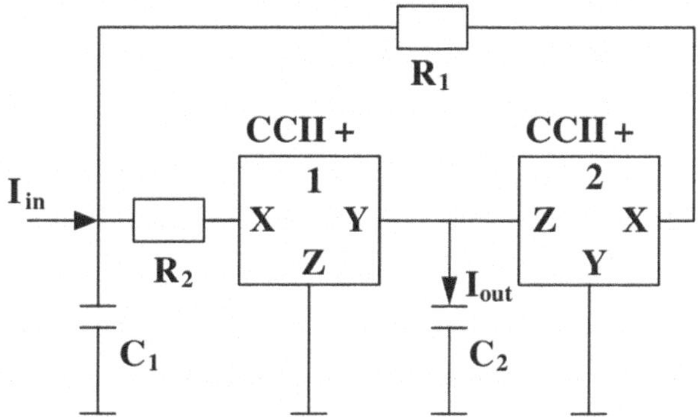

Fig. 11.13 CCII-based band-pass filter. (Uygur and Kuntman 2014)

TSMC 0.18 μm process parameters are used in the design. The change of V_X voltage versus V_Y voltage is shown in Fig. 11.12.

The input stage of the CCII is formed by a pseudodifferential low-voltage amplifier consisting of M_1–M_4. The output stage is the class AB stage from Ferri and Guerrini (2003) with the modification of DTMOS transistor usage for ultra-low-voltage operation. M_1–M_2 and M_5, M_7 are the p-MOS DTMOS transistors. There is feedback in the topology including M6 transistor which helps to decrease the resistance at the X input terminal which should be zero ideally. However, this is very difficult to achieve under subthreshold mode of operation where transistor transconductances are significantly low, which severely affects the resistance seen at the X terminal.

The transistor aspect ratios of the circuit in Fig. 11.11 are tabulated in Table 11.3. From Table 11.3, it can be claimed that relatively large transistors are used in the design. The reason is that they are chosen for correct operation of the circuit under ultra-low supply voltage for the desired frequency range. This is generally the expected case for any MOS circuits operating in the subthreshold region with such low supply voltages where transistor dimensions have been chosen large to enable the flow of targeted current under ultra-low-voltage operation. The circuit is simulated by the program SPICE. Figure 11.13 shows the input range where the V_X voltage follows V_Y voltage. The figure is obtained for a DC voltage source applied to the *Y* terminal, while the *X* terminal is loaded with a 100 kΩ load resistance. From Fig. 11.13, it can be seen that the input voltage range is ±60 mV with a small following error. A relative error term is defined as (Figs. 11.14 and 11.15)

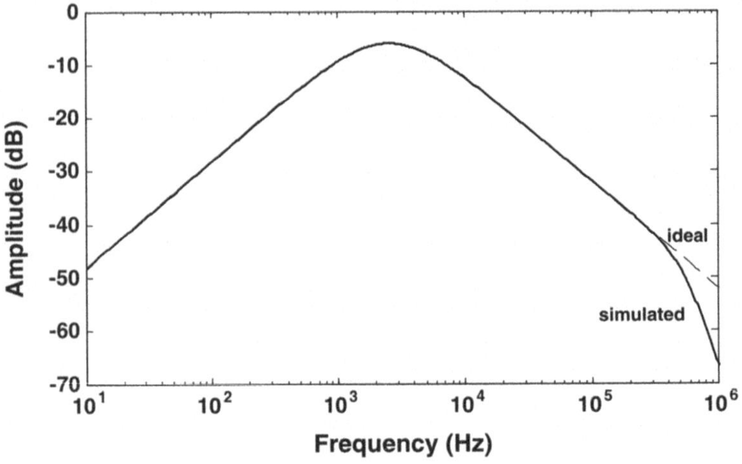

Fig. 11.14 Ideal and simulated filter amplitude response with respect to frequency. (Uygur and Kuntman 2014)

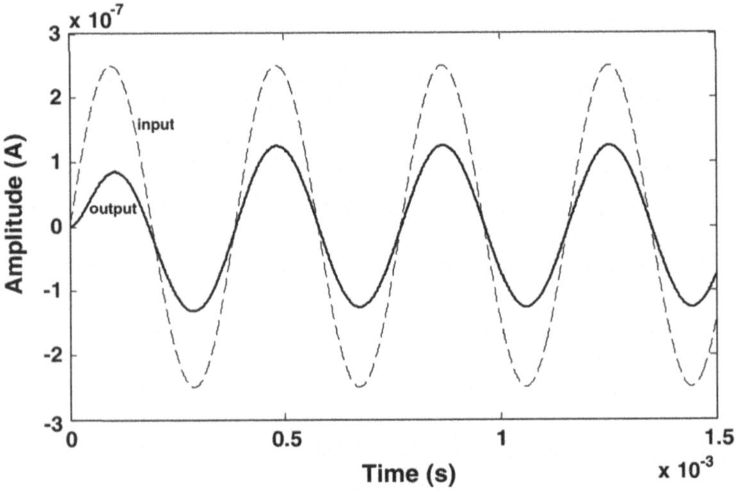

Fig. 11.15 Sinusoidal characteristics of input and output of the filter. (Uygur and Kuntman 2014)

$$\varepsilon = \frac{\left|v_x - v_y\right|}{v_x} \tag{11.7}$$

The proposed DTMOS-based subthreshold current conveyor is used in a band-pass filter configuration which is given in Fabre et al. (1996). This filter configuration is shown in Fig. 11.13. The reason for choosing a band-pass filter type is that in analog hearing aid systems, there are analog filter banks that consist of band-pass filters (Salthouse and Sarpeshkar 2003). These filters should be capable of operating

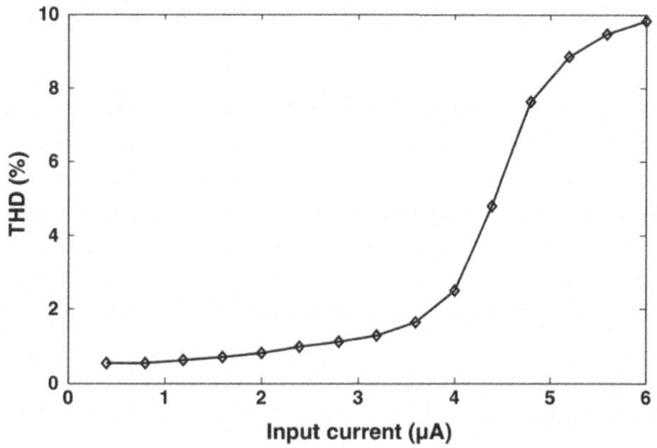

Fig. 11.16 Total harmonic distortion of the filter. (Uygur and Kuntman 2014)

at sound frequencies with very little power consumption to save the life of the battery used in the hearing aid system. The filter circuit employing the proposed CCII is simulated by the program SPICE. The frequency response in Fig. 11.13 and phase response in Fig. 11.13 are obtained when the passive element values of $C_1 = 628$ pF, $C_2 = 628$ pF, $R_1 = 100$ kΩ, and $R_2 = 100$ kΩ are used. The transfer function, pole frequency, and quality factor of the filter are given by

$$
\begin{aligned}
H(s) &= -\frac{sC_2R_2}{s^2C_1C_2R_1R_2 + sC_2(R_1 + R_2) + 1} \\
\omega &= \sqrt{\frac{1}{R_1R_2C_1C_2}} \\
Q &= \frac{\sqrt{R_1R_2}}{R_1 + R_2}\sqrt{\frac{C_1}{C_2}}
\end{aligned}
\tag{11.8}
$$

They are adjusted to example values close to 2.5 kHz pole frequency and Q value of 1/2 which shows that the circuit behaves close to the ideal one throughout the sound frequency range and it might be employed in low-power and low-voltage analog hearing aid filter banks.

The total harmonic distortion of the filter with respect to peak-to-peak input current signal is depicted in Fig. 11.16. From the figure, it is seen that for input current signals not exceeding peak-to-peak 4 μA, the THD is less than 2.5%. To note that, this value has been achieved using only ±0.2 V supply voltage while consuming only 428 nW total power for overall filter circuit.

Concluding Remarks

In this paper, using DTMOS transistors, having close to ideal subthreshold swings, an ultra-low-voltage and ultra-low-power subthreshold CCII is proposed. Its performance is investigated with SPICE simulations. The proposed current conveyor is then employed in a second-order band-pass filter circuit to show its suitability for sound frequency range of operation and speech processing applications. Theoretical and simulated results are found in close agreement. The importance of the design is that, under only ±0.2 V supply voltage and consuming very low power, half of MHz range operation has become possible with using only eight transistors, thanks to the usage of DTMOS transistors and current-mode approach. The circuit is compared with another sub-1 V ultra-low-voltage CCII, available in the literature, and the proposed circuit is found performing significantly better than its bulk-driven alternative. To the best knowledge of the authors, the proposed circuit in this study consists of the least number of transistors in a DTMOS-based CCII circuit in the literature that is able to operate under ±0.2 V supply voltages using standard CMOS process technology.

References

Assaderaghi, F., Parke, S., Sinitsky, D., Bokor, J., Ko, P. K. and Hu, C. (1994). A dynamic threshold voltage MOSFET (DTMOS) for very low voltage operation. IEEE Electron Device Letters, 15(12), 510–512.

Assaderaghi, F., Sinitsky, D., Parke, S. A., Bokor, J., Ko, P. K. and Hu, C. (1997). Dynamic threshold-voltage MOSFET (DTMOS) for ultra-low voltage VLSI. IEEE Transactions on Electron Devices, 44(3), 414–422.

Berg, Y. and Mirmotahari, O. (2011). Ultra low-voltage CMOS current mirrors. Analog Integrated Circuits and Signal Processing, 68(2), 219–232.

Biolek, D. (2003). CDTA-building block for current-mode analog signal processing. In Proceedings of the ECCTD (Vol. 3, pp. 397–400).

Biolek, D., Senani, R., Biolkova, V. and Kolka, Z. (2008). Active elements for analog signal processing: classification, review, and new proposals. Radioengineering, 17(4), 15–32.

Chatterjee, S., Tsividis, Y. and Kinget, P. (2005). 0.5-V analog circuit techniques and their application in OTA and filter design. IEEE journal of solid-state circuits, 40(12), 2373–2387.

Fabre, A., Amrani, H. and Saaid, O. (1996). Current-mode bandpass filters with Q-magnification. IEEE Transactions on Circuits and Systems-II: Analog and Digital Signal Processing, 43(12), 839–842. Analog Integr Circ Sig Process (2014) 81:89–98 97 123

Ferri, G. and Guerrini, N. C. (2003). Low-voltage low-power CMOS current conveyors. Springer Science & Business Media.

İşcan, Z., Özkaya, Ö. and Dokur, Z. (2011, April). Classification of EEG in a steady state visual evoked potential based brain computer interface experiment. In International Conference on Adaptive and Natural Computing Algorithms (pp. 81–88). Springer, Berlin, Heidelberg.

Lee, K. J., Choi, J. N., Yu, C. G., Coligne, J. P. and Park, J. T. (2004). Temperature dependence of DTMOS transistor characteristics. Solid State Electronics, 48(1), 183–187.

Lee, S. C., Lee, S. Y. and Chiang, C. H. (2008). 0.9 V low-power switched-opamp switched-capacitor bandpass filter for electroneurography acquisition systems. IET circuits, devices & systems, 2(2), 257–263.

Ozcan, S., Kuntman, H. H. and Cicekoglu, O. (1998, December). Realization of inductorless RF bandpass amplifiers using immittance simulators employing CCIIs. In Proceedings of the Tenth International Conference on Microelectronics (Cat. No. 98EX186) (pp. 141–144). IEEE.

Salthouse, C. D. and Sarpeshkar, R. (2003). A practical micropower programmable bandpass filter for use in bionic ears. IEEE Journal of Solid-State Circuits, 38(1), 63–70.

Uygur, A. and Kuntman, H. (2013a). DTMOS-based 0.4 V ultra low-voltage low-power VDTA design and its application to EEG data processing. Radioengineering, 22(2), 458–466.

Uygur, A. and Kuntman, H. (2013b, November). An ultra low-voltage, ultra low-power DTMOS-based CCII design for speech processing filters. In 2013 8th international conference on electrical and electronics engineering (ELECO) (pp. 31–35). IEEE.

Uygur, A. and Kuntman, H. (2014). A very compact, 0.4 V DTMOS CCII employed in an audio-frequency filter. Analog Integrated Circuits and Signal Processing, 81(1), 89–98.

Yesil, A., Kacar, F. and Kuntman, H. (2011). New simple CMOS realization of voltage differencing transconductance amplifier and its RF filter application. Radioengineering, 20(3), 632–637.

Zhang, C., Srivastava, A. and Ajmera, P. K. (2006). A 0.8 V CMOS amplifier design. Analog Integrated Circuits and Signal Processing, 47(3), 315–321.

Chapter 12
High-Precision Current-Mode Circuits Based on the MOS Translinear Principle

Basic Concept

It is well known that the decrease in dimensions of MOSFETs in IC fabrication technology affects MOSFET performance and the MOSFET V–I relationship changes from square law to linear. Therefore, errors may occur in the output function of the circuits employing small-sized transistors. Translinear (TL) circuit principle which was originally formulated for loops of bipolar transistors is generalized and the MOS translinear (MTL) principle is derived by Seevinck and Wiegerink (1991). MTL circuits are used in a lot of analog building blocks such as square-root domain integrators, differentiators, filters, oscillators, multipliers, and RMS-DC converters. In this section, a method to reduce the errors generated by the second-order effects in the current-mode circuits employing MTL loop is proposed, and also high-precision circuits are designed using this method (Menekay et al. 2009a, b). The proposed method enables the use of much smaller transistors and the circuits to be designed need smaller chip size than their counterparts. Thus, they may be operated at much higher frequencies. They are appropriate for filtering in square-root domain, modulators, phase discriminator, RMS-DC converters, etc.

Current-mode circuits which are designed by applying the MTL principle are used in synthesizing many nonlinear signal processing functions.

Considering a loop of MOS transistors as indicated in Fig. 12.1, according to the MTL principle Eq. (12.1) can be written as

$$\sum_{\mathrm{cw}} V_{\mathrm{GS}} = \sum_{\mathrm{ccw}} V_{\mathrm{GS}} \tag{12.1}$$

where the subscripts cw and ccw indicate the devices connected clockwise and counterclockwise in the loop, respectively.

H. Kuntman, D. Özenli, *Trends in Circuit Design for Analog Signal Processing*,
Analog Circuits and Signal Processing,
https://doi.org/10.1007/978-3-030-96836-6_12

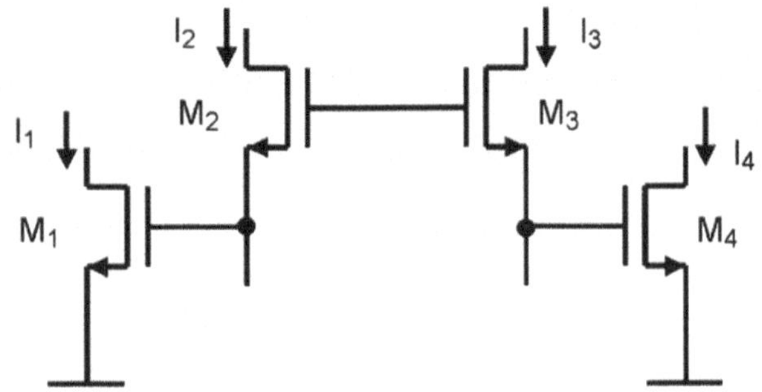

Fig. 12.1 An MTL loop consisting of four transistors

If the second-order effects are negligible, the drain current function for saturation region is

$$I_D = \frac{\beta}{2}(V_{\mathrm{GS}} - V_{\mathrm{TH}})^2 \tag{12.2}$$

From Eq. (12.2) the following expression for V_{GS} can be obtained:

$$V_{\mathrm{GS}} = V_{\mathrm{TH}} + \sqrt{\frac{2I_D}{\beta}} \tag{12.3}$$

Substituting Eq. (12.3) into Eq. (12.1) results in

$$\sum_{\mathrm{cw}}\left(V_{\mathrm{TH}} + \sqrt{\frac{2I_D}{\beta}}\right) = \sum_{\mathrm{ccw}}\left(V_{\mathrm{TH}} + \sqrt{\frac{2I_D}{\beta}}\right) \tag{12.4}$$

Assuming well-matched threshold voltages and neglecting body effect allow the threshold voltages to be dropped as

$$\sum_{\mathrm{cw}}\sqrt{\frac{2I_D}{\beta}} = \sum_{\mathrm{ccw}}\sqrt{\frac{2I_D}{\beta}} \tag{12.5}$$

Relation Eq. (12.5) is a statement of the MTL circuit principle. Square-root and squarer/divider circuits are some of the MTL circuits. Figure 12.2a, b shows a possible way of designing the square-root and squarer/divider circuit. From Eq. (12.5), if the transconductance parameters (β) of the MOS transistors are $2\,\beta_1 = 2\,\beta_2 = \beta_3 = \beta_4 = 2\beta$, the following is obtained:

$$\sqrt{\frac{2I_1}{\beta}} + \sqrt{\frac{2I_2}{\beta}} = \sqrt{\frac{2I_3}{2\beta}} + \sqrt{\frac{2I_4}{2\beta}} \tag{12.6}$$

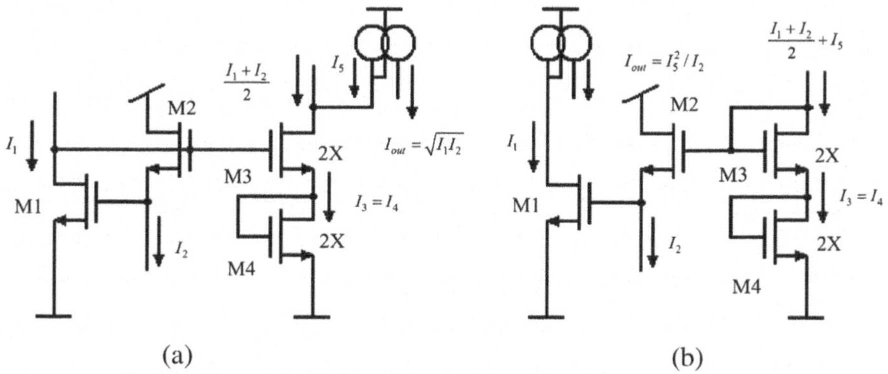

Fig. 12.2 Applications of the stacked VTL loop: (**a**) square-root cell and (**b**) squarer/divider cell

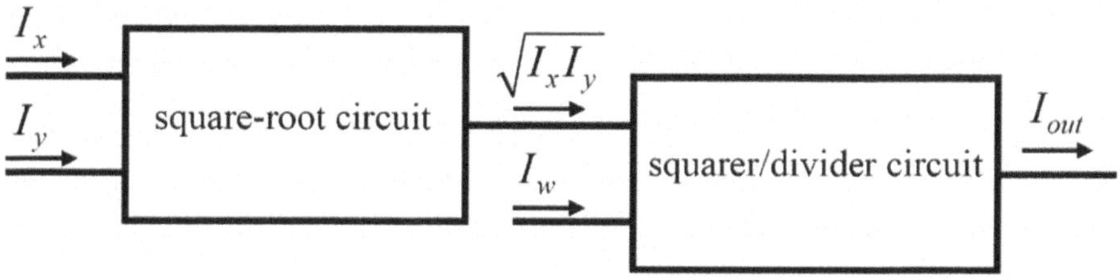

Fig. 12.3 Simplified diagram of the multiplier–divider circuit

According to Fig. 12.2, the drain currents of M_3 and M_4 are the same and taking the square on both sides of Eq. (12.6), the equation can be rewritten as

$$I_3 = \sqrt{I_1 I_2} + \frac{1}{2}(I_1 + I_2) \tag{12.7}$$

Using the KCL equation at the output node, Eq. (12.7) can be obtained as

$$I_{\text{out}} = I_5 = I_3 - \frac{1}{2}(I_1 + I_2) = \sqrt{I_1 I_2} \tag{12.8}$$

Hence, a square-root circuit can be obtained if I_1 and I_2 are the input currents and the output current is a copy of I_5. Alternatively, a squarer/divider circuit is obtained if the output is a copy of either I_1 or I_2 and the inputs are the remaining two currents. A multiplier/divider circuit can be obtained by using these circuits as shown in Fig. 12.3. Since *Ix* and *Iy* are the input currents of the square-root circuit, the output current of the whole circuit can be expressed as in

$$I_{\text{out}} = \frac{I_{\text{sr}}^2}{I_w} = \frac{\left(\sqrt{I_x I_y}\right)^2}{I_w} = \frac{I_x I_y}{I_w} \tag{12.9}$$

The proposed current-mode circuit MTL principle is derived for MOS transistors with an ideal square law characteristic in Eq. (12.5), but various effects which are called second-order effects cause a deviation from the ideal square law behavior and MTL principle. These effects are the body effect, mobility reduction, weak inversion, channel length modulation, component mismatches, errors caused by parasitic resistance and capacitance, and temperature dependence. The mobility reduction affects the statement of the MTL circuit principle that is obtained in Eq. (12.5) as in

$$\sum_{\mathrm{cw}}\sqrt{\frac{2I_D[1+\theta(V_{\mathrm{GS}}-V_{\mathrm{TH}})]}{\beta}} = \sum_{\mathrm{ccw}}\sqrt{\frac{2I_D[1+\theta(V_{\mathrm{GS}}-V_{\mathrm{TH}})]}{\beta}} \tag{12.10}$$

where θ is the mobility reduction parameter. Nowadays, especially the mobility reduction caused by short-channel effects becomes more important in MOSFETs at channel lengths of about 1 μm or less, and therefore errors may occur in the output function of the circuits employing these MOSFETs (Menekay et al. 2009a). Thus, square-root and squarer/divider cells are modified as shown in Fig. 12.4. The error of the output current function caused by second-order effects can be reduced using the resistance R between the gates of the MOS transistors M_2, M_3 and letting the output current Iout flow through this resistance. In this case, a new voltage term is added to the MTL loop function which eliminates the error of the output current function. Assuming that the aspect ratios of the transistors satisfy the $\beta_1 = \beta_2 = \beta$ and $\beta_3 = \beta_4 = 2\beta$, the following is obtained using the MTL principle:

$$\sum_{i=1,2} V_{\mathrm{GS}i} - \sum_{j=1,2} V_{\mathrm{TH}j} = I_{\mathrm{out}}\cdot R + \sum_{k=1,2} V_{\mathrm{GS}k} - \sum_{i=1,2} V_{\mathrm{TH}l} \tag{12.11}$$

If the mobility reduction is taken into consideration, the drain current of a MOS transistor operated in saturation is given by

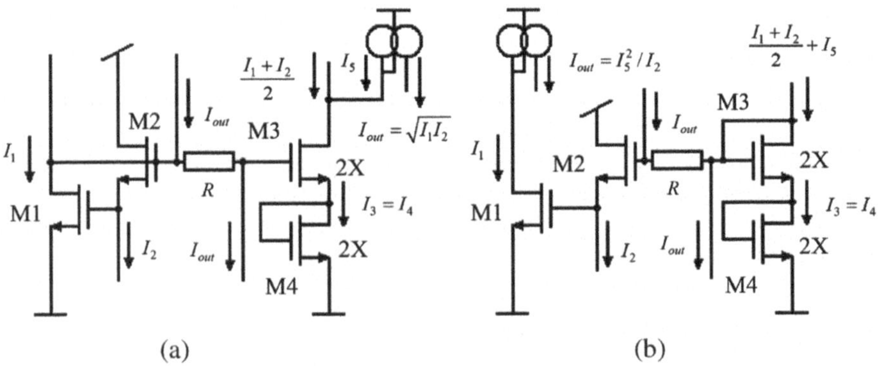

Fig. 12.4 Proposed simplified current-mode circuits: (**a**) square-root cell and (**b**) squarer/divider cell. (Menekay et al. 2009a)

$$I_D = \frac{\beta}{2} \frac{(V_{GS} - V_{TH})^2}{1 + \theta'(V_{GS} - V_{TH})} \tag{12.12}$$

θ' is the new mobility reduction parameter that can be expressed by

$$\theta' = \theta + \frac{\mu}{v_{max} L_{eff}(1 + F_B)} \tag{12.13}$$

where θ' is the mobility reduction parameter, μ is the carrier mobility, v_{max} is the maximum carrier velocity, L_{eff} is the channel length, and F_B is the bulk charge parameter. According to Eq. (12.12), $(V_{GS} - V_{TH})$ voltage can be written as

$$V_{GS} - V_{TH} = \frac{I_D \theta'}{\beta} + \sqrt{\left(\frac{I_D \theta'}{\beta}\right)^2 + \frac{2 I_D}{\beta}} \tag{12.14}$$

Assuming that the drain current is small as follows:

$$\left(\frac{I_D \theta'}{\beta}\right)^2 \ll \left(\frac{2 I_D}{\beta}\right)$$

$(V_{GS} - V_{TH})$ voltage equation can be expressed by

$$V_{GS} - V_{TH} \approx \frac{I_D \theta'}{\beta} + \sqrt{\frac{2 I_D}{\beta}} \tag{12.15}$$

Combining the equations

$$\begin{aligned} &\frac{I_1 \theta'}{\beta} + \sqrt{\frac{2 I_1}{\beta}} + \frac{I_2 \theta'}{\beta} + \sqrt{\frac{2 I_2}{\beta}} \\ &= I_{out} \cdot R + \frac{I_3 \theta'}{2\beta} + \sqrt{\frac{I_3}{\beta}} + \frac{I_4 \theta'}{2\beta} + \sqrt{\frac{I_4}{\beta}} \end{aligned} \tag{12.16}$$

Since the drain currents of the M_3 and M_4 transistors are the same, the equation can be expressed by

$$\begin{aligned} &\frac{I_1 \theta'}{\beta} + \sqrt{\frac{2 I_1}{\beta}} + \frac{I_2 \theta'}{\beta} + \sqrt{\frac{2 I_2}{\beta}} \\ &= I_{out} \cdot R + \frac{I_3 \theta'}{\beta} + 2\sqrt{\frac{I_3}{\beta}} \end{aligned} \tag{12.17}$$

Using the KCL equation at the output node of the modified square-root cell in Fig. 12.4a, the following can be obtained:

$$I_3 = 0.5(I_1 - I_2) + \sqrt{I_1 I_2} \tag{12.18}$$

Multiplying each term with $4/\beta$ and taking the square-root on both sides of equation, the equation can be written as

$$\sqrt{\frac{2I_1}{\beta}} + \sqrt{\frac{2I_2}{\beta}} = 2\sqrt{\frac{I_3}{\beta}} \tag{12.19}$$

Combining the equations, the following can be obtained:

$$I_{\text{out}} \cdot R = \frac{\theta'}{\beta}(I_1 + I_2 - I_3) \tag{12.20}$$

Taking the Iout current as the copy of the output current of the circuit in Fig. 12.4a and considering the mobility reduction effect, output current function can be expressed as

$$\sqrt{I_1 I_2} = (I_1 + I_2)\frac{\theta'}{2(\beta \cdot R + \theta')}E \tag{12.21}$$

E is defined as

$$E = \frac{\sqrt{I_1 I_2}}{(I_1 + I_2)} \frac{2(\beta \cdot R + \theta')}{\theta'} \tag{12.22}$$

If $I_2 = n \times I_1$ error function E-1 is drawn in Fig. 12.5. As shown in Fig. 12.5, when input currents I_1 and I_2 are equal ($n = 1$), output current function error generated by mobility reduction will be zero.

Resistances are realized employing MOSFETs where their value can be controlled by the control voltages Vc in the proposed circuits. By this way the output current functions of the circuits can be controlled and the function errors of the currents can be eliminated. Thus, the proposed circuits can be realized as shown in Figs. 12.6, 12.7, and 12.8.

Simulations

To verify the performance of the circuits proposed, SPICE simulations were performed using TSMC 0.35 μm level 3 CMOS process parameters. The power supply voltage is taken as 3 V. The transistor dimensions are chosen as $W/L = 3$ μm/0.7 μm for M_{a1}, M_{a2}, M_{a9}, M_{a10}, M_{a19}, M_{b10}, and M_{b14}; $W/L = 6$ μm/0.7 μm for M_{a3}–M_{a8}, M_{a11}–M_{a18}, M_{b1}, M_{b2}, M_{b5}–M_{b9}, M_{b11}–M_{b13}, and M_{b15}–M_{b18}; $W/L = 12$ μm/0.7 μm for M_{b3}, and M_{b4}; and $W/L = 15$ μm/0.7 μm for M_{b19}. To obtain time domain SPICE simulation results of the proposed square-root circuit, the input

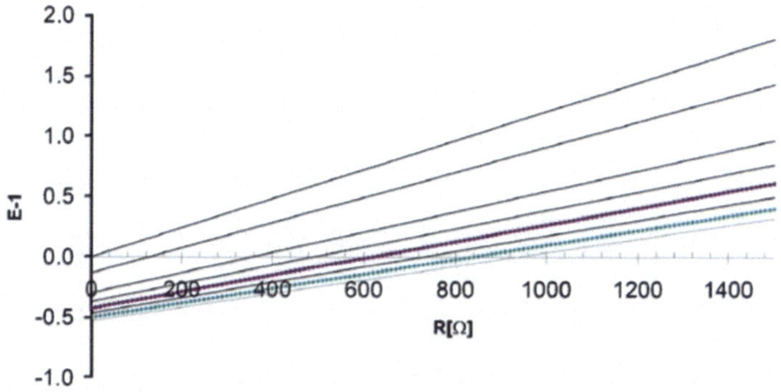

Fig. 12.5 Errors of E-1 for several n (= 0.25 V^{-1}, = 3 × 10^{-4} A/V^2). (Menekay et al. 2009a)

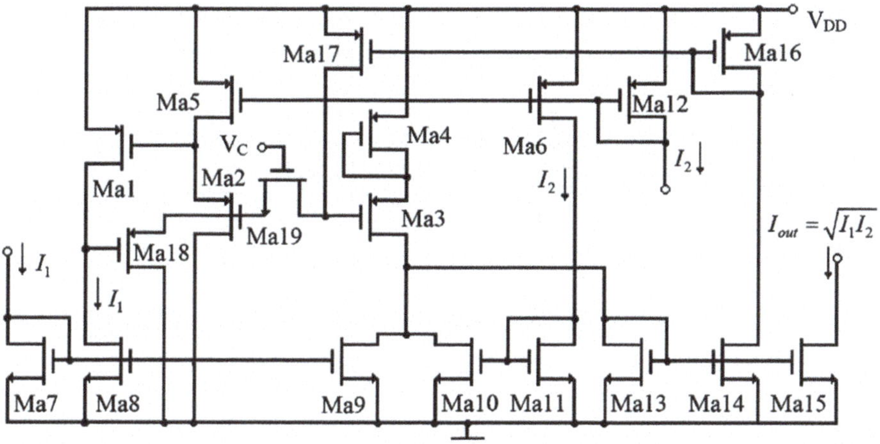

Fig. 12.6 Proposed current-mode square-root circuit. (Menekay et al. 2009a)

current I_1 is constant DC current with a value of 5 μA where the other input current I_2 is applied as a triangular wave with amplitude of 20 μA. The output currents of the proposed square-root circuit and the conventional circuit are simulated with the ideal function as shown in Fig. 12.9. For the proposed squarer/divider circuit, the input current I_1 is applied as a triangular wave with amplitude of 7 μA (initial DC current value of I_1 is 3 μA) where the other input current I_2 is constant DC current with a value of 7 μA. The output currents of the proposed squarer/divider circuit and the conventional circuit are simulated with the ideal function as shown in Fig. 12.10. Similarly for the proposed multiplier/divider circuit, the first input current I_x is applied as a triangular wave with amplitude of 9 μA (initial DC current value of I_x is 1 μA) where the other input currents I_y and I_w are constant DC current with a value of 3 μA. The output currents of the proposed multiplier/divider circuit and the conventional circuit are simulated with the ideal function as shown in Fig. 12.11.

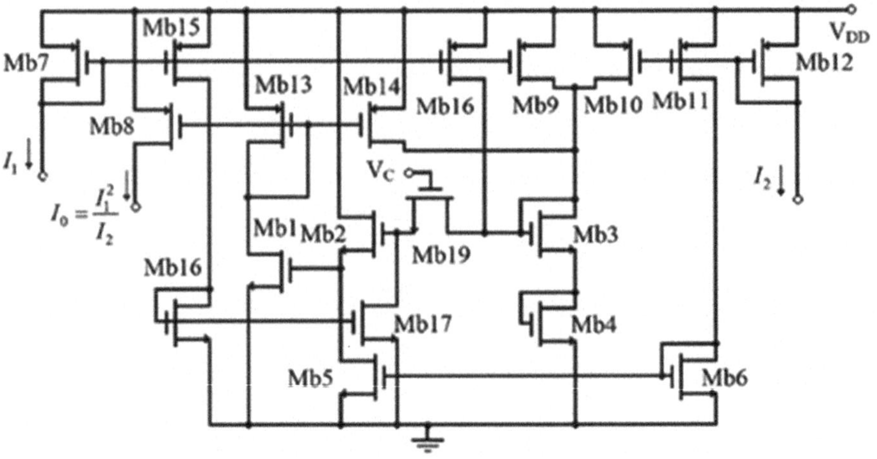

Fig. 12.7 Proposed current-mode squarer/divider circuit. (Menekay et al. 2009a)

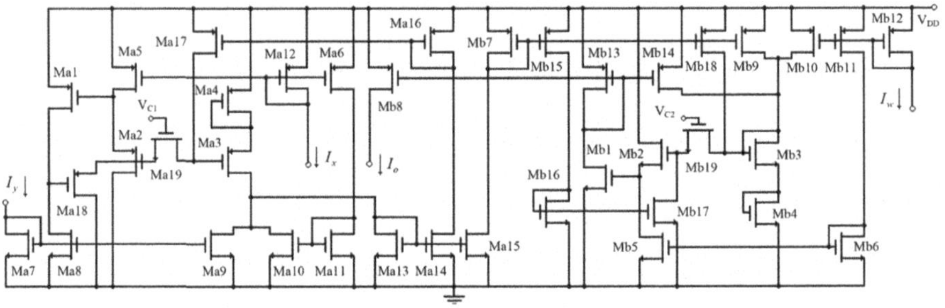

Fig. 12.8 Proposed current-mode multiplier/divider circuit. (Menekay et al. 2009a)

As expected, the characteristics of the output currents of the proposed circuits show approximately the ideal square-root functions. The small-signal bandwidths of the proposed square-root, squarer/divider, and multiplier/divider circuits measured are about 360, 200, and 75 MHz, respectively.

Concluding Remarks

In this part, a method to reduce the errors generated by the second-order effects in the current-mode circuits employing MOS translinear loop is proposed and this method is applied to the square-root, squarer/divider, and the multiplier/divider circuit. The main advantages of the proposed circuits are reducing the errors of the output current function, able to control the output current with the control voltage, operating at higher frequencies, and providing more efficient power consumption. The proposed

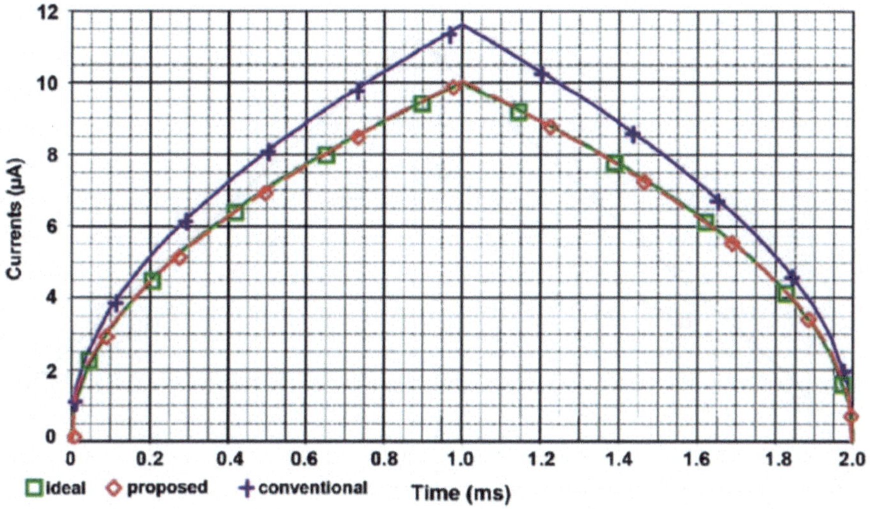

Fig. 12.9 The output currents and ideal curve of the square-root circuit. (Menekay et al. 2009a)

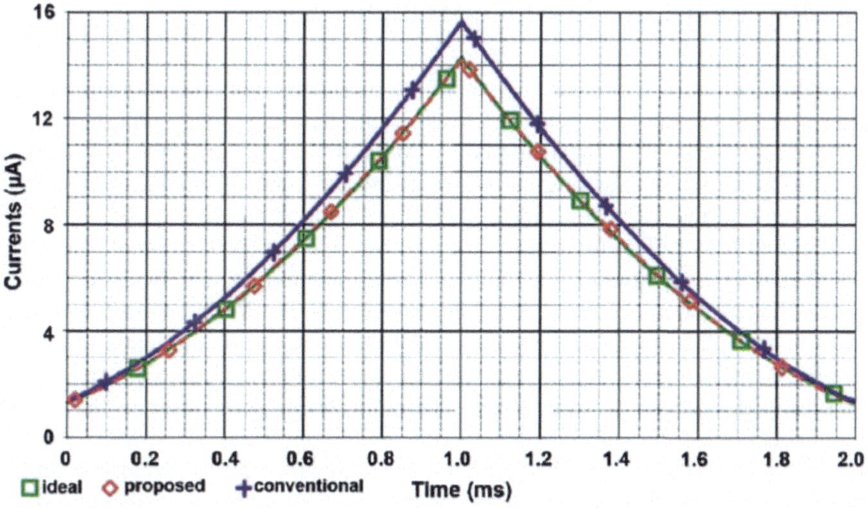

Fig. 12.10 The output currents and ideal curve of the squarer–divider. (Menekay et al. 2009a)

multiplier/divider circuit that is consisting of a square-root and a squarer/divider circuit is suitable for fuzzy logic controllers, artificial neural networks, modulators, phase discriminators, adaptive filters, cryptography systems, RMS-DC converters, sine/cosine synthesizers, etc.

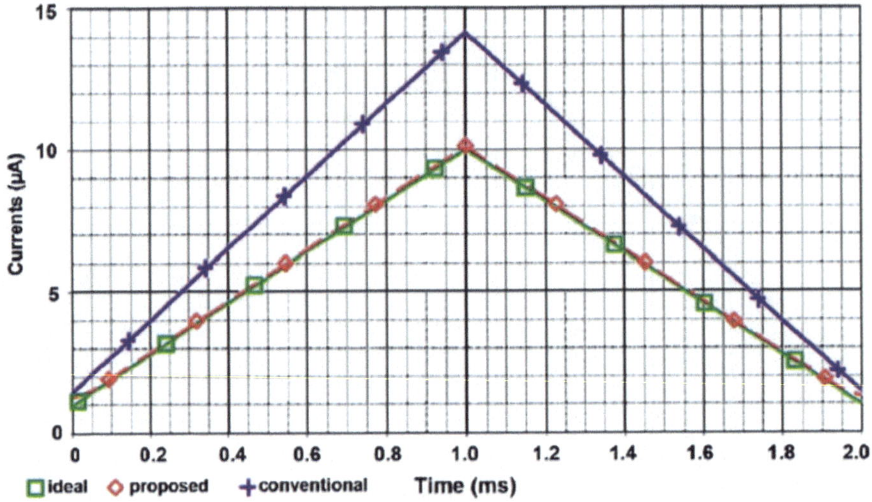

Fig. 12.11 The output currents and ideal curve of the multiplier–divider. (Menekay et al. 2009a)

High-Precision Current-Mode Multiplier/Divider

The multiplication of two signals is one of the most important operations in analog signal processing. Recently several CMOS multipliers have been reported in the literature and some of them are based on MOS translinear principle. Translinear circuit principle which was originally formulated for loops of bipolar transistors is generalized and the MOS translinear (MTL) principle is derived by Seevinck and Wiegerink (1991). MTL circuits are designed by applying MTL principle and used in synthesizing many nonlinear signal processing functions. Square-root circuit and squarer/divider circuit are two important structures of the MTL circuits. It is well known that the decrease in dimensions of MOS transistor in IC fabrication technology affects MOS transistor performance and the MOS transistor voltage–current relationship deviates from square law. Therefore, errors may occur in the output current function of the current-mode circuits employing small-sized MOS transistors. Nowadays, due to the decrease in dimensions of MOS transistors in IC fabrication technologies, second-order effects cause more errors in the MOSFET performance. Actually drain current expression includes effects of W, L, and VDS on effective threshold, velocity saturation effects, effective mobility dependence on normal field, and channel length modulation. Especially, the short-channel effects become more important in MOS transistors at channel lengths of about 1 μm or less and require modifications to the MOS models as well as the circuits that are designed using these MOS transistors (Tarcan and Kuntman 2003).

Current-Mode Multiplier/Divider Circuit

Due to the second-order effects small-sized MOS transistors do not operate properly according to the square characteristic, and therefore, errors may occur in the output current function of the current-mode circuits employing these transistors.

The conventional current-mode multiplier/divider circuit based on MTL principle (Lopez-Martin and Carlosena 2000, 2001) and the proposed circuit (Menekay et al. 2009b) are shown in Fig. 12.12. The output current function error of the square-root circuit that forms the left part of the circuit shown in Fig. 12.12 is reduced by the current flowing over the resistor *R*1. This process is carried out by flowing output current over the transistors M_{a14}, M_{a16}, M_{a17}, and M_{a18}. Similarly, the output current function error of the squarer/divider circuit that forms the right part of the circuit shown in Fig. 12.12 is reduced by the current flowing over the resistor *R*2. The transistors M_{b15}, M_{b16}, M_{b17}, and M_{b18} are used to provide the correction current flowing over the R_2 resistor. By this way, the output current function error of the proposed current-mode multiplier/divider circuit can be eliminated. Thus, the proposed circuit reduces the output current function errors which are not only the second-order effect resulting from the mobility reduction but also the other second-order effects such as mismatches which are examined in the simulation results. If the resistances in Fig. 12.12b are realized with the transistors, the resistance values can be controlled by the V_{C1} and V_{C2} control voltages, and the proposed

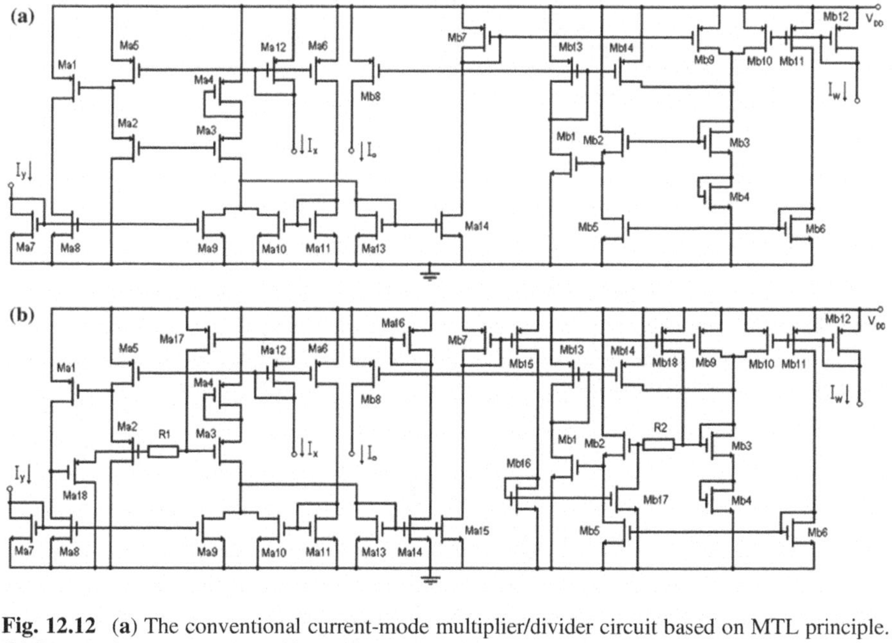

Fig. 12.12 (**a**) The conventional current-mode multiplier/divider circuit based on MTL principle. (**b**) Proposed current-mode multiplier/divider circuit. (Menekay et al. 2009b)

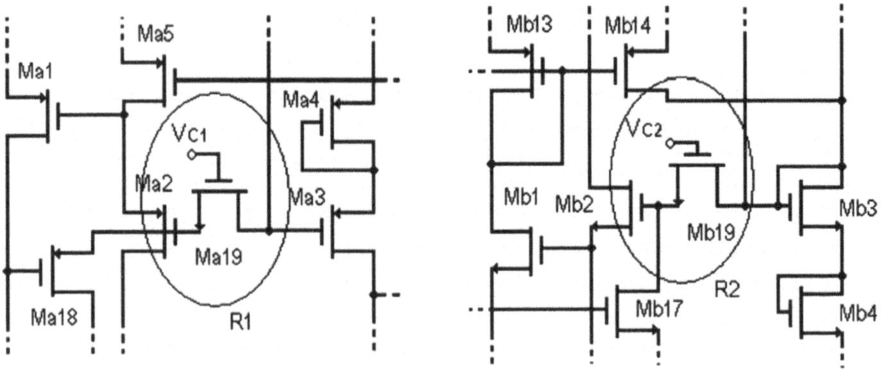

Fig. 12.13 Realization of resistances by transistors in the proposed circuit. (Menekay et al. 2009b)

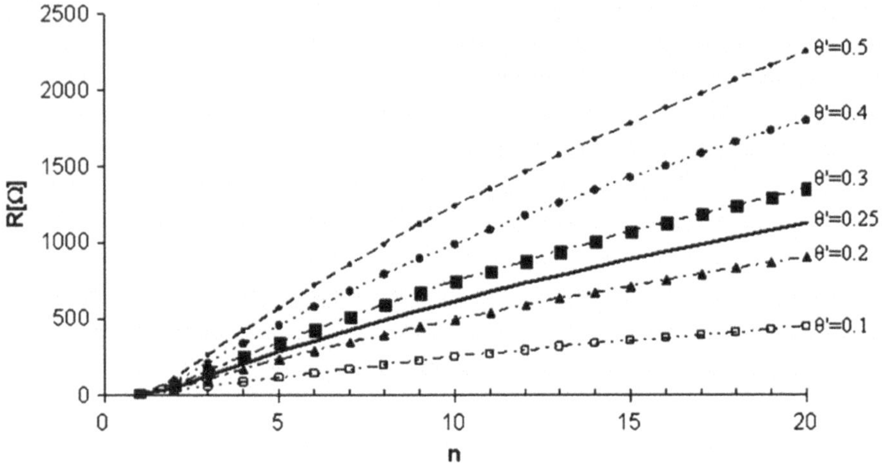

Fig. 12.14 Variation in resistance R value with respect to n corresponding to different θ'. (Menekay et al. 2009b)

circuit becomes appropriate for VLSI fabrication (Fig. 12.13). So the output current function of the proposed circuit can be controlled by these control voltages.

Taking the I_{out} current as the copy of the output current of the circuit, the resistance R can be obtained by

$$R = \frac{\theta'}{\beta}\left[\frac{I_1 + I_2}{2\sqrt{I_1 I_2}} - 1\right] \tag{12.23}$$

Considering the mobility reduction effect, this equation indicated that the value of the resistance R decreases the output current function error of the circuit. It can be expressed as shown previously in Eqs. (12.21) and (12.22).

The variation in resistance R value with respect to n corresponding to different θ' is illustrated in Fig. 12.14.

Simulation Results

Reduction of the output current function error simulation: To verify the proposed circuit, SPICE simulations were performed using TSMC 0.35 lm level 3 CMOS process parameters. The device dimensions of transistors used in the proposed circuit are shown in Table 12.1. The power supply voltage is 3 V. To obtain time domain SPICE simulation results of the proposed multiplier/divider circuit, a triangular wave with an amplitude of 5 μA is applied as the input current *Ix* where the other input currents *Iy* and *Iw* are constant DC currents with a value of 8 μA, as shown in Fig. 12.15. The output currents of the conventional and the proposed circuits illustrated in Fig. 12.16 are simulated for different R_1 and R_2 values. The results are shown with the ideal function in Figs. 12.17 and 12.18. As expected, for different values of the resistances R_1 and R_2 the output current function varies around the ideal curve. Because of the second-order effects the output current function of the conventional circuit is different from the ideal curve. It is apparent that the error of the output current function can be reduced by choosing the appropriate resistance values.

As expected, the characteristic of the output current function of the proposed multiplier/divider circuit shows approximately the ideal current function and the output current function of the conventional circuit is very different from the ideal curve. The output current function of the proposed circuit can be modified by changing V_{C1} and V_{C2} control voltage.

AC and DC analysis of the proposed circuit. The small-signal bandwidth measured from the input I_x to the output (with $I_y = I_w = 15$ μA) is 44 MHz. The DC transfer characteristics of the multiplier/divider circuit are shown in Fig. 12.18. Input currents were $I_w = 3$ μA, *Iy* values ranging from 2 μA to 10 μA in 2 μA steps.

Using the proposed circuit as an analog amplitude modulator, Fig. 12.19 shows the input and output waveforms of the multiplier operating as an analog amplitude modulator. Input currents are taken as $I_x = 5 \times (1 + \sin 2\pi f t)$ μA with $f = 0.2$ MHz, $I_y = 5 \times (1 + \sin 2\pi f t)$ μA with $f = 2$ MHz, and I_w is taken as a fix current of 5 μA. The modulated AC output current of the circuit is given in Fig. 12.19a. The

Table 12.1 Transistor dimensions

	W/*L* (μm)		*W*/*L* (μm)		*W*/*L* (μm)		*W*/*L* (μm)
Ma1	6/.7	Ma11	12/.7	Mb1	12/.7	Mb11	12/.7
Ma2	6/.7	Ma12	12/.7	Mb2	12/.7	Mb12	12/.7
Ma3	12/.7	Ma13	12/.7	Mb3	24/.7	Mb13	12/.7
Ma4	12/.7	Ma14	12/.7	Mb4	12/.7	Mb14	6/.7
Ma5	12/.7	Ma15	12/.7	Mb5	12/.7	Mb15	12/.7
Ma6	12/.7	Ma16	12/.7	Mb6	12/.7	Mb16	12/.7
Ma7	12/.7	Ma17	12/.7	Mb7	12/.7	Mb17	12/.7
Ma8	12/.7	Ma18	12/.7	Mb8	12/.7	Mb18	12/.7
Ma9	6/.7	Ma19	12/.7	Mb9	12/.7	Mb19	80/.7
Ma10	6/.7		10/.7	Mb10	6/.7		

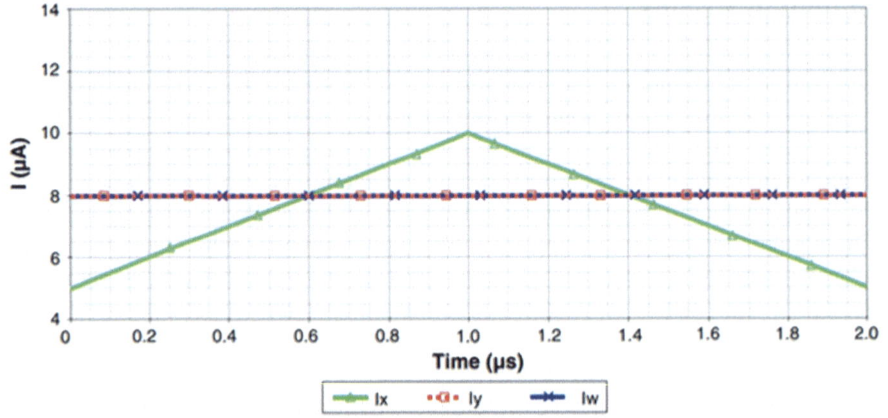

Fig. 12.15 Input currents *Ix*, *Iy*, and *Iw* of the proposed multiplier/divider. (Menekay et al. 2009b)

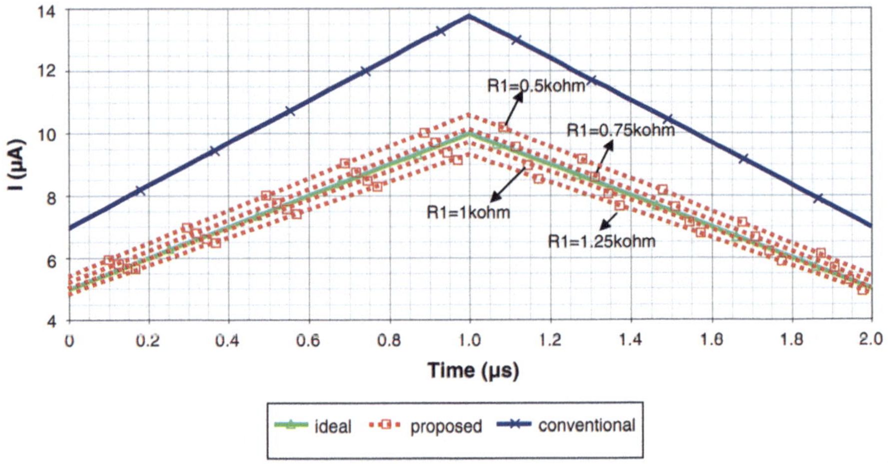

Fig. 12.16 Comparison of the output currents and the ideal curve for the different *R*1 values (*R*2 = 50 ohm). (Menekay et al. 2009b)

waveform obtained is for a one-quadrant multiplier. In our case, the desired amplitude modulator output can be achieved simply by subtracting the signal I_x from the output current I_o. The resulting output current is illustrated in Fig. 12.19b. Note that the circuit can be easily converted to a four-quadrant multiplier. Applying the well-known methods for this type of multiplier/divider circuits realizes the function $I_o = I_x I_y / I_w$. A comparison of the main parameters of the proposed multiplier/divider circuit and already published circuits is given in Table 12.2. The circuit proposed provides to the IC designers advantages such as smaller chip size and lower distortion.

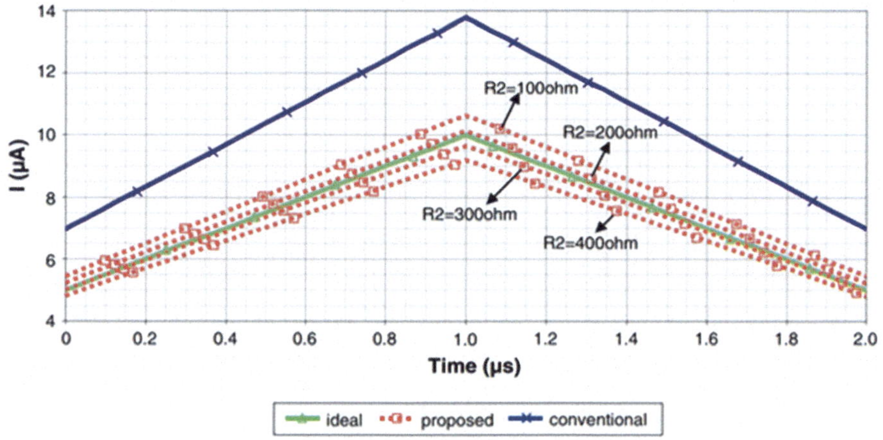

Fig. 12.17 Comparison of the output currents and the ideal curve for the different *R*2 values (*R*1 = 500 ohm). (Menekay et al. 2009b)

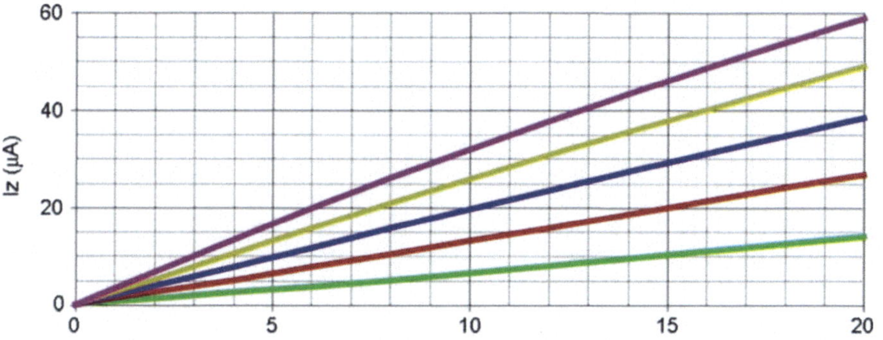

Fig. 12.18 Simulated DC transfer characteristics of the multiplier/divider. (Menekay et al. 2009b)

Concluding Remarks

This section describes a method to reduce the errors generated by the second-order effects such as mobility reduction, transistor mismatches, etc. in the current-mode circuits employing MOS translinear loops. In addition, a high-precision multiplier/divider circuit is also designed using the introduced method. It presents interesting features such as very small area, power efficiency, and precision. The circuit response is tunable over control voltages; therefore, it provides the possibility of reducing the error caused by secondary effects. Hence, it is shown that the proposed multiplier/divider circuit is suitable for fuzzy logic controllers, artificial neural networks, modulators, phase discriminators, adaptive filters, cryptography systems, RMS-DC converters, sine/cosine synthesizers, etc. As a result, it can be considered as a useful building block for IC designer.

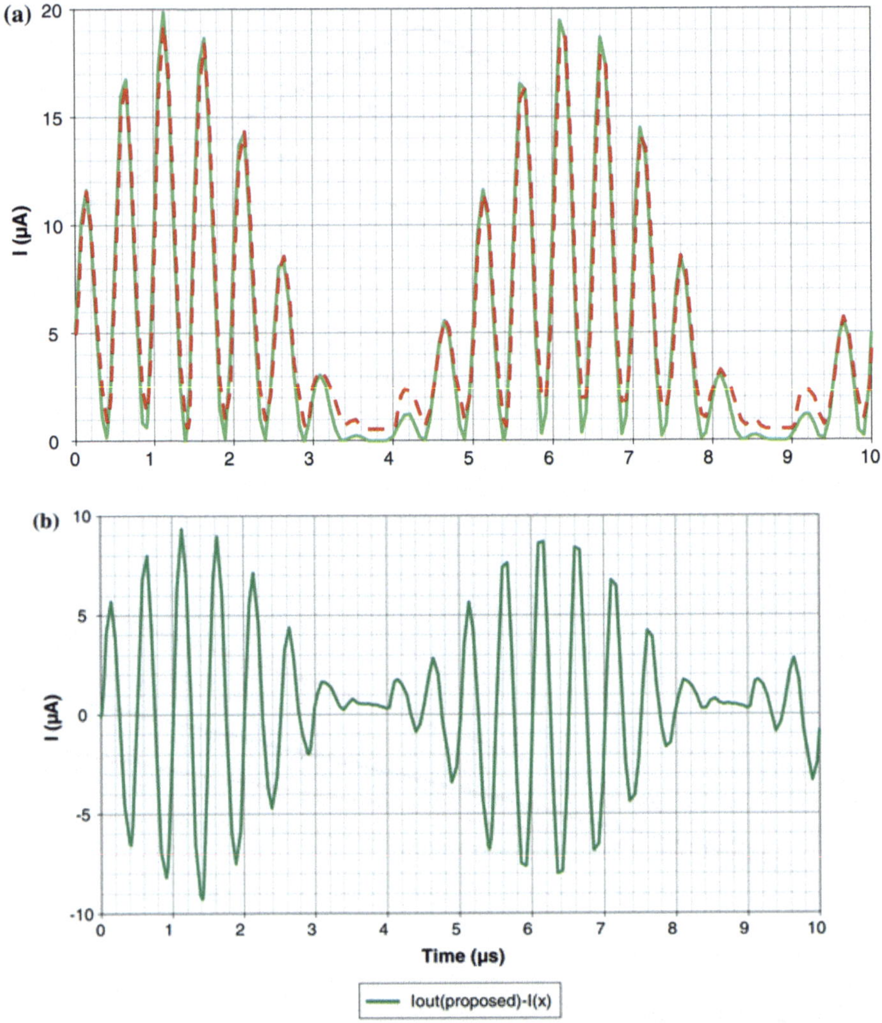

Fig. 12.19 Modulated AC output current of the multiplier/divider as an amplitude modulator: (**a**) output waveform and (**b**) AM output obtained by subtracting I_x from output current. (Menekay et al. 2009b)

Table 12.2 Performance parameters of the multiplier/dividers

	Value		
Parameters	Proposed	De La Cruz-Blas et al., 2003	Oliveira and Oki, 2007
Technology	0.35 μm CMOS	0.8 μm CMOS	0.35 μm CMOS
Supply voltage	3 V	1.5 V	1.5 V
THD	0.14% (*input* : 10 μA_{pp} at 1 MHz)	0.90% (*input* : 24 μA_{pp} at 1 kHz 5.5 MHz	0.83% (*input* : 20 μA_{pp} at 10 MHz)
BW (simulated)	44 MHz	5.5 MHz	140 MHz
Area	0.01 mm^2	0.10 mm^2	0.02 mm^2

References

De La Cruz-Blas, C. A., Lopez-Martin, A. J. and Carlosena, A. (2003). 1.5 V four-quadrant CMOS current multiplier/divider. Electronics Letters, 39(5), 434–436.

Lopez-Martin, A. J. and Carlosena, A. (2000). Design of MOS-translinear multiplier/dividers in analog VLSI. VLSI Design, 11(4), 321–329.

Lopez-Martin, A. J. and Carlosena, A. (2000). Design of MOS-translinear multiplier/dividers in analog VLSI. VLSI Design Journal, 11(4), 321–329.

Lopez-Martin, A. J. and Carlosena, A. (2001). A versatile 1.5 V current-mode CMOS analog multiplier/divider circuit. In European conference on circuit theory and design, ECCTD'01 (Vol. 2, pp. 89–92).

Lopez-Martin, A. J. and Carlosena, A. (2001). Systematic design of companding systems by component substitution. Analog Integrated Circuits and Signal Processing, 28(1), 91–106.

Lopez-Martin, A. J. and Carlosena, A. (2001, September). A current-mode CMOS RMS-DC converter for very low-voltage applications. In ICECS 2001. 8th IEEE International Conference on Electronics, Circuits and Systems (Cat. No. 01EX483) (Vol. 1, pp. 425–428). IEEE.

Menekay, S., Tarcan, R. C. and Kuntman, H. (2009a). Novel high-precision current-mode circuits based on the MOS-translinear principle. AEU-International Journal of Electronics and Communications, 63(11), 992–997.

Menekay, S., Tarcan, R. C. and Kuntman, H. (2009b). Novel high-precision current-mode multiplier/divider. Analog Integrated Circuits and Signal Processing, 60(3), 237–248.

Oliveira, V. J. S. and Oki, N. (2007). Low voltage four-quadrant current multiplier: An improved topology for n-well CMOS process. In Design & technology of integrated systems in nanoscale era, DTIS. International conference (pp. 52–55).

Seevinck, E. and Wiegerink, R. J. (1991). Generalized translinear circuit principle. IEEE journal of solid-state circuits, 26(8), 1098–1102.

Tarcan, R. C. and Kuntman, H. (2003). A new low distortion analog multiplier. AEU-International Journal of Electronics and Communications, 57(6), 365–371.

Index

A
Abuelmaatti configuration, 162
Active-only circuits
  gm-C filters, 181
  higher-order band-pass and band-stop filter functions, 182
  implement LPN and HPN functions, 182
  SPICE parameters, 182
Analog circuits
  amplifier types, 2
  overview, 2
  unavoided building blocks, 5
Analog continuous-time active filters, 181
Analog signal processing circuits built, 161
Angular resonant frequency, 185
AP function, 192
Arbel–Goldminz output stage, 143, 146
Audio-frequency filter
  band-pass filter configuration, 280
  CCII-based band-pass filter, 279
  circuit topology, 277
  ideal and simulated filter amplitude response, 280
  power consumption efficiency, 278
  simulation, 277
  sinusoidal characteristics, 280
  total harmonic distortion, 281
  transistor dimensions, 279
  ultra-low-power, 278
  ultra-low-voltage, 278
  ultra-low-voltage operation, 279

B
Band-pass filter, 249
Band-pass function, 255
BP filter configuration, 150
Broadband wireless access (BWA), 253
Butterworth filter, 258

C
CDTA-based biquadratic filter structure, 257, 258
CFOA-based synthetic inductor, 134
CMOS
  realization methods, 254
  VDTA-based design, 259, 262
COA-based band-stop filter, 156
COA-based filter realizations, 147–148
COA-based high-pass filter topology, 147
COA-based low-pass filter topology, 147
COA-based notch filter (band-stop), 154
Compensation capacitance, 153
Current conveyor
  active sensitivities, 54
  CCII-based oscillator circuit, 73
  circuit symbols, 40, 41
  classification, 39
  concept, 39
  current-controlled conveyor (CCCII), 65
    bipolar realization circuit, 66, 68
    electrical symbol, 65
  current-mode multifunction filters, 49

H. Kuntman, D. Özenli, *Trends in Circuit Design for Analog Signal Processing*, Analog Circuits and Signal Processing, https://doi.org/10.1007/978-3-030-96836-6

Current conveyor (*cont.*)
dual-output, 49
ECCII
BP response, 63
CMOS realization, 59
electrical symbol, 57
LP response, 62
nth-order current TF, 58, 60
second-order filter, 60
sensitivity analysis, 61
filter topologies, 50
filter transfer function (TF), 56, 68
inductance simulator, 70
CCCII bias current, 71
floating inductance, 70
frequency response, 72
input signal level, 42
matrix form, 40
nonlinear behavior, 42
nonlinear distortion, 45
pole angular frequencies, 51
results and verification, 53–55
second-order low-pass filter, 46
simulation result, 74
transfer function, 47
using CCIIs, 47
Current differencing buffered amplifier (CDBA)
circuit symbol, 226
experimental verification, 232, 234
non-ideal CDBA, 231
Current differencing transconductance amplifier (CDTA), 226, 270
CMOS realization, 237, 238, 241
filter applications, 235–237
Current feedback operational amplifiers
analog signal processing circuits, 161
applications and simulation, 172, 175
CFOA n-nodes plus ground network, 164
conventional operational amplifier, 161
inductance simulator, 171, 172
lossy series inductance, 171
n-nodes plus ground general configuration, 164
noninverting (infinite impedance) and inverting (zero impedance) input, 163
oscillation frequency and oscillation condition, 166
oscillator topology, 165
redundant admittances, 165
Current-mode circuits, 119
Current-mode operational amplifier (COA)
DC values of, 145
differential output transconductance stage, 144
input-output transcharacteristics, 155
open-loop frequency, 149
output resistance and current gain, 143
performance parameters, 153
simulated and ideal band-pass filter responses, 150
simulation results of, 150
total harmonic distortion (THD) values, 151, 158
zero input resistance and infinite output resistance, 143
Current-output-based active devices (COBADs), 25

D

Difference current conveyor (DDCC), 76
Differential input voltage (*VP–VN)*, 247
Differential voltage current conveyor (DVCC), 77
CMOS implementation, 77
electrical symbol, 77
KHN-biquad circuit, 78
transfer functions, 78
using Kirchhoff's current law (KCL), 80
Direct-sequence spread spectrum (DSSS), 251
DO-DDCC
CMOS implementation, 76
electrical symbol, 76
DO-OTA-C oscillator
realization, 33
topologies, 35
Double tuned amplifier realization, 249
DTMOS-based circuits
biomedical applications, 271
circuit symbol, 268
conventional analog circuits, 267
digital circuits, 268
double-tuned band-pass filter, 273
double-tuned circuit, 272
filter circuit, 273
gain factor, 273
high-transconductance MOSFET, 268
ideal and simulated frequency responses, 274
mobile carrier, 269
MOS transistor's drain current, 271
standard MOS drain currents, 269
supply voltage, 269
supply voltage levels, 267

time response, 274
transistor, 268, 269
ultra-low supply voltage, 270
ultra-low-voltage analog circuits, 268
VDTA-based band-pass filters, 272
VDTA circuit transistor dimensions, 272
VDTA design, 270
voltage difference, 270
Dual-output current conveyors (DO-CCIIs), 76

E
Electroencephalography (EEG)
brain–computer interface applications, 276
characteristics, 274
data, 275
data processing, 277
digital filtering approach, 277
filter application, 275
frequency component, 276
frequency signals, 275
input data signal amplitudes, 275
MATLAB program, 274
measurement setup, 275
post-filter frequency spectrum, 276
pre-filter frequency spectrum, 276
time response, 275
Electronically tunable current conveyor (ECCII), 56
Equation-based sizing, 200

F
Feedback technique, 255–263
Four-terminal floating nullor (FTFN)
circuit description, 131
common-source and common-gate amplifiers, 121
critical op-amp specifications, 119
current-mode band-pass filter, 124, 125
FTFN-based Chua's Circuit, 134, 135, 138
implementations of, 120
op-amp and cross-coupled current mirrors of, 119
open-loop transconductance, 121, 123
passive components, 126
PSPICE simulation program, 123
realization of, 127
series immittance simulator, 131
sinusoidal oscillator circuit, 129, 130
SPICE simulation, 129
symbol of, 120
transconductance amplifier, 121
universal parallel immittance simulator, 128
Frequency agile filter, 255
Frequency-hopping pattern, 253
Frequency-hopping spread spectrum (FHSS), 251
Frequency responses, 190
Frequency stability analysis, 167
Fully differential current conveyor (FDCCII), 87
CMOS implementation, 90
current-mode MISO biquad filters, 91
current transfer accuracy, 88
equivalent circuit, 87
SIFO biquad filter, 90
simulation results, 93

G
Global positioning systems, 256

H
HAVE QUICK (HQ), 254
High-pass notch, 191

I
Inductance simulator, 171, 172

K
Knowledge-based sizing, 200

L
Linear operation conditions, 5
Lower trace w-output, 234
Low-pass notch, 192

M
Monte-Carlo analysis, 207
MOS capacitances (MOS-cap), 215
MOS translinear principle
amplitude modulator, 300
analog amplitude modulator, 297
analog building blocks, 285
analog signal processing, 294
bipolar transistors, 294
conventional current-mode multiplier/divider circuit, 295
current-mode circuits, 285, 288, 295
current-mode multiplier/divider circuit, 292

MOS translinear principle (*cont.*)
current-mode squarer/divider circuit, 292
current-mode square-root circuit, 291
DC transfer characteristics, 299
drain current, 289
IC designers, 298
IC fabrication technology, 285
input currents, 298
KCL equation, 287, 289
mobility reduction, 288, 290
mobility reduction effect, 296
mobility reduction parameter, 288
multiplier/divider circuit, 287
output current function, 288
output currents, 298, 299
resistance R value, 296
resistances, 290
resistance values, 295
second-order effects, 286
short-channel effects, 294
simulations, 290, 291
SPICE simulations, 297
squarer–divider, 293
squarer/divider cell, 287, 288
squarer/divider circuit, 294
square-root cell, 287, 288
square-root circuit, 287, 293, 294
threshold voltages, 286
triangular wave, 297
voltage equation, 289
MOS-C circuits, *see* MOS-only circuits
MOS-only circuits
AC small signal equivalent circuit, 215
analog design, 203
analog IC, 199
AP filter, 216
automation flow algorithm, 203, 204
automation flowchart, 204, 205
automation model, 205
bias transistors, 220
bottom-plate and sidewall capacitances, 201
capacitance, 212
categories, 199
center frequency, 212
chip area, 217
circuit components, 206
CM-BP filter, 207
CM filter, 207, 211, 216
CM single-input multi-output filter, 215
core transistor currents, 205
current-mode phase shifters, 219
design constraints, 205, 206
design cycles, 201
feedback, 211
filter designs, 199
filter gains, 218
first-order AP architectures' layouts, 221
floating capacitor, 221
frequency response, 217, 218
gain expressions, 220
innovative solutions, 199
linear interpolation, 201
Monte-Carlo analysis, 208, 211, 213, 214
Monte-Carlo simulation, 223
NMOS device, 202
NMOS transistors, 214
output current, 210
peak-to-peak current, 213
pole frequency, 219
polynomial regression models, 201–203, 224
simulation, 209
small signal model, 200
small signal parameters, 222
temperature behavior, 208
temperatures, 219
transconductance gain, 216
transfer functions, 206
transistors, 199, 217
tunability, 209, 218

N
Natural angular frequency, 147

O
Op-amps
effects of non-idealities, 187
frequency response dependence, 196
gain-bandwidth product dependence, 195
open-loop frequency response, 189
Open-loop frequency, 149
Open-loop transconductance gains, 196
Operational amplifier-based circuits, 161
Operational transconductance amplifier (OTA), 11, 181
CMOS topology, 187
biomedical applications, 17–19
BPF circuits, 31
circuit symbol, 11
CMOS OTA realization, 12
CMOS OTA structure, 15
dimensions of MOS transistors in, 188
DO-OTA, 20
active-feedback, 27

CMOS circuits, 27–30
COBADs, 25
conventional MOS, 24
drawbacks, 22
input stages, 22
linearity regions, 28
electroencephalogram (EEG), 17
filter circuits, 191
input signal level, 14
LP filter design, 15
measured frequency response, 21
and Op-amp models, 183
in open-loop mode, 185
new voltage-mode and current-mode multifunction filters, 190
realization circuit, 13
second proposed multifunction type filter, 185
total harmonic distortion (THD), 30
transconductances, 22
transfer functions, 12, 13
VLSI building blocks, 17
Operational transresistance amplifier (OTRA)
circuit symbol, 100
CMOS 65 nm technology, 99
CMOS implementation, 101
concept, 99
current-mode dual-function filter, 104
current-mode filter, 105
differential OTRA, 107
circuit symbol, 107
CMOS topology, 108
gain stage, 109
input stage, 108
simulation result, 111
transistor dimensions, 110
immittance simulator topologies, 103
input terminals, 100
PSPICE simulation results, 106
single-output, 100–102
universal MOS-C filter structure
filter frequency response, 116
MOS structure, 114
simulation result, 112
universal MOS-C filter structure, 113
OTA-C oscillator networks, 32–33

P
Pole angular frequency, 243
Pole quality factor, 227
PSPICE simulation, 133, 138, 245
program, 194, 241
result, 75

Q
Quality factor, 255

S
Simulation-based sizing, 200
Single Channel Ground and Airborne Radio System (SINCGARS), 254
Single-input four-outputs (SIFO) biquad filter, 88
Single-input three-outputs (SITO) current-mode multifunction filter, 132
Sinusoidal oscillator circuit, 129
SPICE, 148
SPICE simulations, 175
Supply current sensing method (SCSM), 119
Systematic circuit generation method, 162

T
Terminal phasor voltage, 44
Third-generation current conveyor (CCIII), 80
circuit symbol, 82
inductor topologies, 84
simulation results, 86
Total harmonic distortion (THD), 246
analysis, 207
values, 151
Transistor dimensions, 145, 153

U
Upper trace z-output, 234

V
Voltage differencing buffered amplifier, 226
Voltage differencing transconductance amplifier (VDTA), 226, 270
CMOS implementation, 248
Voltage operational amplifier (VOA), 143

Z
Zero-IF receivers, 254

The manufacturer's authorised representative in the EU is Springer Nature Customer Service Centre GmbH, Europaplatz 3, 69115 Heidelberg, Germany. If you have any concerns regarding our products, please contact ProductSafety@springernature.com

Printed and bound by CPI Group (UK) Ltd, Croydon, CR0 4YY
30/09/2026
02225496-0009